Epitope Mapping

The Practical Approach Series

Related **Practical Approach** Series Titles

Protein Purifications 1
Protein Purifications 2
Protein–Ligand Interactions: structure and spectroscopy
Protein–Ligand Interactions: hydrodynamic and calorimetry
Bioinformatics: sequence, structure and databanks
Functional Genomics
Essential Molecular Biology 2/e
Immunoassay
Monoclonal Antibodies
Fmoc Solid Phase Peptide Synthesis
Lymphocytes 2/e
Protein Localization by Fluorescence Microscopy
Immunodiagnostics
High Resolution Chromatography
Post-Translational Modification
Protein Expression
Mutation Detection
Immunochemistry 1
Immunochemistry 2
Protein Function 2/e
Protein Structure Prediction
Antibody Engineering
Gene Probes 1
Gene Probes 2
DNA Cloning 2: Expression Systems
Glycobiology
Immunochemistry

Please see the **Practical Approach** series website at
http://www.oup.co.uk/pas
for full contents lists of all Practical Approach titles.

Epitope Mapping

A Practical Approach

Edited by

Olwyn M. R. Westwood

Senior Lecturer in Immunology
School of Life Sciences University of Surrey
Roehampton and St. George's Hospital Medical
School, London, U.K.

and

Frank C. Hay

Professor of Immunology,
St George's Hospital Medical School,
London, U.K.

OXFORD
UNIVERSITY PRESS

OXFORD
UNIVERSITY PRESS

Great Clarendon Street, Oxford OX2 6DP

Oxford University Press is a department of the University of Oxford.
It furthers the University's objective of excellence in research,
scholarship, and education by publishing worldwide in

Oxford New York

Athens Auckland Bangkok Bogotá Buenos Aires Cape Town Chennai
Dar es Salaam Delhi Florence Hong Kong Istanbul Karachi Kolkata
Kuala Lumpur Madrid Melbourne Mexico City Mumbai Nairobi Paris
São Paulo Shanghai Singapore Taipei Tokyo Toronto Warsaw

with associated companies in Berlin Ibadan

Oxford is a registered trade mark of Oxford University Press in the UK
and in certain other countries

Published in the United States by Oxford University Press Inc., New York

First published 2001

A catalogue record for this title is available from the British Library

Library of Congress Cataloging in Publication Data
Epitote mapping : a practical approach / edited by Olwyn Westwood,
Frank Hay.
(Practical approach series ; 248)
Includes bibliographical references and index.
1. Antigenic determinants–Laboratory manuals. I. Westwood, Olwyn
M. R. II. Hay,
Frank C. III. Series.

QR186.3 .E65 2001 616.07'92–dc21 00–048305

1 3 5 7 9 10 8 6 4 2

ISBN 0 19 963653 2 (Hbk.)
ISBN 0 19 963652 4 (Pbk.)

Typeset in Swift by Footnote Graphics, Warminster, Wilts
Printed in Great Britain on acid-free paper
by The Bath Press, Avon

Preface

Epitope Mapping: A Practical Approach is intended for the researcher who needs to define the regions on antigens which make contact with the antigen binding-site of antibodies or T cell receptors. The same procedures can also be easily adapted to identify regions on other ligands in their interaction with receptors. Certainly, the pharmaceutical industry is ever mindful of the need to produce drugs of minimum toxicity that are easy to manufacture and can be rapidly screened for biological activity. The synthesis of organic compounds that mimic bioactive molecules has often been exploited in drug design and the techniques of epitope mapping have led to a novel approach to this problem. Protocols have been provided by Dr. Ian Matthews (Chapter 6) including the design, preparation and characterization of peptoids as peptidomimetics.

An innovative method for identifying linear epitopes on macromolecules, e.g. proteins, is the use of multiple pin peptide scanning ('Pepscan') that utilizes a modified ELISA technique for the rapid screening of individual or panels of antibodies, i.e. B-cell epitope analysis with a vast number of peptides. The methods and trouble-shooting tips for the analysis of data is clearly covered in the Chapter 2 by Dr. Nazira Sumar. With certain adaptations, the same technology may be employed for delineating T cell epitopes, both T helper cells (see Chapter 3 by Dr. D. Mutch & Dr. O. M. R. Westwood) and cytotoxic T cell epitope analysis (see Chapter 5 by Dr. T. Elliott & Dr. J. S. Haurum) as well as overlapping T- and B-cell epitope analysis, an important concept in vaccine design (see Chapter 4 by Dr. Sowsan Atabani).

The use of molecular biology techniques in site-directed mutagenesis has allowed the researcher to investigate the role of individual amino acids in immune complex formation. Dr. S. Perdue (Chapter 10) has written an excellent overview with examples of these concepts, thus providing a general body of knowledge from which to obtain further information. The methods given are straightforward and trouble-shooting points clear for those wishing to use these techniques. Likewise with the phage display libraries, this brings together two methodologies that link the genetic and amino acid sequence of a peptide. In Chapter 9, Dr. S. Williams and co-authors have shown how this library approach may be applied in looking for immunodominant epitopes of polyclonal antisera as well as in monoclonal antibodies. It should be noted that for those wishing to prepare monoclonal antibodies against antigens of interest, Dr. P. N. Nelson, who has considerable experience in this field, has written a straightforward method for their preparation and screening (see Chapter 7).

We were pleased to be able to persuade an excellent team of specialist researchers to form the authors for this text. Our grateful thanks for their willingness to write for this book is freely acknowledged, especially in the ethos where the research assessment exercise appears to place more value in the writing of research papers than other scholarly activities, such as the preparation of under-graduate and post-graduate textbooks. Though it is with sorrow that we record the recent death of one of our authors, Professor G. Magnusson (Lund Institute of Technology, Sweden) who made a significant contribution to our understanding of carbohydrate biochemistry in relation to epitope analysis (see Chapter 8 that he co-authored with Dr. U. J. Nilsson).

No previous knowledge of epitope mapping is assumed when reading *Epitope Mapping—A Practical Approach*. Although most readers are likely to have some familiarity with immunological techniques, care has been taken to ensure that guidance to the basic literature has been included. It is our hope that this text will stimulate interest and encourage the researcher to explore the literature, experiment and thus contribute to further great understanding of epitope analysis in relation to health, disease diagnosis, treatment and prevention.

Dr. Olwyn M. R. Westwood
Professor Frank C. Hay

Contents

Protocol list

Case study: identification of dominant peptide epitopes

General approaches to mutagenesis in epitope mapping techniques in site-directed mutagenesis

Abbreviations

BLG	beta-lactoglobulin
BrdU	Bromodeoxiuridine
BSA	bovine serum albumin
CDR	complementarity determining regions
CFA	complete Freund's adjuvant
CMV	cytomegalovirus
CTL	cytotoxic T lymphocytes
DAB	diaminobenzidine
DIC	1,3-diisopropylcarbodiimide
DIPEA	*N,N*-diisopropylethylamine
DMF	*N,N*-dimethylacetamide
DMSO	methyl sulfoxide (dimethyl sulfoxide)
EBV	Epstein Barr Virus
EDC	ethyl 3-(dimethylamino)propylcarbodiimide
EDT	ethanedithiol
ELISA	enzyme-linked immunosorbent assay
FBS	fetal bovine serum
FCA	Freund's complete adjuvant
FCS	fetal calf serum
FF	forward flanking
FITC	fluorescein isothiocyanate
FM	forward mutagenic
GAM	goat anti-mouse
HA	haemagglutination
HAI	haemagglutination inhibition
HC	heavy chain
HF	hydrofluoric acid
HGPRT	hypoxanthine-guanine phosphoribosyl transferase
HIV	Human immunodeficiency virus
HLA	Human leukocyte antigen
HOBt	hydroxybenzotriazole hydrate
HPLC	high performance liquid chromatography
HRP	horseradish peroxidase
[^{3}H]-TdR	tritiated thymidine

IC	inhibitor concentration
IC_{50}	inhibitor concentration causing 50% inhibition
IEF	iso-electric focusing
IF	internal flanking
IFN	interferon
IFA	incomplete Freund's adjuvant
IFN-gamma	interferon gamma
i.p.	intraperitoneal
KLH	keyhole limpet haemocyanin
LISS	low ionic strength saline
Lf	Limit of flocculation
mAb	monoclonal antibody
MAP	multiple antigenic peptide
MHC	major histocompatibility complex
MP	megaprimer
MS	mass spectroscopy
NCP	non-cleavable peptides
NHS	*N*-hydroxysuccinimide
NMR	nuclear magnetic resonance
NOE	nuclear Overhauser effect
OD	optical density
OD_{50}	optical density at 50% inhibition
PAGE	polyacrylamide gel electrophoresis
PAS	protein A Sepharose
PBMC	peripheral blood mononuclear cells
PBS	phosphate-buffered saline
PCR	polymerase chain reaction
PHS	pooled human serum
PPD	purified protein derivative
PVP	polyvinylpyrrolidone
RM	reverse mutagenic
RSV	respiratory syncital virus
RT	room temperature
s.c.	subcutaneous
SD	standard deviation
SDS	sodium dodecyl sulfate
SDS–PAGE	SDS–polyacrylamide gel electrophoresis
SI	stimulation index
SPOTs	simple precision original test system
SPPS	solid phase peptide synthesis
SRBC	sheep red blood cells
TAP	transporter associated with antigen processing
TFA	trifluoroacetic acid
TcR	T cell receptor
TNF-α	tumour recess factor-alpha
TT	tetanus toxoid

Chapter 1

An introduction to epitope mapping

Olwyn M. R. Westwood
School of Life Sciences, University of Surrey Roehampton, and Division of Immunology, St. George's Hospital Medical School, London, UK.

Frank C. Hay
Division of Immunology, St George's Hospital Medical School, London, UK.

Antigens are highly diverse, varying in size as well as the composition of their primary sequence. In the case of proteins, post-translational processing with the addition of moieties, such as carbohydrates during glycosylation, may alter secondary structure. The relative location of epitopes on the surface of an antigen demands ample consideration in relation to the structure when defining and predicting antigenic sites. Although, their identity can only really be found secondarily to the examination of the cellular and humoral products of specific immunity, i.e. lymphocytes and antibodies. Interest in defining epitopes has been borne out, at least in part, from a need for producing non-pathogenic vaccines. At a more technical level, the cloning and sequencing of DNA has allowed the construction of genes for the synthesis of precise antigens in non-human cells. Such advances have obviated the risk of infections, such as hepatitis B and human immunodeficiency virus (HIV) that are potential hazards when preparing antigens from body tissues and fluids.

Antigenic sites for antibody binding are only found on the surface of the molecule and conform to two possible architectural options. An epitope that constitutes part of a linear amino acid sequence on a polypeptide chain is known as a continuous or linear epitope. Where the epitope is formed from two stretches of the primary sequence which are distant from one another, but brought together in the folded molecule's secondary or tertiary structure, then it is known as a conformational epitope.

1 Environmental conditions can influence protein structure

The environment of the whole antigen will affect its molecular folding and thus the nature of epitopes exposed on the molecule's surface. Moreover, the location of the antigen demands consideration since the intracellular conformation differs from the extracellular counterpart, for a protein passing through the cellular organelles, e.g. endoplasmic reticulum, is required to adopt a loose or open conformation (1). Most naturally occurring proteins are stable, held together in secondary and tertiary structures, i.e. their native conformation, by intramolecular bonds such as hydrogen bonding, hydrophobic, and electrostatic forces. Accordingly these forces may be altered by their environmental conditions, such as pH, temperature, as well as the nature of the solvent in which they are immersed. Treatment with denaturing agents can lead to loss of α-helix or β-pleated structures, converting the protein to a random coil formation and losing its compact structure, for example, breaking disulfide linkage using 2-mercaptoethanol, or the use of urea or guanidinium hydrochloride to alter hydrogen and hydrophobic interactions. When the disulfide and other intramolecule bonds apart from the peptide linkages are broken, the protein is said to be denatured. As a consequence, the molecule may be rendered non-functional if it is less globular, and its physicochemical properties such as solubility and sedimentation rate are altered. The study by Michaelsen *et al.* (1975) is a classic example of how the reduction and alkylation of rabbit immunoglobulins resulted in their diminished capacity to induce cytotoxicity (2). Thus the common use of reducing agents in SDS–polyacrylamide gel electrophoresis (SDS–PAGE) may result in the detection of a different set of epitopes from those exposed on the native protein. For instance, an antibody may not bind to a denatured protein in a Western blot analysis because its structure or shape is different from the native protein. Obviously such factors are of little clinical significance, but such data may be critical for preparing effective diagnostic reagents. Hence it can sometimes be useful to deliberately immunize with the denatured antigen to produce reagents capable of recognizing antigens in SDS–PAGE or in fixed tissue sections.

Ionic strength of the medium affects antibody–antigen interactions *in vitro*, for lower ionic strength solutions will promote antigen–antibody binding. Blood group serologists have exploited this idea for developing shorter incubation times in blood group and cross-matching techniques (3). In this situation, erythrocytes washed and suspended in low ionic strength saline (LISS), are incubated with serum, to ascertain whether atypical blood group antigens are present on the surface of cells that could react with serum antibodies.

1.1 Locating the epitope of the molecule

Any accessible part of a molecule may be considered to be potentially antigenic, some inherently produce a greater response than others. It is a debatable point, but some workers believe that the immune response is largely dependent on the ability of the specific host (4), whereas other suggest that the nature of the antigen is independent of the host being immunized (5). When comparing the immunogenicity of native versus denatured proteins, this is generally a problem for evaluating B cell epitopes rather than for T cell epitopes. It has been well established that immunoglobulins recognize conformation rather than sequence. This somewhat explains why some antibodies will only work in certain assay or analytical systems.

For antibody-binding to an antigen, the specific epitope must be exposed on the molecular surface. There are a number of criteria that might affect position of the epitope, such as:

(a) The mobility of the region of the protein.

(b) Nature of the primary sequence (in the case of protein antigens) where there is the potential to form a loop or turn, e.g. presence of proline residues.

(c) The hydrophilicity.

With reference to molecular mobility of an antigen, X-rays and NMR studies may indicate surface amino acid residues with higher mobility, and areas on the molecule where there are rigid structures, such as α-helices and β-sheets. As a general rule, the hydrophobic stretches of sequence will reside in the interior of the molecule, with the hydrophilic stretches found on the surface. Therefore highly hydrophilic sequences are likely to be on the surface of a molecular fold and thus putative epitopes. When known, the amino acid sequence data is valuable for predicting epitopes, particularly when used in conjunction with hydrophilicity parameters for each residue (6). As far as primary sequences are concerned, the sequence of the amino acids in the peptide chain specify the ultimate three-dimensional (3D) structure. Therefore, from a knowledge of X-ray crystallographic data from other protein sequences that are similar to the protein of interest, its 3D structure may be predicted, and thus so could the exposure of potential epitopes. Moreover, it is a simplistic, but accepted, view that proteins with comparable intermediate or transitional folding during their manufacture and intracellular transport also may have similar native structures (7).

With respect to T cell epitopes it has been generally acknowledged that the T cell receptor recognized amino acid sequence within the peptide groove of the major histocompatibility complex. In this context, it was the nature of the antigen, rather than its shape that was recognized by the T cell arm of the immune system. However, this has since become a subject of some debate. Bach *et al.* (1998) have refuted this evidence in their work on mimic sequence motifs of the pancreatic antigen in insulin-dependent diabetes mellitus, i.e. the 65 kDa glutamic acid decarboxylase (GAD65). They found T cells specific for GAD65 could be stimulated by conformational peptide mimic that had little sequence homology with the defined autoantigenic epitope (8). Their findings have suggested that knowledge of the primary residue contacts with the T cell receptor within the epitope and the MHC class II binding motifs is particularly significant.

The pharmaceutical industry is ever mindful of the need to produce drugs that are easy to produce; having minimum toxicity and that can be rapidly screened for biological activity. This natural concept of mimicry is also being exploited for the design and synthesis of organic compounds that mimic bioactive molecules. Chapter 6 provides protocols that cover the design, preparation, and characterization of peptoids as peptidomimetics.

2 Lessons from an historical perspective

Traditional methods for defining linear epitopes have included fragmentation of proteins either by chemical cleavage, e.g. using cyanogen bromide or by enzymatic digestion. If proteolytic enzymes are to be used, then prior knowledge of the primary sequence is generally required. Where the protein of interest is glycosylated, the removal of post-translational moieties with endoglycosidases may be advisable. Yet post-translational modifications such as oligosaccharide chains of glycoproteins, can form part of an epitope. Alternatively the carbohydrate moiety might even mask an epitope that was present on the native non-glycosylated protein.

2.1 Enzymatic cleavage for epitope mapping

The relative position of an epitope on an intact protein may be detected by limited digestion of the antigen followed by immunoblotting of the protein fragments (9). This may be of clinical relevance when investigating a defect with a protein such as ankyrin—a structural protein with a membrane skeleton where defects may be inherited or induced. In either case, structural changes are liable to result in abnormalities in cellular deformability or fragility (8). Competitive inhibition of epitope binding by synthetic peptides is also used in mapping. But it is expedient for the researcher to be aware that biological activity may be inhibited. Furthermore data may not necessarily allow genuine 'competition' to be distinguished from steric hindrance owing to 'crowding' of the site because of the sheer size of the antibody molecule.

When investigating the nature of putative epitopes on a protein, several enzymatic methods are available, e.g. digestion of the protein into smaller peptide fragments, deglycosylation. The classic work of Rodney Porter (1959) who rationalized the structure of immunoglobulins into their relative areas of activity relied on the hydrolysis of the polypeptide chain at specific positions. By hydrolysis of IgG with papain he produced fragments that bound antigen, i.e. Fab region, and an intact peptide fragment that could be crystallized, i.e. Fc (10).

The purity of both the enzyme preparation as well as the antigen to be digested requires consideration, as does the nature of the protein structure being hydrolysed. Immunoglobulins do not only bind antigen, but also may be considered as antigens. The globular structure of immunoglobulin is a classic example of a molecule that is relatively resistant to enzymatic hydrolysis. The activities of some enzymes are restricted to the hinge region of the molecule where there is some degree of flexibility. Conversely within the Fab region there is a relative absence of movement that can in turn affect hydrolysis. Moreover, the different subclasses of IgG differ with respect to their hinge regions; thus IgG2 and IgG4 that have restricted hinge lengths are resistant to papain hydrolysis (although IgG4 may be hydrolysed in the presence of cysteine). In contrast, IgG1 and IgG3 molecules have flexible hinge regions and are more sensitive to hydrolysis (10).

Carboxypeptidase for C-terminal amino acid sequence analysis has been used in the investigation of the role of MHC class I molecules in selecting antigenic peptides. These peptides, which are imported into the endoplasmic reticulum for assembly in the MHC class I groove, appear to be protected from cleavage by proteases. One proposal has been that MHC class I molecules are involved in peptide selection by sampling their affinity before protease degradation or efflux into the cytosol (11).

2.2 Protein sequence analysis

To determine the primary sequence of a polypeptide any disulfide bonds must be cleaved, which may be performed using dithiothreitol, or by oxidation with peroxy-formic acid which converts cysteine to cysteic acid (13). Protein that has been purified to homogeneity may then be sequenced from the carboxy (C-) or amino (N-) terminus. C-terminal sequence analysis utilizes the naturally occurring enzyme carboxypeptidase. N-terminal amino acid sequencing is performed by the Edman degradation where the peptide is reacted with phenylisothiocyanate to produce the phenylthiocarbamol derivative. This is subsequently treated with anhydrous hydrochloric acid in an organic solvent to produce a phenylthiohydantoin derivative. The amino acid is separated from the N-terminus of the peptide and identified by chromatography. Partial degradation by chemical cleavage of peptides of proteins has traditionally been used to determine the primary structure. The peptides cleaved may be chromatographically separated, sequenced, and may provide a complete primary sequence (14). To ensure precise data, more than one chemical cleavage protocol may need to be used to produce peptides of differing lengths because they were hydrolysed at specific residue sites. Then the primary sequence may be defined by comparing the overlapping sequences deduced from data obtained from the individual peptide fragments.

Peptides of varying lengths are produced and sequenced: cyanogen bromide selectively cleaves peptides at the carbonyl group of methionine residues, converting the methionine to a C-terminal homoserine lactone unit (15). Epitope mapping using this method was employed by Suphioglu *et al.* (1993) to identify IgE-binding epitopes on the grass pollen allergens of *Pooidae*, *Chloridoideae*, and *Panicoideae* (16). When evaluating the peptide fragments produced from treating the Fc region of IgG with cyanogen bromide, the different allotypes must be considered. For instance G1m(a) has methionine residues at positions 252 and 428, whereas the non-Gm(a) allotype has a methionine at residue 358. Thus cyanogen bromide digestion of these two allotypes produces very different peptide fragments. Other reagents used for the fragmentation of proteins include hydroxyamine that cleaves asparagine–glycine bonds, and 2-nitro-5-thiocyanobenzoate that cleaves peptide chains at the amine side of cysteine residues (17).

Using chemical cleavage produces peptides with different sequences and analysis of the overlapping sequences can produce precise data. Such methods have distinct advantages over enzymatic hydrolysis in that they are not liable to steric hindrance. Moreover, since chemicals are generally smaller molecules than enzymes, they can penetrate the protein to reach target sites deep within the tertiary structure. Only a limited number of residues can be sequenced at any one cycle owing to 'noise' when evaluating chromatographic peaks. Nevertheless, such information has proved invaluable for extrapolating genetic sequences to synthesize cDNA probes used for identification and isolation of clones from genomic and cDNA libraries.

2.3 Chemical characteristics of sequences

The recognition and binding of an antigen to a lymphocyte receptor is highly dependent on the chemical characteristics of the residues constituting the epitope. Since antigenic epitopes are an integral part of the whole antigen, overall an isolated peptide fragment or synthesized peptide may well have a different conformation from that found when it forms part of the whole antigen. Some amino acids appear to be more immunogenic than others, since they are seen in high frequency as part of antigenic determinants, these include: His, Lys, Ala, Leu, Asp, Arg (in order of decreasing antigenicity). Thus possible antigenic sites may be located by calculating regions of immunogenic potential (18). Knowledge of these parameters would be useful in conjunction with X-ray crystallography data of a named antigen to locate an epitope.

Certain amino acid residues within antigenic epitopes have been shown to be highly significant for binding immunoglobulins. For example, Artandi *et al.* (1992) who analysed specificity of rheumatoid factors, natural autoantibodies whose antigen is the Fc region of IgG, found that the His-435 in the Fc region of the IgG was an important residue for binding (19). A cross-species comparative study of primary sequence data for a named protein could indicate sites of antigenicity, and may improve our understanding of the relative roles of key residues within the protein. From an evolutionary perspective, sites of reactivity may be enhanced, diminished, or totally eliminated.

Optical configuration will also affect immunogenicity; L-forms generally produce a better immune response. One reason is that D-forms of bacterial antigens are not as easily processed by antigen-presenting cells, so are not as effectively presented to T lymphocytes (20).

3 Synthetic peptide technologies for epitope mapping

Geysen *et al.* (1987) investigated possible ways of identifying epitopes on macromolecules like proteins and developed the Pepscan technique, but this requires prior knowledge of amino acid sequence for the protein of interest. Briefly, multiple peptides are synthesized on polystyrene pins; these sequences overlap and span the complete length of the protein. The polystyrene pins are arranged in blocks of 9×12 (96 pins) that may be immersed into the wells of a microtitre plate format. A modified ELISA may be used to screen individual or panels of antibodies for distinguishing linear B cell epitopes (21). The technology and possible troubleshooting for the analysis of data is clearly covered in Chapter 2. This technique may also be adapted for the delineation of T cell epitopes, both T helper cells (see Chapter 3) and cytotoxic T cell epitope analysis (see Chapter 5). But these authors have also advocated the use of other methodologies including the purification of antigens presented by MHC class I molecules and expression cloning for identifying CTL epitopes following transfection of plasmid DNA into Chinese hamster ovary cells (COS-7 cells) or HeLa cells.

T and B cell epitopes may be overlapping or adjacent within a single sequence, for within a peptide as short as a 10-mer, the respective T and B cell sites may be unique (22). Since B cells can also act as antigen-presenting cells, they are capable of presenting antigen for T helper cell recognition. Accordingly, the cytokine network operates whereby these two immune cell types may augment or dampen the immune response. Thus a single peptide may be able to stimulate both the humoral and cellular arm of immunity, and this has significant implication for possible therapies, e.g. vaccine development. Chapter 4 includes protocols for the identification of combined B and T cell epitopes within a stretch of amino acid sequence using synthetic peptide technologies and complements the chapter on the Pepscan technique (Chapter 2).

Attempts have been made to mimic the shape of conformational epitopes with peptides by building up the structure in iterative stages. First all possible 400 dimeric peptides are synthesized and those showing best binding to antibody are then lengthened at either end with each of the 20 amino acids in turn. Again the best antibody binders are selected and further amino acids added. In this way, with the making of relatively few peptides, mimotopes that mimic conformational structures may sometimes be selected. The technique has had limited success as it has a tendency to select for similar sequences that bind to many different antibodies. The technique with random phage sequences, described in Chapter 9, where full-length peptides are used at the first selection, has proved more useful to generate unique structures for individual antibodies.

4 Chemical modification of antigens

Chemical modifying reagents with the ability to penetrate reaction sites have been strategically employed to investigate potential active interaction sites. It is well documented that changes to side chains on macromolecules, e.g. proteins, immunoglobulins, can alter antigenicity and/or biological function. For instance, Boc-Cys(Npys)-OH blocks the thiol groups of cysteine residues and can render T cell epitopes inactive (23).

Immunoglobulins are excellent candidates for investigating antigenicity since they have a number of active sites, e.g. binding sites for the Fc receptor, complement. The IgG molecule is the antigen for naturally occurring antibodies, called rheumatoid factors (RFs) that are detected in a number of different connective tissue disorders, e.g. rheumatoid arthritis, systemic lupus erythematosus (24, 25). To examine the possible binding sites for RFs, IgG has been chemically modified by carbamylation: this involves the treatment of IgG with 2 M potassium cyanate and the subsequent modification of α and ε amino groups, particularly affecting lysine residues. Conflicting evidence has been obtained regarding changes in antigenicity of carbamylated IgG; whilst some workers have not found changes in polyclonal antisera reactivity (26), others have suggested that changes in reactivity of the Fc region could be detected (27). These discrepancies are not surprising, since changes in antigenicity would be dependent on the extent of chemical modification and whether the carbamylation had provoked changes in the epitope *per se* and/or the molecular conformation. This is not simply a chemical artefact that happens *in vitro*, for the addition of isocyanic acid (a degradation product of urea), i.e. carbamylation of proteins, happens *in vivo*, e.g. erythrocyte membrane proteins of uraemic patients (28).

Nitration is another example of how the use of one technique is very much antigen-dependent. For example, nitrated human IgG failed to react with rheumatoid factors, whereas nitrated rabbit IgG maintained reactivity. Such discrepancies were attributed to differences in amino acid sequences between rabbit and human IgG that led to changes in conformation post-treatment. This in turn affected the antigenic site where there are tyrosine residues that are directly involved in the rheumatoid factor-binding of human IgG (26). Tetranitromethane modifies tyrosine residues but there are some limitations:

(a) The extent of treatment is governed by the concentration of the reagent used in the reaction mixture.

(b) Possible phylogenic variation in the structure of proteins.

5 Site-directed mutagenesis as a tool for epitope mapping

With knowledge of the amino acid sequence data for a protein of interest, the precise residues that form the epitope involved in immune recognition may be determined. Where a protein exhibits evolutionary variation or where there is the possibility of mutants, this divergence will inevitably govern its 'foreignness' to the host, and thus the subsequent immune response. When the protein being studied is conserved across species, often the antigen produces a limited immune response. The most probable rationale is that antigen is not perceived as 'dangerous', instead it is tolerated like so many other 'self'-antigens.

From a purely academic perspective, the study of natural variants has demonstrated some success in mapping epitopes on protein antigens, but the technique has its limitations:

(a) The number of different proteins available for study is restricted to ones where there are multiple species isolates available, moreover, are these proteins of any real interest?

(b) Researchers are limited to studying the few amino acid variations that occur naturally rather than numerous strategic changes along the primary sequence.

(c) Single amino acid substitutions rarely occur in isolation. Even where a protein has been conserved across species there is likely to be a number of sequence differences and this may result in variation within protein conformation, and thus the nature of the epitope exposed on the antigen surface. Therefore when analysing the between-species variation of a named antigen and its binding to a specific antibody, any discrepancy may not be the contribution of a single residue on the antigen surface, but also the effects of other amino acid substitutions.

Site-directed mutagenesis is a powerful tool that solves all of these problems for it allows any single amino acid within a protein sequence to be substituted with another. When an individual residue is changed, the effect of this change on antibody binding can be assessed by a variety of techniques (29). Unlike the case with natural variants or mutants, with site-directed mutagenesis, the role of individual amino acids within the immune complex formed may be evaluated. For example, an amino acid may be substituted to produce a variation in side chain, electrostatic charge (smaller size residue), or one that produces a definite change in conformation, e.g. proline. This allows an in-depth analysis of the contribution of specific amino acids to the energy of complex formation.

To determine the residues involved in complex formation, the technique of alanine-scanning mutagenesis may be employed. In this instance, single point mutants of a protein are systematically generated to assess the molecular interactions at several protein–protein interfaces (30). Alanine is an amino acid with a very simple structure, i.e. only has a single methyl group as the side chain (which compares to the carbon in the side chains of all amino acids except glycine and proline). So alanine substitution has the advantage of effectively removing from the antigen–antibody interface the energetic contributions of all side chain atoms beyond the C position, including intramolecular forces, e.g. side chain hydrogen bonds. Accordingly, this technique affords the researcher the ability to evaluate the contribution of individual amino acids at the antigen–antibody interface. Perdue (Chapter 10) has given an excellent overview and examples of the concept of site-directed mutagenesis, providing the researcher with a general body of knowledge from which to obtain further information. The methods given are straightforward and troubleshooting points clear for those wishing to use these techniques.

6 Hybridoma technology and epitope analysis

Chapter 7 provides detailed protocols for production of monoclonal antibody probes. Monoclonal antibodies (mAbs) are themselves powerful reagents in diagnostic applications and as tools for the investigation of macromolecules and cells (31, 32), and permit standardization and production of an unlimited supply of a reagent. This in turn may be used to set up a technique where the antibody used is of a high titre and the protocol is highly reproducible. Conversely, polyclonal reagents are limited in supply and generally of low titre, and there is likely to be between-batch variation.

6.1 Protein footprinting in epitope analysis

Hybridoma technology for the production of monoclonal antibodies has the added advantage that being epitope-specific, these antibodies in turn may be used in the study of protein structure. The technique of protein footprinting relies on the fact that the epitope is protected from proteolytic or chemical cleavage when bound as an antibody–antigen complex, and the protected epitope may then be eluted (33). Albeit data on the primary structure are not needed, ultimately knowledge of the tertiary conformation is useful if the location of the epitope on the whole molecule is required. Obviously with a larger antigen, the physicochemical interaction that stabilizes the epitope and its interaction with the antibody binding site are also not available. Much information on the interface of antigen–antibody interactions has been derived from X-ray crystallographic studies, and again, lysozyme as the model antigen in association with the monoclonal antibody has been extensively studied (34).

7 Generating monoclonal antibody

Nelson (Chapter 7) has commented on the reactivity of monoclonal antibodies to desired target antigens, and looks at how this may be applied to the development of secure assay systems and reliable functional studies. The more traditional techniques (enzymatic or chemical degradation studies) for localizing target epitope may be used in conjunction with analysing mAb reactivity in assay systems, e.g. enzyme-linked immunosorbent assay, haemagglutination, and slot-blotting. Likewise *in situ* localization of cell surface or intracellular antigenic determinants may be achieved by immunocytochemical techniques. Again it is essential to point out that mAbs, like polyclonal antisera, only bind if the epitope conformation is maintained. Conversely, if the assay system or the technique in which it is utilized, e.g. antigen denaturation in SDS–PAGE prior to Western blot analysis, modifies the epitope, then a positive antigen–antibody complex may not be detected. Evidently, other methodologies, e.g. for isolating cell membrane components and protein purification steps are required, and it is suggested that these protocols are obtained from other sources (35–37). Ultimately, epitope localization can be further refined by epitope mapping studies which specifically highlight salient residues (38). Nelson has provided information on where techniques have been applied as well as useful information on company addresses and financial costing for researchers considering developing monoclonal reagents.

8 Phage display libraries

Phage display libraries bring together two methodologies, linking the genetic and amino acid sequence of a peptide. The technique involves the cloning of oligonucleotide sequences that encode peptides via their ligation into a vector DNA whilst preserving the reading frame of the phage fusion protein. The filamentous bacteriophage then displays the gene product, i.e. peptide encoded by the oligonucleotide, as a fused protein on the bacteriophage coat protein. Accordingly, transfection of bacteriophage into a known recipient bacterial strain means distinct peptides that mimic native epitopes, i.e. 'mimotopes' are displayed. These random phage display libraries may be screened using an antibody, then the nature of the peptide elucidated by DNA sequencing to confirm the amino acid sequence of specific epitope. Such libraries have the distinct advantage that positive bacteriophage particles (with respect to antibody binding) may be amplified in bacterial culture. In addition, there is the opportunity for further analysis of individual phage clones such as phage ELISAs. Chapter 9 shows how a peptide library made up of nine random amino acids could be used to delineate the epitopes recognized by anti-β-lactoglobulin antibodies. Although the preliminary studies using phage display libraries evaluated the epitopes of monoclonal antibodies, recently, Williams *et al.* have shown how the same technique may be applied to looking for immunodominant epitopes of polyclonal antisera, and they were also able to compare their technique directly with data obtained using a Pepscan assay (synthetic peptide library on plastic pins) that they used to screen the same antiserum. Moreover, they have suggested possible uses including the generation of human antisera for use as pharmaceutical agents. Certainly biological libraries such as those with phage display have the advantage of:

(a) Being able to be propagated almost *ad infinitum*.

(b) Numerous selection cycles may be performed to purify and enhance a rare ligand.

(c) Since sequence analysis of the isolated bacteriophage particles is straightforward, the selection for binding activity also means selecting for the encoding gene.

(d) Synthetic libraries do not have the 'biological selection' bias towards a particular amino acid sequence that may occur *in vivo*. Moreover the peptides used are not individually synthesized, therefore their conformation is not affected by the rest of the bacteriopnage coat protein.

9 Carbohydrates and their significance when epitope mapping

It was Landsteiner whose interest in blood group serology led to the investigation of cell surface carbohydrate antigens (ABO blood group system) on erythrocytes, long before the structure of immunoglobulin molecules had been determined. Glycosylation of physiologically active molecules is often essential for their functionality. Oligosaccharide moieties fulfil a number of biological roles, including the protection of the core protein from proteolytic attack, induction and maintenance of the active site configuration, and may or may not decrease immunogenicity. Deglycosylation may lead to loss of antigenicity if the sugars form part of the epitope, or alternatively if the charge and/or conformation of the active epitope is modified following their removal. For instance, aglycosylated IgG has a reduced capacity for complement- and FcγR-mediated processes. Changes in IgG glycosylation have been associated with some autoimmune diseases, e.g. rheumatoid arthritis, where the terminal galactose is missing in up to 60% of IgG molecules—the so-called %$G_{(o)}$ which can be determined by lectin affinity studies (39). This methodology relies on the binding of galactose molecules to *Ricinus communis* agglutinin, and *N*-acetyl glucosamine to *Bandeiraea simplicifolia* II. The ratio of binding to these two lectins may be used to calculate the $G_{(o)}$ in IgG. Glycosylation changes in IgG have been implicated as contributing to immune complex formation. Westwood *et al.* (1994) proposed that putative epitopes that are partially masked by the oligosaccharides in normal IgG and exposed agalactosyl IgG epitopes are therefore available as targets for rheumatoid factor-binding (40).

Some physiological molecules exist naturally in glycosylated and non-glycosylated forms. Epitope mapping to assess any possible minor differences in configuration may be profitable when investigating receptor–ligand interactions and drug design, as well as (where appropriate) evaluating physiological function. Prolactin, a polypeptide hormone produced and secreted by the anterior pituitary gland and endometrium, is one such example. The glycosylated form is the major form produced by the endometrium during the menstrual cycle. The release of prolactin by the pituitary is stimulated by thyrotrophin-releasing factor, but inhibited by the drug bromocriptene, yet both have no effect on the release by the endometrium (41). Although the role of this protein is not fully understood, it is possible that minor changes in molecular structure that delineates the different isoforms may augment or diminish biological activity (42).

Broadly there are two methods available for the removal of oligosaccharides from proteins: chemical or enzymatic hydrolysis. Endoglycosidases such as neuraminidase (that hydrolyses sialic acid) may be utilized, but they are notoriously inefficient with limited activity, depending on the source of the enzyme (43). Chemical deglycosylation involves using trifluormethane sulfonic acid. This is a highly aggressive method that again is not necessarily 100% effective at removing the carbohydrate, and can also affect the integrity of the protein core (44).

In the assessment and characterization of carbohydrate moieties, the researcher may need knowledge of:

(a) The oligosaccharide sequence.

(b) The nature of the binding of carbohydrate to the protein (N-linked or O-linked).

(c) The intramolecular interactions between the protein and the carbohydrate that can in turn affect conformation and charge.

Chapter 8 gives information and protocols on how carbohydrate libraries (collections of chemically modified carbohydrates) may be employed in the epitope mapping of carbohydrate binding proteins, as well as their use as probes. In addition they have also given examples of how data may be interpreted with respect to molecular interactions.

10 Approaches to epitope mapping

The best approach to defining antigenic epitopes depends mainly on the resources available but some factors merit consideration before beginning analysis.

10.1 Polyclonal or monoclonal antibody?

In most circumstances the end-result of experimentation is to understand how whole animals or individuals are responding to antigens. In order to do this polyclonal systems require investigation. The results obtained from epitope mapping with antisera can, however, be horrendously complicated to analyse. Even supposedly non-immune sera can give multiple positive spikes when analysed in Pepscan assays. Monoclonal antibodies are much cleaner to work with but have the disadvantage that they may be non-representative of the antibody population as a whole.

10.2 Whole antigen available?

A great deal can be learned about the likely position of epitopes, in their native conformation, by carrying out initial studies to try to localize and chemically characterize principal sites. Enzymic or chemical fragmentation to identify key areas; denaturation to reveal linear versus conformational epitopes; deglycosylation to pin-point carbohydrate-dependent sites; specific amino acid modification to highlight key residues; crystallography to enable prediction from likely surface structures.

10.3 Amino acid sequence known?

Algorithmic predictions of likely epitopes can cut down the number of peptides that need to be synthesized; the full sequence then allows the power of Pepscan to be applied.

10.4 Nucleotide sequence available?

The ease and comparative cheapness of phage display makes this a favoured approach especially as the epitope is displayed in the context of a whole protein. Site-directed mutagenesis then allows the key residues in the identified epitope to be delineated.

References

1. Austen, B. M. and Westwood, O. M. R. (1991). *Protein targeting and secretion*. In Focus series, IRL Press, Oxford, New York, Tokyo.
2. Michaelsen, T. E., Wisloff, F., and Natwig, J. B. (1975). *Scand. J. Immunol.*, **4**, 71.
3. Judd, W. J., Steiner, E. A., and Capps, R. D. (1982). *Transfusion*, **22**, 185.
4. Benjamin, D. C., Berzofsky, J. A., East, I. J., *et al.* (1984). *Annu. Rev. Immunol.*, **2**, 67.

5. Berzofsky, J. A. (1985). *Science*, **229**, 932.
6. Hopp, T. P. and Woods, K. R. (1983). *Mol. Immunol.*, **20**, 483.
7. Baker, P. (2000). *Nature*, **405**, 39.
8. Bach, J. M., Otto, H., Jung, G., Cohen, H., Boitard, C., Bach, J. F., *et al.* (1998). *Eur. J. Immunol.*, **28**, 1902.
9. Willardson, B. M., Thevenin, B. J., Harrison, M. J., Kuster, W. M., Benson, M. D., and Low, P. S. (1989). *J. Biol. Chem.*, **264**, 15893.
10. Porter, R. R. (1959). *Biochem. J.*, **73**, 119.
11. Hunneyball, I. M. and Stanworth, D. R. (1975). *Immunology*, **29**, 921.
12. Ojcius, D. M., Langlade-Demoyen, P., Gachelin, G., and Kourilsky, P. (1994). *J. Immunol.*, **152**, 2798.
13. Carne, A. F. (1997). *Methods Mol. Biol.*, **64**, 271.
14. Spackman, D. H., Stein, W. H., and Moore, S. (1958). *Anal. Chem.*, **30**, 1190.
15. Gross, E. (1967). In *Methods in enzymology* Vol. 11, p. 238.
16. Suphiolu, C., Singh, M. B., and Knox, R. B. (1993). *Intl. Arch. Allergy Immunol.*, **102**, 144.
17. Jacobson, G. R., Schaffer, M. H., Stark, G. R., and Vanaman, T. C. (1973). *J. Biol. Chem.*, **248**, 6583.
18. Welling, G. W., Weijer, W. J., van der Zee, R., and Welling-Wester, S. (1985). *FEBS Lett.*, **188**, 215.
19. Artandi, S. E., Calame, K. L., Morrison, S. L., and Bonagura, V. R. (1992). *Proc. Natl. Acad. Sci. USA*, **89**, 94.
20. Jaton, J. C. and Sela, M. (1968). *J. Biol. Chem.*, **243**, 5616.
21. Geysen, H. M., Rodda, S. J., Mason, T. J., Tribbick, G., and Schoofs, P. G. (1987). *J. Immunol. Methods*, **102**, 259.
22. Lehner, T., Walker, P., Smerdon, R., Childerstone, A., Bergmeier, L. A., and Haron, J. (1990). *Arch. Oral Biol.*, **35** Suppl, 39S.
23. Mourier, G., Maillere, B., Cotton, J., Herve, M., Leroy, S., Leonettic, M., *et al.* (1994). *J. Immunol. Methods*, **171**, 65.
24. Hay, F. C. (1988). *Br. J. Rheumatol.*, **27**, 157.
25. Isenberg, D. A., Williams, W., Le Page, S., Swana, G., Feldman, R., Addison, I., *et al.* (1988). *Br. J. Rheumatol.*, **27**, 431.
26. Hunneyball, I. M. and Stanworth, D. R. (1976). *Immunology*, **30**, 881.
27. Jefferis, R. and Hodgeson, L. F. (1987). *Biochem. Soc. Trans.*, 621st Meeting, pp. 472–3.
28. Trepanier, D. J., Thibert, R. J., Draisey, T. F., and Caines, P. S. (1996). *Clin. Biochem.*, **29**, 347.
29. Benjamin, D. C. and Perdue, S. S. (1996). Methods: a companion. *Methods in enzymology*, Vol. 9, pp. 508–15.
30. McCarthy-Troke, M. H., Harrison, P. T., Campbell, I., and Allen, J. M. (1997). *Biochem. Soc. Trans.*, **25**, 361S.
31. Blottiere, H. M., Daculsi, G., Anegon, I., Pouezat, J. A., Nelson, P. N., and Passuti, N. (1995). *Biomaterials*, **16**, 497.
32. Nelson, P. N., Fletcher, S. M., De Lange, G. G., Van Leeuwen, A. M., Goodall, M., and Jefferis R. (1990). *Vox. Sang.*, **59**, 190.
33. Sheshberadaran, H. and Payne, L. G. (1989). In *Methods in enzymology* Vol. 178, pp. 746–64.
34. Atassi, M. Z. (1979). *CRC Crit. Rev. Biochem.*, **6**, 371.
35. Rose, N. R., de Macario, E. C., Folds, J. D., Lane, H. C., and Nakamura, R. M. (1997). *Manual of clinical laboratory immunology* (5th edn). American Society for Microbiology, Washington DC.
36. Johnstone, A. and Thorpe, R. (1987). *Immunochemistry in practice* (2nd edn). Blackwell Scientific Publications, Oxford.

37. Hay, F. C. and Westwood, O. M. R. (2001). *Practical immunology* (4th edn). Blackwell Science, Oxford (in press).
38. Nelson, P. N., Westwood, O. M., Jefferis, R., Goodall, M., and Hay, F. C. (1997). *Biochem. Soc. Trans.*, **25**, 373S.
39. Sumar, N., Bodman, K. B., Rademacher, T. W., Dwek, R. A., Williams, P., Parekh, R. B., *et al.* (1990). *J. Immunol. Methods*, **20**, 127.
40. Westwood, O. M. R., Soltys, A. J., Austen, B. M., and Hay, F. C. (1994). *Clin. Exp. Rheum.*, **12**, S-110.
41. Healy, D. L. (1991). *Baillieres Clin. Obstet. Gynaecol.*, **1**, 95.
42. Sinha, Y. N. and Sorenson, R. L. (1993). *Proc. Soc. Exp. Biol. Med.*, **203**, 123.
43. Fleit, H. B. and Kuhnle, M. (1988). *J. Immunol.*, **140**, 3120.
44. Edge, A. S. B., Faltynek, C. R., Liselotte, H., Reichert Jr. L. E., and Weber, P. (1981). *Anal. Biochem.*, **118**, 113.

Chapter 2
Multiple Pin Peptide Scanning ("Pepscan")

Nazira Sumar
Department of Surgery, St. George's Hospital Medical School, Cranmer Terrace, London SW17 ORE, UK.

1 Introduction

The identification of regions of interaction between an antigen and antibody is an important area of research in molecular immunology. The correct identification of epitopes not only allows one to map where the important regions of an antigen are located in its three-dimensional structure but more importantly, it is instrumental in the diagnosis and prognosis of disease, in immunointervention, and in the design of drugs and generation of vaccines where specific peptides are used to induce antibodies to pathogenic organisms.

Various methods have been used to identify epitopes, these include predictive algorithms, which identify possible epitopes, these are then synthesized and screened. One of the prerequisites to use algorithms for epitope prediction is that the amino acid sequence of the protein needs to be known and the major draw back is that the predictions are not always accurate.

Enzymatic and chemical cleavage using cyanogen bromide have also been used to generate peptide fragments which are then screened to locate epitopes on the fragments. Once the fragment is identified, it is sequenced and the epitope characterized. The amino acid sequence of the protein is not required initially for this method.

A popular method for identifying epitopes is epitope mapping using synthetic overlapping peptides spanning the entire sequence of the protein of interest.

The synthesis of peptides can be carried out using resin-based technologies where the peptide is synthesized on the resin and then cleaved. Using this conventional system of protein synthesis, milligram amounts of protein are produced but only one peptide can be synthesized at a time. The bottleneck in epitope analysis therefore arose at the peptide synthesis level. For an epitope mapping method the two considerations of importance were that the method would allow the parallel synthesis of a large number of peptides and subsequent testing of large numbers of samples.

The ability to identify epitopes speeded up enormously by a novel development in solid phase peptide synthesis and testing known as Pepscan technology developed by Geysen *et al.* (1). The Pepscan method utilizes solid phase synthesis of peptides on polystyrene pins. In this method (1), multiple peptides are synthesized essentially by Merrifield's solid phase protein synthesis method (2) on specially designed polystyrene pins in a 9 × 12, 96-well microtitre plate format which are then screened against sera or antibodies of interest using ELISA to identify linear B cell epitopes. This novel development of solid phase peptide synthesis on plastic pins as the support for the synthesis in a 96-well format has permitted easy and efficient synthesis and subsequent screening of large numbers of peptides by ELISA.

The original Geysen method (1) where the peptides were pin-bound has been adapted for T cell assays by developing chemistries that allow cleavage of peptides from the pins (3). Hence, Pepscan can be used for both B and T cell epitope mapping by its versatility in the synthesis of both pin-bound and cleaved peptides. The peptides can also be further modified to incorporate different endings, for example, biotinylated for use in solution phase B cell epitope mapping. The basic technology of the multipin peptide synthesis has also been adapted to allow multi-milligram quantities to be synthesized without sacrificing the advantages of the 96-well format (4). Several other methods of simultaneous multiple peptide synthesis have been reported (5–7).

In this chapter, I shall describe synthesis of peptides, methods for epitope mapping using pin-bound and cleaved biotinylated peptides, applications of Pepscan, use of 'T' (tea-bag) peptide synthesis, synthesis and screening of peptides on membranes and peptide libraries.

1.1 Brief outline of Pepscan

The amino acid sequence of the protein is required. The amino acid sequence of the protein is determined mostly through the DNA sequence. Linear peptides of a given length spanning the entire amino acid sequence of the protein are synthesized on pins which are attached to a plastic support. The pins are incubated with sera or antibodies of interest. The pins are then incubated with species-specific secondary antibody conjugated with enzyme. The complex is visualized using an appropriate substrate and the colour developed read in an ELISA reader. Pepscan has the potential therefore to identify linear epitopes on an antigen (8).

For T cells, peptides are synthesized on pins then cleaved and used in T cell assays for identification of Th cell epitopes and cytotoxic T cell epitopes with Th clones, cytotoxic T cell clones, and PBMC (8).

2 Solid phase peptide synthesis

Before going on to multiple synthesis on pins, I shall describe briefly peptide synthesis on solid phase. In solid phase peptide synthesis (SPPS), peptides are built on an insoluble polymeric support by the sequential addition of amino acids.

Two types of chemistry are available for peptide synthesis, these are the t-Boc and the Fmoc chemistries which use butyloxycarbonyl (t-Boc) and flurenylmethyloxycarbonyl (Fmoc) as the α-*N* amino protecting groups.

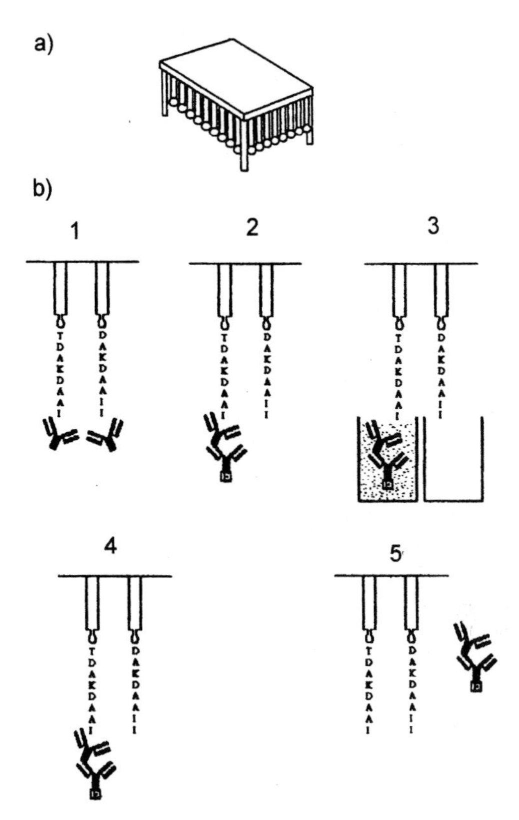

Figure 1 Pepscan assay. (a) Figure showing the plastic block with pins attached to it in a 96-well format. (b) Diagram showing the steps in the Pepscan assay using pins. Step 1: binding of primary antibody (monoclonal or polyclonal antibody or serum) to the peptide on the pin. Step 2: binding of anti-species enzyme-conjugated secondary antibody to the primary antibody. Step 3: visualization of enzyme–antibody complex by immersion of pins in substrate solution. Step 4: regeneration of pins. Removal of antibody complex from the peptide on the pin by sonication. Step 5: regenerated pins ready for the next assay.

Schematically, SPPS is represented in *Figure 2*, showing the attachment, deprotection, chain elongation, and cleavage steps in the synthesis of the peptide.

2.1 Attachment

The peptide chain is synthesized on an insoluble polymeric support by the attachment of the first Boc/Fmoc α-*N*-protected amino acid via the C-terminus. Synthesis takes place from the C-terminus to N-terminus of the peptide. The linker can vary according to whether a peptide acid or amide is being synthesized. The side chain amino acid also needs protection to prevent unwanted polymerization or side chain reactions. Various side chain protecting groups are used depending on the functional group of the amino acid and the type of chemistry used for the synthesis, i.e. Boc or Fmoc. For example, for serine, threonine, and tyrosine, the functional group being protected is -OH, the protecting group used is t-But for Fmoc chemistry, and benzyl for Boc chemistry.

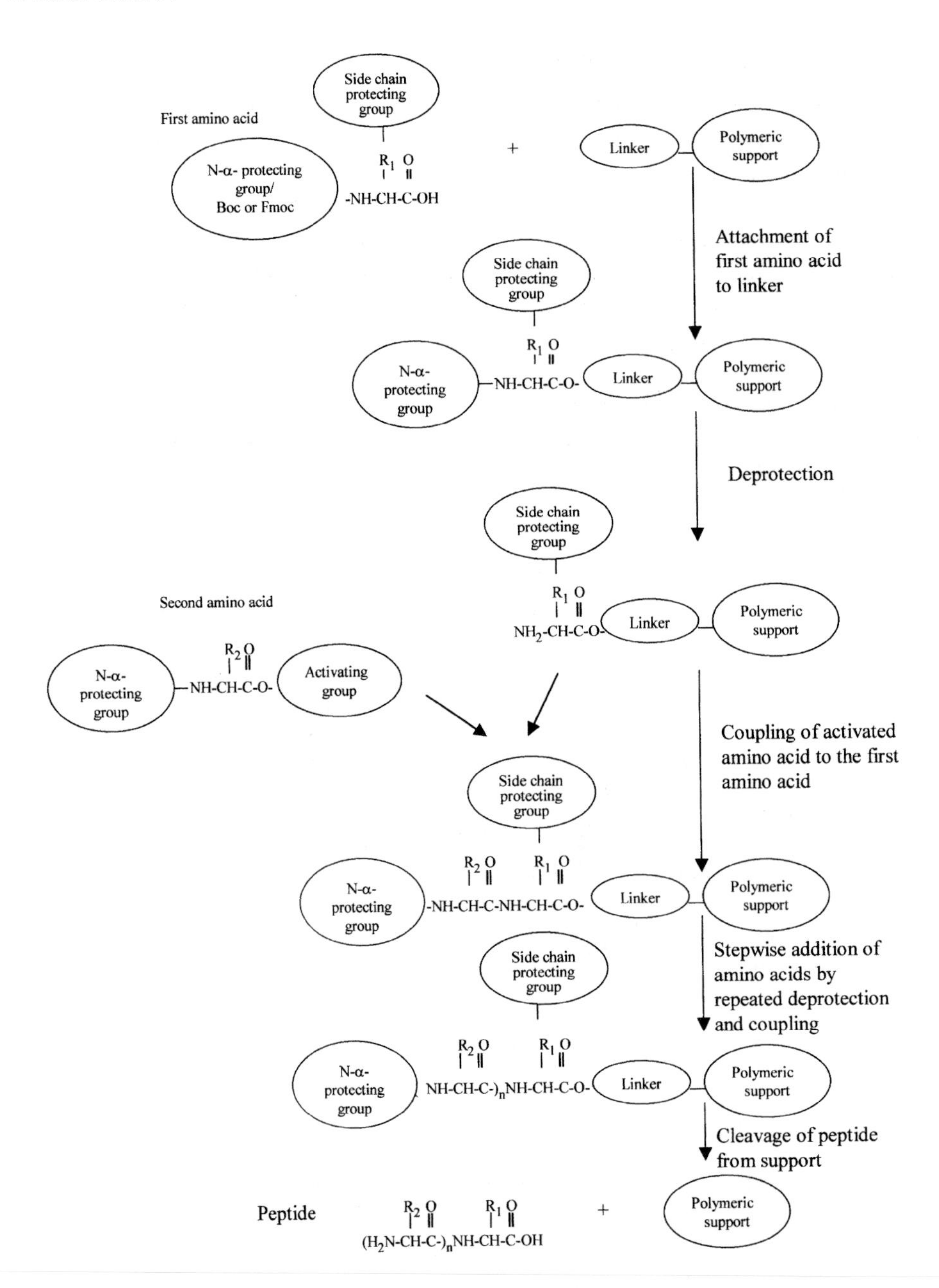

Figure 2 Diagram of solid phase peptide synthesis showing the attachment, *N*-α-deprotection, coupling, repeated deprotection, and coupling steps in peptide synthesis, and final cleavage of peptide from the support.

2.2 Deprotection and coupling

The α-N-protecting group is removed before the next amino acid is added. Cleavage of the Boc protecting group requires TFA whereas removal of the Fmoc protecting group requires piperidine.

After deprotection the next protected, activated amino acid is coupled to the first amino acid forming a peptide. The coupling process requires the incoming amino acid to be activated at the α carboxyl group so that it can react with the amino group of the growing peptide chain. For resin deprotection, coupling, and washing, dichloromethane and dimethyl formamide are used.

2.3 Elongation of peptide chain

Sequential amino acids are added by repeated deprotection and coupling steps until the required length is achieved.

2.4 Cleavage

The peptide is then cleaved from the resin. Final cleavage of the peptide from the resin and side chain deprotection for Boc chemistry requires strong acid—

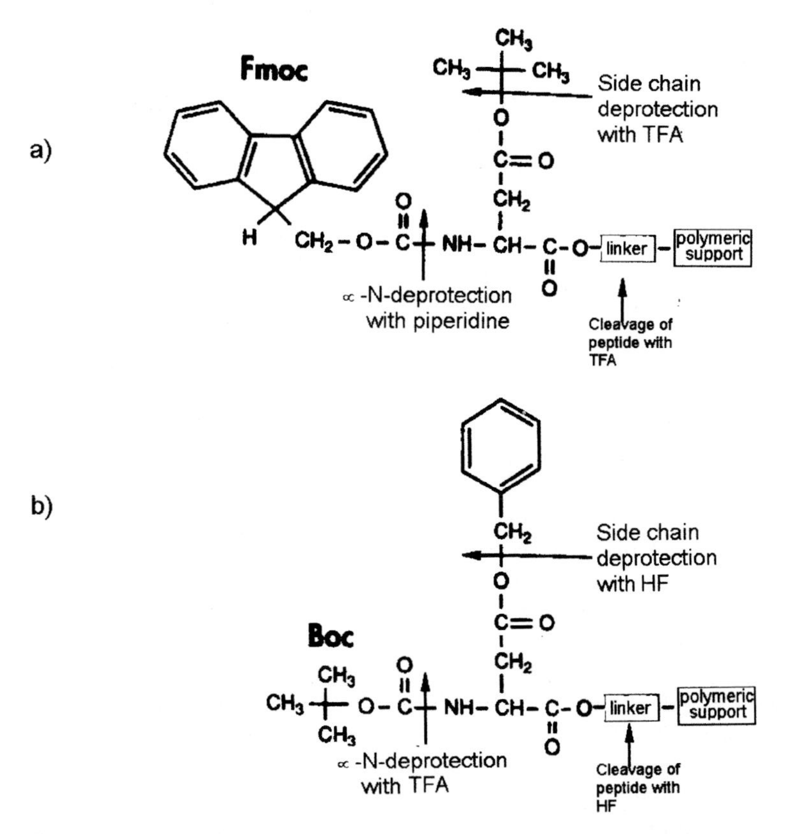

Figure 3 Cleavage of side chain and α-N-deprotecting groups and final cleavage of peptide from the support. (a) Cleavage strategies for Fmoc chemistry. (b) Cleavage strategies for Boc chemistry.

hydrofluoric acid (HF) or trifluoromethanesulfonic acid whereas Fmoc chemistry uses TFA. In general then, Fmoc chemistry is gentler in terms of use of mild base treatment during Fmoc group deprotection and TFA is only used in the final cleavage. Boc chemistry utilizes rather harsh conditions, TFA is used in the Boc group deprotection which is a repetitive cycle and during the final cleavage and deprotection of the final Boc, HF is used. Therefore, Fmoc chemistry is preferred as it uses milder conditions.

3 Multiple peptide synthesis on pins

There are some excellent protocols on multiple peptide synthesis and synthetic peptides in epitope mapping mainly by Rodda, Tribbick, Maeji, and Bray (8–12).

Custom-made peptides can also be purchased from various suppliers. Chiron Technologies Pty supply custom-made peptides or customers can purchase peptide synthesis kits with full instructions for the users to synthesize their own peptides. According to Geysen *et al.* (1) peptide synthesis on pins is simple and does not require an extensive knowledge of organic chemistry.

The kind of peptides one uses depends very much on what assay the peptides are being used for; for B cell epitope mapping, both pin-bound and cleaved peptides can be used, in the latter case it is preferable to use biotinylated cleaved peptides.

For T cell epitope mapping, cleaved peptides are used, again the peptide endings can vary depending on whether one is looking at Tc cells, Th, cells or T cell clones.

Table 1 shows the different kind of kits that are available from Chiron Technology.

In all cases, the initial synthesis is on pins and then the peptides can stay pin-bound (non-cleaved) or cleaved as shown in *Figure 3*. The pin technology uses the milder Fmoc chemistry.

3.1 Pins

The pins are polyethylene rods onto which 6% acrylic acid has been polymerized as polyacrylic acid. The pins are mounted on a plastic holder with a 8 × 12 format similar to the ELISA plates, thus facilitating not only the synthesis of multiple peptides but also the screening of peptides for epitopes.

3.2 Choice of peptide length

Identification of sequential epitopes can be achieved by using a complete set of overlapping peptides spanning the entire amino acid sequence of the antigen.

Detailed studies of antibody binding peptides have led to the general consensus that the size of sequential epitopes is between five to eight residues long (1, 13). To scan all the peptides through a protein, one needs to consider the cost and effort in making and scanning all the peptides as well as ensuring that no epitopes are missed due to insufficient overlap between successive peptides.

Table 1 Kits for peptide synthesis using the pin technology (9, 10)

Kit name	Amount of peptide	Peptide produced	Used for (10)
Non-cleavable peptides			
Non-cleavable multipin peptide synthesis kit (NCP)	50 nmol	N–peptide–linker–pin[a]	Antibody epitope scanning using pin-bound peptides
Cleavable peptides			
Cleavable peptide kit, diketopiperazine group at the C-termini (DKP)	1 μmol	N–peptide–DKP[b]	(i) Producing biotinylated peptides for antibody scanning (ii) Th cell epitope scanning
Cleavable peptide kit, glycine acid or amide at the C-termini (GAP)	1 μmol	N-peptide–glycine acid/amide[c]	Tc cell epitope scanning
Multipin multiple peptide synthesis kit (MPS)	7–8 μmol	N–peptide–acid[c]	General peptide synthesis
Multipin multiple peptide synthesis kit (MPS)	7–8 μmol	N–peptide–amide[c]	General peptide synthesis

[a] Acetyl group at the N-termini.

[b] Can have biotin group at the N-termini.

[c] Can have acetyl, biotin, or free amine at the N-termini.

One approach would be to make overlapping octapeptides stepping through the sequence one residue at a time, i.e. octamers offset by 1. Therefore for a protein of n residues, one would require the synthesis and testing of (n − 7) octapeptides (1). A protein of 400 residues would require the synthesis and screening of (400 − 7) 393 octamers.

However, as a starting point Rodda *et al.* (8) suggest that a set of overlapping peptides should not drop below 8-mers as the longest peptide length for which all possible sequences are present in the scan such that 9-mers offset by 2, 10-mers offset by 3, 11-mers offset by 4, and so on up to 20-mers offset by 13 can be used. The number of peptides that need to be synthesized is affected by the peptide length and offset, e.g. for a protein of 100 amino acids, synthesis of 8-mer peptides offset by 1 would require 93 peptides, whereas synthesis of 9-mers offset by 2 would reduce this number to 47, and for 10-mers offset by 3, the number of peptides required to be synthesized would be 31. For 11-mers, offset by 4, the number of peptides would fall to 24, approximately a quarter of those for 8-mers offset by 1 (14).

Similar considerations apply to designing a peptide set for T cell epitope mapping, i.e. peptide length, offset, quantity, and purity of peptide. Solubility of the peptides is important, hence the peptides need to be screened for hydrophobicity, this can be done using various computer programs and the peptide length adjusted to make the peptides more soluble (8).

For T cell epitopes the maximum continuous sequence length for which no sequence within the protein is missing is 9 residues or more. For Th epitopes the peptide set consists of 16-mers offset by 4, hence the number of peptides needed to be synthesized is around a quarter of the number of amino acid residues in the protein. For preliminary Tc epitope location use 12-mers and upwards, however for accurate Tc mapping, 9-mers offset by 1 are used and in this case, the number of peptides approximates the number of amino acids in the protein.

Naturally processed Tc peptides are usually 8, 9, or 10 amino acids whereas naturally processed Th peptides are between 13–18 amino acids.

Protocol 1

Peptide synthesis on pins[a]

Having decided on the type of peptide ending, order the appropriate pin synthesis kit. The synthesis kit from Chiron comes with pins, plastic baths, reaction trays, technical manual, software, pins with pre-synthesized control positive and negative peptides, and corresponding monoclonal antibody for the positive control.

Equipment and reagents

Make sure all reagents are AR or the highest quality.

- Computer
- Computer controlled display unit called Pin Aid which aids in the dispensing of reagents (Chiron or its distributors): consists of an 8 × 12 array with LED display which lights up the wells that require a particular activated amino acid solution to be added at the coupling step in peptide synthesis
- Pipettors (solvent resistant)
- Activating agent (diisopropylcarbodiimide, DIC), catalyst (1-hydroxybenzotriazole, HOBt) (Sigma, Fluka, Merck, Aldrich)
- Fmoc protected amino acids with the appropriate side chain protection (Chiron Mimotopes, Bachem, Novabiochem, Sigma)
- Acetic anhydride, diisopropylethylamine
- Piperidine for deprotection (Aldrich, Fluka, Merck)
- Solvents: dimethyl formamide (DMF) tested to have only low levels of amine, methanol (Merck, Fluka, Sigma, Aldrich)
- Indicator: 0.7% bromophenol blue in DMF (Merck)
- Side chain deprotection: TFA (Merck, Aldrich, Fluka)
- Scavengers and reducing agents: ethanedithiol (EDT), anisole, mercaptoethanol (Merck, Aldrich, Fluka)
- Biotin or long chain biotin if biotinylation of peptides if required
- Appropriate solvents for cleavage or washing dried peptide include HPLC grade acetonitrile, ether, petroleum ether

A. Peptide synthesis

1. Select appropriate kit type and design of peptide set.
2. Generate a peptide synthesis schedule by entering the peptide sequence into the kit software. This will produce a printed schedule showing:
   (a) The layout of the peptides including controls on each 'block' which holds 96 pins.
   (b) The location and amount of amino acids to be added for each coupling cycle of synthesis and the amount of catalysts and activation reagent required.

Protocol 1 continued

3 Following the synthesis schedule produced in step 2, put the required pins onto the holder. All 96 pins may not be required depending on the number of peptides being synthesized and whether the peptides are the same length or not.

B. Deprotect the pins in readiness to accept the first amino acid

1 To Fmoc deprotect the α amino acid on the pins, add 20% piperidine/DMF to the bath and place the block of pins in the bath ensuring the tips are covered. Cover and leave for 20 min at room temperature (RT).

2 After 20 min, wash the pins in DMF for 2 min, air dry for 2 min followed by a 2 min methanol wash, and air dry for 2 min.

3 Repeat methanol washes three times and air dry the block.

4 Prepare and activate the required amount of each activated Fmoc protected amino acid.

5 Put the required volume of each amino acid into the correct wells of the reaction tray, checking the computer generated synthesis print-out for the well positions. It may help if two people performed this step, one person reads the position and the other dispenses. Alternatively, use the Pin Aid if available, the appropriate well position is shown by the lit up LEDs.

C. Coupling

1 Lower the pins into the solutions in the wells of the reaction tray. Make sure the block of pins is in the right orientation. Incubate for 2 h or longer at 20–25 °C in a sealed polythene bag or polyethylene box. Coupling is complete when the blue staining from the reactive pin surface disappears.

2 Wash the pins in DMF and methanol and air dry.

3 Start the next cycle of amino acid addition. Repeat part B, and part C, steps 1 and 2 until all the peptides are synthesized.

4 Deprotect the final Fmoc amino acid by following part B and part C, steps 1 and 2. Then proceed to part F for side chain deprotection unless N-terminal acetylation or biotinylation are required. For pin-bound peptides, N-terminal acetylation is recommended (9). Carry out N-terminal acetylation according to part D and biotinylation according to part E.

D. N-terminal acetylation (9)

1 Place the pins in a mixture of DMF/acetic anhydride/diisopropylethylamine in a ratio of 193:6:1 in 200 ml.

2 Leave for 90 min at RT.

3 Wash the pins in a methanol bath and air dry.

4 Proceed to part F.

Protocol 1 continued

E. N-terminal biotinylation (9)

1. Make a 125 mM solution of biotin in DMF and activate it with 10 × concentrate solutions of activating agent (158 mg DIC in 1 ml DMF) and catalyst (192 mg HOBt in 1 ml DMF) in a ratio of 80 biotin :10 DIC:10 HOBt (by vol.).
2. Dispense appropriate amount of activated biotin (150 μl for 1 μmole scale and 450 μl for 5 μmole scale) into the wells in the reaction tray and incubate the pins (after part C) in the tray for 2 h or more.
3. Wash the pins in methanol.
4. Proceed to part F.

F. Side chain deprotection and cleavage

1. Immerse the pins in TFA/ EDT/ anisole solution (38:1:1) for 2.5 h at RT. For 1 μmole scale and 5 μmole scale use 0.3 ml and 1.5 ml respectively per pin. At this stage the side chain deprotection also simultaneously cleaves off the peptides from the pins on the MPS kits while for the non-cleavable peptides (NCP), DKP, and GAP kits, the peptides are still attached on the pins. For MPS kits go to part F, step 5.
2. For the DKP, GAP, and NCP kits, wash the pins in methanol (0.5% acetic acid in 1:1 methanol/water) for 1 h. Follow with two more washes in methanol. The NCP peptides (non-cleavable) on pins are ready for testing.
3. Cleavage of peptides from the GAP kit. Cleave the peptides directly into a rack of 1 ml polypropylene tubes. Add 0.7 ml of 0.1 M NaOH (or 0.1 M NaOH in 40% (w/v) acetonitrile to solubilize hydrophobic residues) to each tube and place the pins in the tubes for 0.5–1 h. Sonication reduces the cleavage time. Neutralize the peptides immediately after cleavage. The peptides are ready for use.
4. Cleavage of peptides from DKP kit. Cleave the peptides in a neutral or alkali buffer. Immerse each pin in 0.8 ml of cleavage buffer like 0.1 M sodium phosphate pH 7.6 or 0.05 M Hepes pH 7.6 (40% (w/v) acetonitrile/water can be included to solubilize hydrophobic residues) for 16 h or 1 h if the pins in buffer are sonicated during cleavage. After cleavage, the peptides are ready for use.
5. Further processing of cleaved peptides from the MPS kits from part F, step 1. The cleaved peptides are in TFA solution. Dry down the TFA solution containing the peptide to ~ 0.1 ml using a gentle stream of dry nitrogen in a good chemical fume hood. Extract each peptide with 8 ml of cold ether/petroleum ether/mercaptoethanol[b] in a ratio of 1:2:0.003 for 30 min. Spin at 3200 g in a flame-proof centrifuge and collect the precipitate by decanting the supernatant. Wash the precipitated peptide with 4 ml of cold 1:2 ether/petroleum ether[b] and collect precipitate as above. Dry the pellet with a gentle stream of nitrogen.

[a] Modified from ref. 8 by kind permission from Oxford University Press.

[b] Take great care as these are highly flammable.

In a conventional day three amino acid couplings can be done, making it possible to synthesize a set of 15-mer peptides in two weeks.

Once the peptides have been synthesized, the non-cleavable peptides on pins are stored dry in the refrigerator, in the presence of desiccant, the peptides are stable for months (9). Cleaved peptides, can be stored frozen in aliquots at −20 °C or lower or as lyophilized powders.

4 Testing of antibody epitopes

4.1 Pin-bound non-cleavable peptides

Pin-bound peptides have a peptide coating of around 50 nmol/pin enough for 50 tests. They can be used with monoclonal and polyclonal antibodies and allow high sensitivity of detection of antibody epitopes. However, one block of peptides can only be used for one test/day after which the pins have to be regenerated (see protocol for regeneration). The reproducibility of the assay decreases with the use of the pins and the quality of synthesis of the peptide cannot be verified easily although control peptides are included in the synthesis cycle. Rodda *et al.* (8) recommend pin-bound peptides be made in duplicate to test for reproducibility and to allow for test and control sera to be run in parallel if required. Although control peptides are synthesized to check the efficiency of synthesis and purity, the individual pin-bound peptides cannot be checked for purity, therefore once a peptide has been identified as an epitope, it should be re-synthesized by conventional methods and screened to validate the result.

Protocol 2

ELISA using pin-bound peptides[a]

Equipment and reagents

- Non-cleavable peptides on pins
- Epitope mapping kit: includes a control antibody which binds the control peptide PLAQ and does not bind the negative control peptide with the sequence GLAQ
- 96-well flat-bottom microtitre plates
- Shallow plastic baths when all the pins are incubated in the same solution
- Horizontal shaker run at 80–100 r.p.m.
- Elisa reader (e.g. Anthos, Lab systems Multiskan)
- Computer and software for storing and analysing data
- Phosphate-buffered saline (PBS)
- PBST: PBS containing 0.1% Tween 20
- PBSTA (sample diluent): PBST with 0.1% sodium azide
- Substrate solution: 50 mg ABTS (Sigma tablets, Cat. No. A-9941) in 100 ml citrate phosphate buffer (citric acid anhydrous 2.16 g and sodium phosphate 4.59 g disodium hydrogen orthophosphate 12 H_2O in 500 ml), pH 4.0 and 30 μl of 30% hydrogen peroxide
- Conjugate diluent: PBST with 1% sheep, goat, or rabbit serum (v/v), depending on the source of the antibody and 0.1% sodium caseinate (w/v)
- Anti-species horseradish peroxidase (HRP)-conjugated IgG antibody (Sigma, Dako)

Protocol 2 continued

Method

1 Check that there is no binding of conjugate to the pin-bound peptides, i.e. conjugate blank test, follow step 2 and then 5–8.

2 Blocking of the peptide pins. Fill the wells of a microtitre plate with PBST[b] (0.2 ml/well) and place the pins into the plate. Incubate the pins in the PBST for 1 h at RT. This step reduces non-specific binding to the pins.

3 Incubation with primary antibody. Dilute the test sample (monoclonal antibody, polyclonal antibody, or serum sample) in PBSTA,[c] dilute hyperimmune sera 1/5000 and patient and control sera at 1/000 to start with as it is best not to overload the pins with excess antibody. Add 0.2 ml of diluted sample per well in a microtitre plate. Take the pins out of the blocking solution, shaking off any excess, and place the pins in the sample solution in the plate. Incubate overnight at 4°C on a horizontal shaker at 100 r.p.m.

4 Wash in PBS four times.

5 Secondary antibody. Dilute the secondary antibody in conjugate diluent using the predetermined dilution arrived at in *Protocol 2a* and place 0.2 ml/well in a microtitre plate. Incubate the pins in the secondary antibody for 1 h at RT on a shaking platform at 100 r.p.m.

6 Development of colour. Wash the pins three times in PBST, and finally wash the pins once with PBS to remove traces of Tween from the pins. Place substrate solution in a microtitre plate (0.2 ml/well) and incubate the pins in the substrate for 30–60 min.

7 Read the absorbance of the coloured product in the microtitre plate at 405 nm using a reference filter at 492 nm. The reaction stops as soon as the pins are removed from the substrate, and it re-starts as soon as the pins are put back in the substrate. The pins can be reinserted in the substrate. Store the data for further analysis (Section 5).

8 Regenerate the pins (*Protocol* 3).

[a] Modified from ref. 8 by kind permission from Oxford University Press.

[b] Rodda *et al.* (8) use PBST for blocking and PBSTA for sample dilution. Worthington and Morgan (14) use PBS, 0.1% Tween containing 1% ovalbumin, and 1% BSA as blocking buffer, sera diluent, and conjugate diluent. Follow manufacturer's instructions.

[c] Avoid azide in any solutions for horseradish peroxidase as HRP is sensitive to azide.

The optimum concentration of the conjugate is determined by assaying with different concentrations of the conjugate against different concentrations of the test antibody (14). The sensitivity of the conjugate will tend to plateau with increasing concentration (12).

The antibody complexed to peptide needs to be removed from the pins before the next assay can be performed on the same set of pins.

Protocol 2a

Optimization of secondary antibody conjugate concentration[a]

Equipment and reagents

- See *Protocol 2*
- Test antibody, 0.1 M bicarbonate buffer pH 9.6

Method

1. Coat a microtitre plate with test antibody serially diluted (1/50–1/10 000) in bicarbonate buffer. Incubate at room temperature for 1 h.
2. Wash plates three times in PBST.
3. Add blocking buffer and incubate for 1 h at RT.
4. Take off the blocking buffer and add serially diluted (1/100–1/10 000) secondary antibody. Incubate for 1 h at RT.
5. Wash three times with PBST and add substrate solution. Monitor colour development and read at 405 nm with reference filter at 492 nm.
6. Using this checkered approach where both the test antibody and conjugate are serially diluted, take the dilution of the conjugate before a major drop in OD occurs at a coating antibody dilution where the absorbance is on scale as the working strength of the conjugate (14).

[a] Modified from ref. 14 by kind permission from Oxford University Press.

Protocol 3

Regeneration of pins

Equipment and reagents

- Disruption (sonication) buffer: 0.1 M sodium phosphate buffer pH 7.2, containing 0.1% 2-mercaptoethanol and 0.1% SDS
- Sonication bath (~ 7 kW power)
- Methanol, water

Method

1. Sonicate the pins in sonication buffer at 60 °C for 10 min.
2. Wash the pins in water at 60 °C and the rinse in warm methanol.
3. Air dry the pins. They are now ready for use in a further ELISA test. If not using immediately, store dry in the cold.

4.2 Biotinylated peptides

The peptides are synthesized with biotin at either the carboxy or the amino terminus, a spacer is incorporated between the biotin and the peptide. The general format of the peptide is biotin-SGSS-peptide, SGSS being the spacer. The C-terminal can be amidated or have a diketopiperazine group.

Biotinylated peptides are synthesized as cleaved peptides and stored as lyophilized powders. Once the peptides are reconstituted and diluted they should be aliquoted and frozen at −20 °C or lower to avoid degradation.

The synthesis range is 1 μmol per pin which is enough for thousands of assays. The assays are reproducible as a fresh aliquot of peptides is used each time and many samples can be processed in one day depending on the user. The sensitivity using biotinylated peptides is lower than on pins as the density of peptide on the pin is higher. The biotinylated peptides are best captured on streptavidin coated plates to ensure uniformity of peptide on the plate. Using streptavidin to capture of biotinylated peptide requires less than 1 μg/ml of biotinylated peptide.

Protocol 4

Epitope mapping with biotinylated peptide ELISA

Equipment and reagents

- Microtitre plates (Nunc, maxisorp F96 Cat. 4204-4)
- Computer for data analysis
- ELISA reader (e.g. Anthos)
- Multichannel pipette
- Biotinylated peptide set (biotin-SGSS-peptide-DKP)
- Streptavidin (Sigma, Cat. No. S-4762)
- Bovine serum albumin (BSA) (Sigma, No. A-7030)
- Blocking buffer:[a] PBS, 0.1% Tween, 2% BSA
- Sample diluent: PBS, 0.1% BSA, 0.1% azide
- Secondary antibody: anti-species HRP-conjugated IgG (Sigma)
- Substrate buffer: as in *Protocol 2* (a ready to use TMB (tetramethylbenzidine) substrate for HRP is available from various suppliers

Method

1. Coat microtitre plates with streptavidin at 5 μg/ml (100 μl/well). Leave in a 37 °C oven till the plates are dry. (At this stage, the plates can be stored dry at 4 °C for up to four weeks. Place the coated plates in a plastic bag with desiccant like silica gel and store.)
2. Wash the plates four times with PBST. When processing a lot of plates together it is better to use a well washer (e.g. Well Wash 4). When washing the plates manually, flood the plates with buffer, then flick the plates upside down to remove the buffer, and finally remove excess buffer by slapping the plates down onto a wodge of paper towels on the bench top.
3. Block the plates with blocking buffer. Dispense 200 μl/well and incubate for 1 h at RT on a horizontal shaker.

Protocol 4 continued

4. Wash the plates four times as in step 2.
5. Reconstitute the biotinylated peptides in 200 μl of a pure solvent like DMSO or DMF or solvent/water mixture. Dilute the required amount of peptides 1/1000 in PBS/BSA/azide. The amount of diluted peptide should be about 1 μg/ml. A 1/5000 dilution of peptides (0.2 μg/ml) can be used but a loss in sensitivity may occur. The dilutions are best done in numbered racked polypropylene tubes from Bio-Rad (Cat. No. 223-9390). The racks are numbered in a 9 × 12 ELISA format and can be racked for use with multichannel pipettes which makes it easy to dispense the diluted peptides into the corresponding wells in step 6. Aliquot the rest of the peptide solutions and freeze at −20 °C or lower. Again in our experience, the reconstituted aliquoted peptides are best stored in the racked numbered tubes thus facilitating dispensing for further dilutions.
6. Dispense 100 μl of diluted biotinylated peptides in duplicate into the corresponding wells and incubate for 1 h at RT on a horizontal shaker.
7. Wash the plates four times as in step 2. At this stage, the peptided plates can be used immediately or dried at 37 °C and stored dry at 4 °C. Several batches of plates can be prepared and stored for later use.
8. Dilute the serum/antibody to be tested in sample diluent, the dilution of the sample varies depending on the sample and the amount of antibody present. For sera from hyperimmunized animals and ascitic fluid from hybridomas start with 1/1000 while for human convalescent and control sera start with 1/500.
9. Add 100 μl/well of diluted sample in duplicate. Include a conjugate blank plate, this is where the plate is processed in the same way as the sample plate but the sample is omitted and instead at this stage sample diluent added to the peptide plate. Incubate for 1 h at RT on a horizontal shaker, sensitivity can be increased by incubating overnight at 4 °C.
10. Wash the plates as in step 2.
11. Add 100 μl/well of conjugate solution diluted in blocking buffer. Incubate the plates for 1 h at RT on a horizontal shaker.
12. Wash the plates as in step 2. Then wash the plates twice with PBS to remove the last traces of Tween.
13. Prepare the substrate solution, and dispense 100 μl/well. Incubate at RT for up to 45 min. Read at 405 nm using a reference filter at 492 nm. If using the ready to use TMB substrate solution, decant the required amount of solution in a clean container and bring it to room temperature, keeping it in the dark until use. Dispense 100 μl/well and incubate the plate for up to 45 min at RT. A blue colour will develop. Stop the reaction with 0.5 M sulfuric acid and read the bright yellow colour produced at 450 nm with 620 nm as a reference.

[a] This blocking buffer is used normally, however Rodda *et al.* (8) recommend the use of sodium caseinate as an effective blocker (1% sodium caseinate/0.1% Tween in PBS).

5 Results and data analysis

Figure 4 shows the scan for a polyclonal rabbit antiserum raised against a peptide sequence of p43, a 43 kDa protein from *Mycobacterium paratuberculosis*. The antiserum was screened against 15-mer biotinylated peptides spanning the entire sequence of p43. The antiserum used at 1/1000 dilution showed reactivity with the peptide sequence it was raised against. A parallel run using pre-immune sera was run. The same biotinylated peptides were screened against sera from Crohn's patients (*Figures 5a, 5b*) and non-convalescent sera from normal healthy controls (*Figure 5c*). A peptide sequence at the carboxy terminus of p43 was identified in Crohn's sera by this method (15).

Since not all sequences in the protein are identified as epitopes, the non-reactive sequences can act as control peptides. However, once an epitope is identified, it is better to synthesize a scrambled sequence of the epitope and test it against the sera of interest. Serum samples tend to give higher backgrounds compared to monoclonal and peptide antibodies and the interpretation of data can be difficult in terms of what constitutes an epitope and where the background cut-off should lie. A test background can be established by ranking the absorbance values and averaging the lowest 25% of the values and then adding three times the standard deviation (1, 14). All values above this cut-off are viewed as significant. This does not always work with serum samples as quite high back-

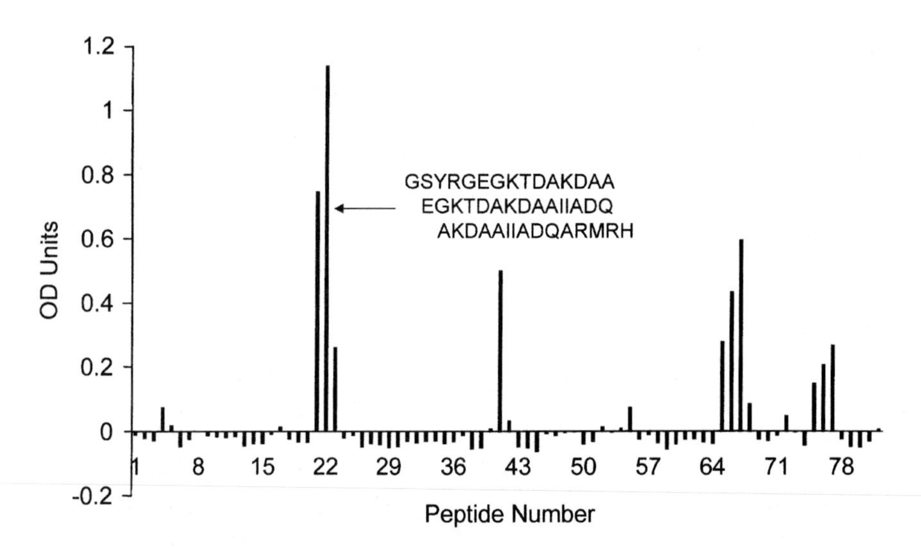

Figure 4 Scan showing the binding of an anti-peptide antibody to 15-mer biotinylated peptides spanning the entire sequence of protein p43 from *Mycobacterium paratuberculosis*. The peptide number is on the horizontal scale while the binding of the antibody is represented on the vertical scale in optical density units (OD) at 405 nm. The antibody was raised to a 18-mer peptide with the sequence, YRGEGKTDAKDAAIIDQ, binding occurs at overlapping peptides 23, 24, and 25. The maximum binding is at peptide 24 which has most of the peptide sequence to which the antibody has been raised.

grounds are often obtained with some sera (cf. *Figures 5a* and *5b*). With serum samples, data can be analysed by plotting a consensus plot for the sera analysed.

(a) The geometric mean antibody titre and range are plotted for the peptides binding antibodies from the sera tested.

(b) A frequency profile is plotted to show the number of sera reacting with each peptide.

(c) The reactivity of each individual serum to each peptide is plotted. In this way, epitopes masked by high background can be revealed (16).

One way to confirm binding to the sequential epitope identified is to elute the serum antibody bound to the pin by using buffers of low or high pH and testing the eluted antibody for specific binding to the peptide by standard ELISA. Tribbick *et al.* (17) have used this elution method to investigate the specificity of many antibodies in polyclonal serum.

6 Further analysis

An epitope identified by a primary scan using overlapping peptides of constant length called General Net, can be further characterized by Window net and Replacement net analysis (*Figure* 6).

6.1 Window net analysis

This allows the minimal (optimal) antibody binding peptide to be defined. Peptides of different lengths representing the epitope are synthesized, the sera re-tested against the peptides, and the minimum length recognized by the antibody defined (1). Amino acid additions and deletions should be carried out at both the amino and carboxy termini.

6.2 Replacement net analysis

Replacement net analysis can be carried out to determine the amino acid residues critical to the binding. This is done by synthesizing a set of peptides comprising the core binding fragment, with each amino acid replaced in turn by the 19 amino acids. Both D- and L-amino acids can be used for replacement. A critical residue is one whose replacement abrogates binding. If binding is reduced or lost compared to the parent residue, then the parent residue is considered to contribute directly to the binding. There are many studies in the literature highlighting the effects of single amino acids changes in the epitope (18–21). Geysen *et al.* (22) using the Pepscan method identified an immunogenic epitope of foot and mouth disease virus to a resolution of a single amino acid. In this study, 205 overlapping hexapeptides were synthesized covering the total 213 amino acid sequence of VP1 of foot and mouth disease virus. Antigenic profiles were plotted

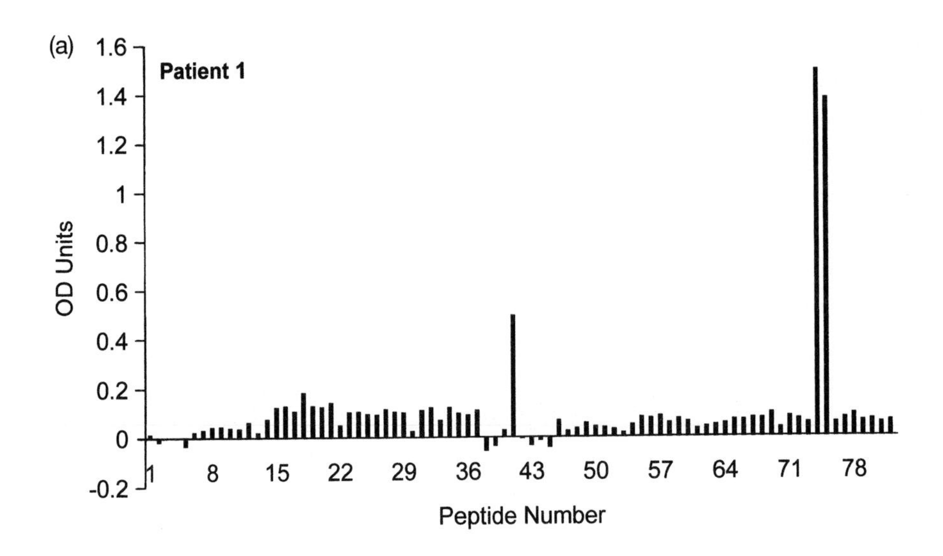
(a)
Patient 1
OD Units
1.6
1.4
1.2
1
0.8
0.6
0.4
0.2
0
-0.2
1
8
15
22
29
36
43
50
57
64
71
78
Peptide Number

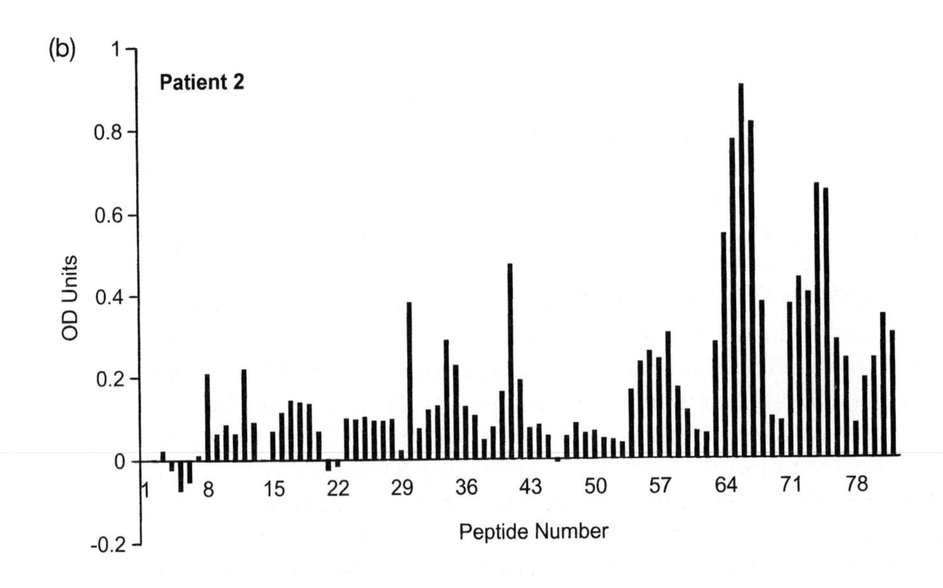
(b)
Patient 2
OD Units
1
0.8
0.6
0.4
0.2
0
-0.2
1
8
15
22
29
36
43
50
57
64
71
78
Peptide Number

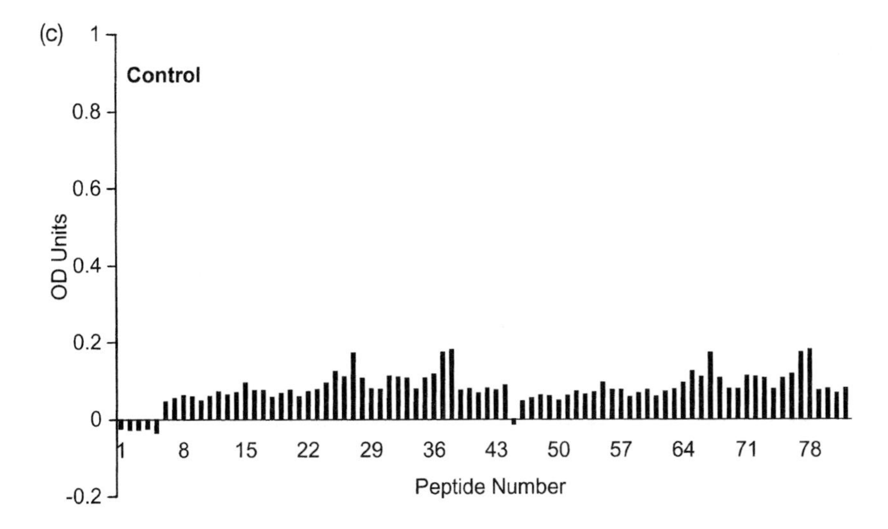

Figure 5 (see ref. 15) Scan showing the binding of two sera from Crohn's patients and one from a normal control to 15-mer biotinylated peptides spanning the entire sequence of protein p43 from *Mycobacterium paratuberculosis.* The peptide number is on the horizontal scale while the binding of the antibody is represented on the vertical scale in optical density units (OD) at 405 nm. (a) Scan of serum from patient 1, showing low background and a clear binding to two carboxy terminal epitopes. (b) Scan of serum from patient 2, showing a much higher background and less defined epitopes. (c) Scan of serum from a normal healthy control.

as peptide number against absorbance (OD). An antigenic peptide was defined as one giving an OD significantly above background level of the test. Peptides 146 (GDLQVL) and 147 (DLQVLA) were identified as epitopes. To further determine whether the epitope was the common sequence DLQVL or GDLQVLA a **replacement net analysis** was carried out. Each of the 6 amino acids in the peptide sequence 146 was substituted in turn by the19 common amino acids. Analysis of binding was carried out with the 120 hexapeptides and the particular amino acids critical for antibody binding were determined. In the testing of replacement nets, the binding was expressed relative to the binding of the parent peptide. Leucine at positions 148 and 151 essential, while glutamine and alanine at 149 and152 were not so important. The contribution of alanine of peptide 147 to the binding was also investigated by synthesizing 20 peptides with the complete sequence of peptide 146 with one of the 19 amino acids added to the carboxy terminus of the peptide. This detailed analysis demonstrates the identification of an epitope at high resolution. Similarly Geysen *et al.* (23) have demonstrated the use of window net and replacement net analysis for the model protein myohaemerythrin. Pepscan analysis has led to a vast number of publications on the identification of epitopes, all of which are too numerous to list. Some of the Pepscan application are given in Section 7.

Identification of epitopes within a protein region.

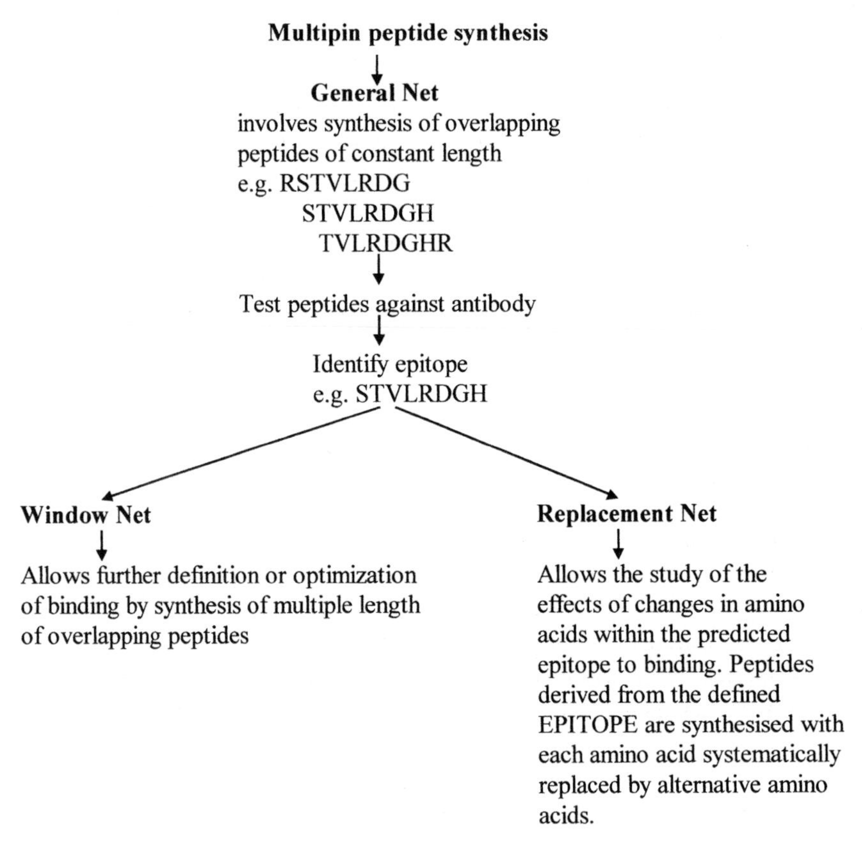

Figure 6 Identification of epitopes within a protein region.

7 Applications of Pepscan

Pepscan has been used in a wide range of applications from the study of antigen–antibody interactions (18, 22) to identification of epitopes on monoclonal antibodies (23–25) and in the study of autoantibodies (19, 26), to detection of epitopes on toxins (20, 21), viruses (27–34), bacteria (35–39), parasites (40), tumour suppressor protein (41), enzymes (42, 43), hormones (44, 45), in the design of vaccines (46–52), and even used for the location of the epitopes/sweet taste determinants (active site) of sweet tasting protein thaumatin (53). Rodda (18) studied the antibody response to myoglobin using Pepscan. A hexapeptide epitope was identified and detailed analysis of the epitope carried out by replacement net analysis showed the N-terminal leucine to be crucial to the binding of antibody to the peptide.

Using the Pepscan method, Koshy *et al.* (51) studied the binding of monoclonal antibody to chicken riboflavin carrier protein (RCP). A novel neutralization site conserved in all known bovine respiratory syncytial virus (BRSV) and human RSV strains was identified by Pepscan analysis of monoclonal antibodies raised against fusion protein complex of BRSV. This linear conserved epitope may be a potential candidate for a peptide based vaccine which can produce neutralizing antibodies to all groups and subgroups of RSV (52).

Cross-reactive idiotypes have been detected on anti-DNA autoantibodies in SLE. Autoantibodies associated with SLE include those that bind to SmB/B' peptides. Overlapping octapeptides of SmB/B' were tested by Pepscan analysis against sera from normal and SLE patients and an octapeptide sequence identified (18). Using a set of overlapping peptides representing the VH and VL regions of monoclonal antibodies derived from mouse models of SLE and reacting these against normal and lupus mice sera, a mosaic of antibody V-region idiotypes were identified (26).

Pepscan has been used to identify a major B cell epitope within the immunodominant nucleoprotein amino subregion of the hepatitis C virus nucleocapsid protein (23).

Epitope mapping of scorpion neurotoxin II from *Androctonus australis* by the Pepscan method identified a new antigenic region for AaH II antibodies, fine analysis showed residues lysine, aspartic acid, and glycine to be important in binding (20).

Using the Pepscan method, the binding of α-bungarotoxin to Torpedo acetylcholine receptor was studied. The epitopes of 12 MCAs against the cytoplasmic side of acetylcholine receptor (AChR) α-subunit were mapped using 300 sequential peptides, a very immunogenic cytoplasmic epitope (VICE-α) was identified (25).

Epitope characterization of monoclonal antibodies against mucin protein core identified a short hydrophilic region in the MUC 1 mucin core as immunodominant in the induction of antibodies (54).

Epitope mapping of a novel fimbrial protein, Pg II from *Porphyromonas gingivalis* revealed seven immunodominant regions which reacted with sera from patients with periodontal diseases (39).

Interesting information was obtained by Pepscan analysis of horseradish peroxidase (HRP isoenzyme C). Epitopes detected in loops and folds of the HRP peptide chain with irregular shapes contained functionally important residues like Arg-38 of the active site of the enzyme and Phe-142 and 143 which form a channel allowing aromatic residues to reach the active site while amino acid residues which formed calcium binding sites did not form part of the epitope (43).

An amino acid sequence (PLITHVLPFEKINE) from alcohol dehydrogenase was studied and a pentamer (HVLPF) binding to a monoclonal antibody identified (42). Potential vaccine candidates have been elucidated by Pepscan studies on the glycoprotein E of *Varicella zoster* virus (47).

Pepscan has also provided the methodology for the analysis of humoral responses induced by vaccination (48, 49) and the study of the fine-specificity of antibodies induced by HIV candidate vaccines (50). Pepscan provides a methodology for the design, evaluation, and selection of candidate vaccines.

8 Other systems for epitope analysis

8.1 Simple precision original test system (SPOTs)

Another method for multiple peptide synthesis and epitope analysis is SPOTs (14) available from Genoysis. In this system, up to 96 peptides are synthesized in parallel using the Fmoc chemistry on a derivatized membrane (instead of pins) and the membrane probed with antibody. The SPOTs has an 8 × 12 cm format with 96 spots allowing 96 peptides to be simultaneously synthesized (one on each spot). The membrane is pre-activated ready for the first amino acid to be added, coupling takes 15–30 minutes, after deprotection the second amino acid is added, each complete cycle taking 90 minutes. Up to a maximum of 15 cycles can be performed.

The membrane with the completed peptides can be probed with antibody. The membrane is blocked, incubated with test antibody, washed, and incubated with species-specific enzyme-conjugated secondary antibody, followed by a wash, and finally incubated with enzyme substrate. Peptide epitopes recognized by the antibody light up as spots on the membrane. As with the pins, the membrane can be cleaned and re-probed allowing for optimization of antibody dilutions. In the SPOTs the peptides can also be cleaved off the membrane, lyophilized, and used in different assays. The coupling efficiencies of three different coupling methods have also been studied by Molina *et al.* (55).

8.2 Tea-bag synthesis

An alternative to Pepscan is the use of T bag synthesis where multiple peptides are simultaneously synthesized on resin (5) or on paper discs (7) as the solid phase contained in separate solvent permeable packets similar to tea-bags. Various methods for multiple peptide synthesis on solid phase have been described. Houghten's T bag synthesis (5) method followed shortly after the peptide synthesis method on pins, this was then followed by synthesis on paper (7, 56), cotton (57–59). In 1990, Krachnak *et al.* (60) described solid phase multiple peptide synthesis on paper discs followed by determination of antibody binding to the disc-bound peptides.

8.2.1 'T' bag (tea-bag) synthesis using resin

In the 'T' bag synthesis method described by Houghten (5), resin is used as the solid phase. The method for peptide synthesis is essentially that described by Merrifield (2) and all resins commonly used for SPPS can be employed.

50–100 mg (0.2–0.8 meq/g) of standard Boc amino acid resin is placed into 15 × 20 mm polypropylene (mesh size 74 μm) pouches. Each of the bags is numbered inside with indelible black ink and the bag permanently sealed. This ensures that each bag is easily identifiable. The bags are now ready for use for simultaneous synthesis of a large number of different peptides as well as for concurrent synthesis of multiple analogues of an individual peptide. More than 100 peptides can be synthesized at a time with quantities of more than 10 mg each.

All the bags are washed, deprotected, and neutralized in a reaction vessel. The bags are then taken out and reacted with individual solutions of protected amino acids for coupling. All the bags requiring a certain amino acid can be processed in the same vessel. After coupling, the bags are put in the reaction vessel for washing, deprotection, and neutralization ready for the addition of the next amino acid. Additional cycles of the washing, deprotection, neutralization, and coupling are carried out until the synthesis is complete. Any variations, e.g. single residue replacement/omissions, shorter chain length can be carried out by removing the bag at the appropriate step completion, modified separately, and added back to the reaction vessel for completion.

After completion of synthesis, protected peptide resins still within the bags are cleaved and the peptides extracted from the resins. The average purity of the peptides obtained was 84% (70–94%). The purity of the peptides synthesized by this method was found to be good or better than the purity of peptides prepared by free resin synthesis.

Using this method a series of 247 replacement analogues of an antigenic peptide sequence (aa 98–110) of influenza HA1 and 13 controls were synthesized (in two weeks). These peptides were probed with monoclonal antibody using a standard ELISA to determine the importance of individual amino acids in antibody binding. This 'T' bag synthesis approach has been used for synthesizing synthetic peptide combinatorial libraries (61, 62) using MBHA resin, t-Boc chemistry, and polypropylene mesh pouches (see Section 8.3.2).

8.2.2 T bag synthesis using paper discs

In this method (7) peptides were simultaneously synthesized on paper discs using Fmoc chemistry. Discs of filter paper (Schleicher and Schull) with 0.6 cm diameter were punched out, activated, and derivatized to get free amino groups on them. 100 of these derivatized discs (200 mg) were then placed in T bags made of 75 μm polypropylene mesh, the bags indelibly marked, and sealed. N-terminal Fmoc protected and side chain protected amino acid was then coupled on the disc, followed by side chain deprotection and washing. Cycles of coupling, deprotection, and washing were carried out until the synthesis was complete. Finally side chain deprotection was carried out. The peptides were ready to be tested for epitope mapping in solid phase immunological procedures without detachment from the paper discs. This method allows the simultaneous synthesis and subsequent immunological testing of large numbers of peptides.

Using this method, epitope mapping was carried out on the feline major allergen Fel d I. A total of 15 000 paper disc-bound peptides (146 nonapeptides overlapping by 8 amino acid residues) were synthesized simultaneously with 100 discs per bag, thereby permitting 100 epitope mapping tests. The tests were carried out using radioimmunoassay and the method was compared to Pepscan where the same peptides were synthesized and binding of each serum tested on pins. Results of paper disc RIA and pin ELISA showed binding of peptides 41–49 to 45–53 and peptides 43–51 and 44–52 in chain 2 of Fel d I respectively.

Studies on the structural requirement for ligand binding to the neuropeptide Y receptor from rat cerebral cortex have been carried out using analogues of the neuropeptide Y model peptide synthesized by the 'T' bag method (63). The T bag peptide synthesis approach using Fmoc chemistry and simultaneous multiple peptide synthesis has been used for antigenic mapping of viral proteins (64). The role of individual amino acids in binding human and macaque antibodies was determined in the human immunodeficiency virus type 1 (HIV-1) gp 41, residues 594–613. Using decapeptides with 9 amino acid overlap, amino acids 599–603 were found to be the main recognition site for 19 human anti-HIV positive sera.

8.3 Peptide epitope libraries

Epitope libraries provide a method for identifying peptide ligands for antibodies, receptors, or other binding proteins and provide a tool to rapidly identify lead ligands in the drug discovery process (65).

An alternative to mapping epitopes via the antigen is to employ combinatorial methods whereby a library of peptides is generated. The library of peptides can be generated by chemical synthesis (1, 61, 62, 66, 67) or biologically by phage display systems (68). Only chemically synthesized peptide libraries will be dealt with here. The library is a vast collection of all theoretically possible peptides consisting of 4, 5, 6, and so on amino acid residues where the total number of variations is calculated as 20^n, n being the number of residues and 20 is the number of optional amino acids. Therefore to synthesize a library of pentapeptides, the total number of variants (i.e. complexity of the library) would be $20^5 = 3.2 \times 10^6$ peptides whereas for a library of hexapeptides the total number of peptides would be $20^6 = 6.4 \times 10^7$ (69).

8.3.1 Peptide libraries on pins

Geysen *et al.* (22, 70), using replacement net analysis showed that for significant binding to an antibody, three amino acid residues within the peptide sequence should have both the correct identity and position and at least two of the three amino acids should be adjacent to one another. If two amino acid positions are fixed, 400 peptide mixtures would be required to synthesize all possible hexapeptides. These 400 hexapeptide mixtures are assayed for binding by a monoclonal antibody. The best binding peptide mixture is identified and subsequent re-synthesis and screening of the peptide mixture with first pair of amino acids extended to three defined residues, followed by four, and so on. Geysen *et al.* (71) using this strategy concluded that what is important is that the peptide should have complementarity between antigen binding site and the surface of the antibody with respect to shape and charge. This led to the term 'Mimotope' which is defined as the optimum binding peptide that mimics the binding of the epitope without necessarily bearing any relationship to the primary sequence. Elucidation of such a binding peptide can be achieved by:

(a) Synthesizing peptide mixtures, e.g. libraries as described above or in Section 8.3.2.

(b) Starting with a dipeptide and building a defined peptide on it, the 'dipeptide strategy' (Section 8.3.3).

8.3.2 Combinatorial libraries using 'T' bags

Houghten *et al.* (61) have developed synthetic peptide combinatorial libraries (which uses an iterative selection and enhancement process to define the most active sequence) composed of free peptides in quantities which can be used in virtually all existing assays). This method essentially uses the same principle as the Geysen *et al.* (71) library approach except that the synthesis is carried out in T bags as described in Section 8.2.1 and the peptides can be used in solution. There is no limitation to the number of peptides that can be synthesized. A hexapeptide library (with N-terminal acetylation and C-terminal amidation) was synthesized, starting with the first two amino acids positions defined in each peptide and the last four positions of equimolar ratios of 18 of the 20 natural L-amino acids (cysteine and tryptophan were left out in the initial library for ease of synthesis). The initial starting sequence can be represented as $Ac\text{-}O_1O_2XXXX\text{-}NH_2$, where O_1 and O_2 are the defined amino acids and X is the equimolar mixture of the 18 amino acids used, hence 324 hexapeptides mixtures were synthesized in the first round. In this method the assay used for testing was the ability of the peptide mixtures to inhibit of binding of a monoclonal antibody to a 13-residue peptide. The optimum defined amino acid mixture that gave maximum inhibition was noted and new peptide mixtures were synthesized with the third amino acid defined (at this stage tryptophan was included in the X position). The iterative process was carried out for the three remaining positions. Using this method a library composed of over 34 million hexapeptides was synthesized and used to identify an antigenic determinant of a monoclonal antibody. The same approach was used in the development of new antimicrobial peptides. A conceptually different approach is the positional scanning synthetic peptide combinatorial library (PS-SPCL) (62) where ten positional decapeptide libraries were synthesized. Each of the ten decapeptide positional peptide libraries is made up of 20 peptide mixtures with a **single** amino acid position defined and the other 9 positions composed of the 18 amino acid mix as described above.

8.3.3 Dipeptide net

In this method, all possible dipeptides are synthesized, each on a separate support, and screened against the antibody. The sequence of the dipeptide that binds best is then extended by re-synthesizing 20 analogues of the dipeptide with only one of the 20 amino acids for the third residue, then taking the best of the tripeptide and synthesizing its analogues with one of the 20 amino acids as the fourth residue and so on. This eventually leads to the identification of an optimum binding peptide. There is a possibility with this approach that a low binding dipeptide which is discarded might actually be as a high binder in a longer peptide.

8.3.4 Peptide libraries on beads

Lam *et al.* (72) have described a method for the synthesis of a peptide library consisting of a set of all possible peptides on resin beads using a 'one-bead, one-peptide' approach. This method involves the synthesis of a large library comprising millions of beads, each bead containing a single representing the universe of possible random peptides in roughly equimolar ratios. Different amino acids have different coupling rates and therefore use of a random mixture of amino acids in a peptide synthesis protocol would lead to unequal representation. To circumvent this, a 'split synthesis' approach was used. A pool of resin beads was distributed into separate reaction vessels containing a single amino acid. This first amino acid was coupled to the resin and after the first coupling cycle, the resins were pooled, split, and distributed into reaction vessels. The process of coupling the second amino acid was carried out. This randomizing and splitting process was continued till the required length of peptides was reached. Using this method, each bead should only contain a single peptide species. Standard solid phase Boc and Fmoc chemistries are applied.

For tripeptides starting with three reaction vessels and three cycles of coupling, 27 tripeptides each with a different sequence would be produced. On a larger scale, starting with 19 reaction vessels and synthesis of pentapeptides, would produce a library of up to 19^5, i.e. 2 476 099 individual peptides of differing sequence. Such a synthesis would take a few days.

Screening of the peptide library, which may seem a daunting task was quite cleverly achieved. The peptide beads were put in a Petri dish and reacted with enzyme labelled or fluorescein labelled acceptor molecule. Where there was binding of the acceptor to the peptide bead, staining was visible which could be easily visualized under a low power microscope. The reactive beads were then removed and microsequenced after removal of the acceptor molecule. Screening of a library of several million beads could be accomplished in an afternoon using 10–15 Petri dishes.

This method has been used to study the binding of a monoclonal antibody to β endorphin specific for a pentapeptide YGGFL. Six reactive beads were retrieved from two million beads screened from the pentapeptide library, the affinities of the ligands were better than those obtained by a phage library method. Using the same pentapeptide library, the peptides binding sequence for streptavidin was found to be have a consensus sequence of HPQ (72).

This method of synthesis and screening of peptide libraries is simple and identifies ligands with affinities virtually identical to the natural ligand. Both D- and L-amino acids can be used and most importantly the peptides sequences on the beads do not need to be predetermined as only those beads that light up need to be sequenced.

Acknowledgements

I would like to thank Barry Hodson of Meltek scientific (Feltham, UK) for his help, Chiron Mimotopes Pty for permission to use information from their technical manual, and Oxford University Press for permission to use information from their books. Thanks also to Natasha Sumar for artwork.

References

1. Geysen, H. M., Rodda, S. J., Mason, T. J., Tribbick, G., and Schoofs, P. G. (1987). *J. Immunol. Methods*, **102**, 259.
2. Merrifield, R. B. (1963). *J. Am. Chem. Soc.*, **85**, 2149.
3. Maeji, N. J., Bray, A. M., and Geysen, H. M. (1990). *J. Immunol. Methods*, **134**, 23.
4. Maeji, N. J., Bray, A. M., Valerio, R. M., and Wang, W. (1995). *Peptide Res.*, **8**, 33.
5. Houghten, R. A. (1985). *Proc. Natl. Acad. Sci. USA*, **82**, 5131.
6. Schnorrenberg, G. and Gerhart, H. (1989). *Tetrahedron*, **45**, 7759.
7. van't Hof, W., Van den Berg, M., and Alaberse, R. C. (1993). *J. Immunol. Methods*, **161**, 177.
8. Rodda, S. J., Tribbick, G., and Maeji, N. J. (1997). In *Immunochemistry: a practical approach* (ed. A. Johnstone and M. Turner), p. 121. Oxford University Press, Oxford.
9. Rodda, S. J. (1997). In *Current protocols in immunology*, suppl. 22, p. 9.7.1. John Wiley and Sons.
10. Bray, A. (1997). In *Immunology methods manual*, p. 809. Academic Press Ltd.
11. Rodda, S. J., Maeji, N. J., and Tribbick, G. (1996). In *Methods in molecular biology* (ed. G. E. Morris), Vol. 66, p. 137. Humana Press.
12. Tribbick, G. (1997). In *Immunology methods manual*, p. 817. Academic Press Ltd.
13. Kabat, E. A. (1970). *Ann. N. Y. Acad. Sci.*, **169**, 43.
14. Worthington, J. and Morgan, K. (1994). In *Peptide antigens: a practical approach* (ed. B. Wisdom), p. 181. Oxford University Press, Oxford.
15. Sumar, N., Tizard, M., Doran, T., Austen, B. M., and Hermon-Taylor, J. (1994). *Proceedings of the fourth international colloquium on paratuberculosis* (ed. R. J. Chiodini, M. T. Collins, and E. O. Bassey).
16. Rodda, S. (ed.) (1992). *Pinnacles*, Vol. 2, No. 1, p. 1. Chiron Mimotopes.
17. Tribbick, G., Triantafyllou, B., Lauricella, R., Rodda, S. J., Mason, T. J., and Geysen, M. H. (1991). *J. Immunol. Methods*, **139**, 155.
18. Rodda, S. J., Geysen, M. H., Mason, T. J., and Schoofs, P. G. (1986). *Mol. Immunol.*, **23**, 603.
19. James, J. and Harley, J. B. (1992). *J. Immunol.*, **148**, 2074.
20. Devaux, C., Mansuelle, P., and Granier, C. (1993). *Mol. Immunol.*, **30**, 1061.
21. Tzartos, S. J. and Remoundos, M. S. (1990). *J. Biol. Chem.*, **265**, 21462.
22. Geysen, M. H., Meleon, R. H., and Barteling, S. J. (1984). *Proc. Natl. Acad. Sci. USA*, **81**, 3998.
23. Geysen, H. M., Tainer, J. A., Rodda, S. J., Mason, T. J., Alexander, H., Getzoff, E. D., *et al.* (1987). *Science*, **235**, 1184.
24. Cerino, A., Boender, P., La Monica, N., Rosa, C., Habets, W., and Mondelli, M. U. (1993). *J. Immunol.*, **151**, 7005.
25. Tzartos, S. J. and Remoundos, M. S. (1992). *Eur. J. Biochem.*, **207**, 915.
26. Ward, F. J., Knies, J. E., Cunningham, C., Harris, W. J., and Staines, N. A. (1997). *Immunology*, **92**, 354.
27. Estepa, A. and Coll, J. M. (1996). *Virology*, **216**, 60.

28. Chassot, S., Lambert, A. K., Godinot, C., Roux, B., Trepo, C., and Cova, L. (1994). *Virology*, **200**, 72.
29. Coursaget, P., Lesage, G., Le, C. P., Mayelo, V., and Bourdil, C. (1991). *Res. Virol.*, **142**, 461.
30. Kohli, E., Maurice, L., Bourgeois, C., Bour, J. B., and Pothier, P. (1993). *Virology*, **194**, 110.
31. Lombardi, S., Garzelli, C., La, R. C., Zaccaro, L., Specter, S., Malvadi, G., *et al.* (1993). *J. Virol.*, **67**, 4742.
32. Pereira, L. G., Torrance, L., Roberts, I. M., and Harrison, B. D. (1994). *Virology*, **203**, 277.
33. Torrance, L. (1992). *Virology*, **191**, 485.
34. Posthumus, W. P., Lenstra, J. A., van, N. A., Schaaper, W. M., van der Zeist Ba, and Meleon, R. H. (1991). *Virology*, **182**, 371.
35. Duim, B., Vogel, L., Puijk, W., Jansen, H. M., Meleon, R. H., Dankert, J., *et al.* (1996). *Infect. Immun.*, **64**, 4673.
36. Peterson, E. M., Cheng, X., Qu, Z., and de la Maza, L. M. (1996). *Mol. Immunol.*, **33**, 335.
37. Wallace, G. R., Ball, A. E., Macfarlane, J., El, S. S., Miles, M. A., and Kelly, J. M. (1992). *Infect. Immun.*, **60**, 2688.
38. Rudin, A. and Svennerholm, A. M. (1996). *Infect. Immun.*, **64**, 4508.
39. Ogawa, T., Yasuda, K., Yamada, K., Mori, H., Ochiai, K., and Haegawa, M. (1995). *FEMS Immunol. Med. Micro.*, **11**, 247.
40. van, A. A., Beckers, P. J., Plasman, H. H., Schaaper, W. M., Sauerwein, R. W., Meuwissen, J. H., *et al.* (1992). *Peptide Res.*, **5**, 269.
41. Lane, D. P., Stephen, C. W., Midgley, C. A., Sparks, A., Hupp, T. R., Daniels, D. A., *et al.* (1996). *Oncogene*, **12**, 2461.
42. Leone, N. A., Whitehouse, D. B., Swallow, D. M., Wallace, G. R., and Adinolfi, A. (1993). *FEBS Lett.*, **335**, 327.
43. Ammosova, T. N., Ouporov, I. V., Rubtsova, My, Ignatenko, O. V., Egorov, A. M., Kolesanova, E. F., *et al.* (1997). *Biochemistry*, **62**, 440.
44. Mol, J. A., van, W. M., Kwant, M., and Meloen, R. (1994). *Neuropeptides*, **27**, 7.
45. van, A. A., Plasman, H. H., Kuperus, D., Kremer, L., Rodriguez, J. P., Albar, R., *et al.* (1994). *Peptide Res.*, **7**, 83.
46. Zhong, G., Berry, J. D., and Choukri, S. (1997). *J. Ind. Microbiol. Biotechnol.*, **19**, 71.
47. Garcia-Valcarcel, M., Fowler, W. J., Harper, D. R., Jefferies, D. J., and Layton, G. T. (1997). *Vaccine*, **15**, 709.
48. Loomis, L. D., Deal, C. D., Kersey, K. S., Burke, D. S., Redfield, R. R., and Birx, D. L. (1995). *J. Acquired Immune Deficiency Syndromes Hum. Retrovirol.*, **10**, 13.
49. Loomis-Price, L. D., Levi, M., Burnett, P. R., van Hamont, J. E., Shafer, R. A., Wahren, B., *et al.* (1997). *J. Ind. Microbiol. Biotechnol.*, **19**, 58.
50. Coeffier, E., Excler, J.-L., Kieny, M. P., Meignier, B., Moste, C., Tartaglia, J., *et al.* (1997). *Aids Res. Hum. Retroviruses*, **13**, 1471.
51. Koshi, T., Karande, A. A., and Radhakantha, A. (1996). *Vaccine*, **14**, 307.
52. Langedijk, J. M. P., Meloen, R. H., and van Oirschot, J. T. (1998). *Arch. Virol.*, **143**, 313.
53. Slootstra, J. W., De Geus, P., Haas, H., Verrips, C. T., and Meloen, R. H. (1995). *Chem. Senses*, **20**, 535.
54. Petrakou, E., Murray, A., and Price, M. R. (1998). *Tumour Biol.*, **19**, 21.
55. Molina, F., Laune, D., Gougat, C., Pau, B., and Granier, C. (1996). *Peptide Res.*, **9**, 151.
56. Frank, R. and Doring, R. (1988). *Tetrahedron*, **44**, 6031.
57. Lebl, M. and Eichler, J. (1989). *Peptide Res.*, **2**, 297.
58. Berg, R. H., Almdal, K., Pedersen, W. B., Holm, A., Tam, J. P., and Merrifield, R. B. (1989). *J. Am. Chem. Soc.*, **111**, 8024.
59. Daniels, S. B., Bernatowicz, M. S., Coull, J. M., and Koster, H. (1989). *Tetrahedron Lett.*, **30**, 4345.

60. Krachnak, V., Vagner, J., Novak, J., Suchanokova, A., and Roubal, J. (1990). *Anal. Biochem.*, **189**, 80.
61. Houghten, R. A., Pinnala, C., Blondelle, S. E., Appel, J. R., Dooley, C. T., and Cuervo, J. H. (1991). *Nature*, **354**, 84.
62. Pinnala, C., Appel, J. R., and Houghten, R. A. (1994). *Biochem. J.*, **301**, 847.
63. Baeza, C. R. and Unden, A. (1990). *FEBS Lett.*, **277**, 23.
64. Salberg, M., Ruden, U., Magnius, L. O., Norrby, E., and Wahren, B. (1991). *Immunol. Lett.*, **30**, 59.
65. Stockman, B. J., Bannow, C. A., Miceli, R. M., Degraaf, M. E., Fischer, H. D., and Smith, C. W. (1995). *Int. J. Peptide Protein Res.*, **45**, 11.
66. Fodor, S. P. A., Read, J. L., Pirrung, M. C., Stryer, L., Lu, A. T., and Solas, D. (1991). *Science*, **251**, 767.
67. Pinnala, C., Appel, R., and Houghten, R. (1996). In *Epitope mapping protocols* (ed. G. E. Morris), p. 171. Humana Press Inc., Tottowa, NJ.
68. Lane, D. P. and Stephen, C. W. (1993). *Curr. Opin. Immunol.*, **5**, 268.
69. Gershoni, J. M., Stern, B., and Denisova, G. (1997). *Immunol. Today*, **18**, 108.
70. Geysen, M. H., Bartleing, S. J., and Meloen, R. H. (1985). *Proc. Natl. Acad. Sci. USA*, **82**, 178.
71. Geysen, M. H., Rodda, S. J., and Mason, T. J. (1986). *Mol. Immunol.*, **23**, 709.
72. Lam, S., Salmon, S. E., Hersh, E. M., Hruby, V. J., Kazmierski, W. M., and Knapp, R. J. (1991). *Nature*, **354**, 82.

Chapter 3

Methodological tips for human T cell epitope mapping when using pin technology peptide arrays

David A. Mutch
Visionary Voyager Corporation Pty Ltd., Fern Road, Upper Ferntree Gully, Victoria 3156 Australia

Olwyn M. R. Westwood
School of Life Sciences, University of Surrey Roehampton and Division of Immunology, St. George's Hospital Medical School, London, UK.

1 Introduction

The conditions for efficient large scale mapping of T cell epitopes within proteins recognized by human T helper cells are discussed. Many of the inherent difficulties in using freshly isolated human peripheral blood mononuclear cells (PBMC) and T helper cells are also described, as well as how such problems may best be overcome. Observation and awareness of such difficulties are crucial to obtaining accurate and meaningful results for epitope mapping studies. Although these methods precisely locate and size T helper cell epitopes, they rely on the availability of high quality, large scale, low volume peptides presented to freshly isolated human peripheral blood lymphocytes or cultured clone cells, in a non-toxic, biologically compatible buffer system. MultiPin Technology type peptides developed at Chiron Mimotopes Pty Ltd. (see also Chapter 2) are used in the procedures described in this chapter. However, other multiple peptide synthesis systems are available but it is important to ensure that similar biologically compatible environments are used.

Throughout the development of the techniques, many strategies have been rigorously investigated. These include the basic optimal culture conditions, and best peptide synthesis/mapping strategies; also important is a suitable computer-based results analysis system based on Poisson distributions ('ALLOC'; see below). This program is used to accurately determine and account for the level of background cellular proliferation within each assay so that responses that can be attributed solely to cells responding to specific peptide stimuli are isolated effectively. This mathematical system was conceived and developed by Dr Mario Geysen.

It is important to note at the outset the major (and possibly controversial) differences between the methods described in this chapter and those traditionally used in T cell epitope mapping studies. Here, the identification of T cell epitopes relies on the estimation of responding cell frequencies rather than the traditional method of relating cellular response estimates directly to the levels of cellular proliferation or radiolabel incorporation. The level of radiolabel incorporation of some cultures is used only to establish a background proliferation cut-off value. Importantly, the level of proliferation (detected by label incorporation) of cells considered to have responded positively to the applied peptide stimulus is not used to gauge epitope strength or prevalence.

2 Historic foundations to large scale T cell epitope mapping

Many methods and reviews have been published to describe assays to detect the antigenic stimulation of cells *in vitro* (see list at end of the Chapter). However, all have been undertaken either using vastly different cell types, equipment, or techniques, or using similar equipment and techniques but with mitogen- or strong recall whole antigen-stimulated cultures. Almost all of these early methods do not allow for effective scale-up to large scale peptide mapping studies. Schellekens and Eijsvoogel (1968) undertook an initial attempt at optimization and standardization of human PBMC proliferation assays. These authors used [^{14}C]-thymidine uptake to measure mitogen stimulation of human PBMC in 4 ml glass culture tubes. Assays were undertaken to determine the kinetics of radiolabel uptake and the correlation with actual incorporation, labelling time, and pulse duration, the influence of culture conditions, incubation times, mitogen concentration, cell viability, and assay reproducibility. The findings of this work were used as a foundation to devise, evaluate, and standardize many subsequent studies (1).

Du Bois *et al.* (1973) modified the early methods to use 1 ml microculture tubes and, as well as mitogen, included an antigen cocktail containing five strong recall antigens. The most noteworthy modifications were the use of glass fibre filter mats to harvest radiolabelled cells, and the demonstration that the specific activity of the radiolabel (i.e. the ratio of labelled thymidine to 'cold' thymidine) affected the rate of radiolabel incorporation. In addition, they observed that:

(a) Stored serum was often much more toxic in proliferation assays than fresh serum.

(b) The presence of erythrocytes did not affect mitogen stimulation but did depress antigen-induced proliferation.

In addition, the peak time of response varied between day 3, 4, or 5, even in mitogen-stimulated assays while certain types of plastic used in these assays were found to be toxic. Overall, these authors concluded that, since the mechanisms determining optimal reactivity were unknown, most variables should be thoroughly investigated for each series of assays before optimal and reliable results can be expected (2).

Quantitative assessment of human peripheral blood helper T cells, specific for tetanus toxoid (TT) or purified protein derivative (PPD), was undertaken by van Oers *et al.* (1978) using limiting dilution analysis and the uptake of [^{3}H]-TdR to assess proliferative responses. Significantly, only three of the 11 donors tested showed moderately constant or reproducible responding cell frequencies over a three month period. Even for these donors the correlation was poor ($R^2 = 0.47$) and varied by between half to double the original estimation (3). It has also been reported that the standard assay for lymphocyte (transformation) responses to the antigens was almost entirely dependent upon the participation of progeny lymphocytes after the initiating events (4).

These authors found that the dilution process inherent in limiting dilution assays probably affected the outcome by diluting other cell types required for optimal expansion of specific T cell clones, and that antigen-induced proliferation may be more susceptible to 'other' factors due to the increased assay time required. Moreover, the comparison of results, even within the same assay, could be influenced considerably by the calculation method used. For instance, whether data was log-transformed or not or whether the calculation of positives was done by subtraction or division of the background responses. This observation highlights the degree of uncertainty that must be addressed when interpreting proliferative data based only on comparisons of peak counts per minute (c.p.m. values).

A study using 20 μl hanging drop cultures showed that cell concentration, incubation period, and the concentration of allogeneic stimulator cells had interrelated effects on the measured proliferation rate of responding PBMC (5). Hensen and Elferink (1984) noted the importance of using a large number of replicates to accurately measure responses to a single whole antigen. Importantly, these authors reported the adverse effects of random proliferation occurring in control cultures and the inadequacy of the stimulation index as a measure of specific proliferation (6).

3 Fundamental considerations

Very few studies have investigated the precision or the validity and meaning of results obtained from T cell mapping assays using peptides. Most have been based on early development work which utilized mitogens or very strong whole antigens such as PPD that provide clearly positive results in stimulation assays. This is not a problem when using similar strong antigens or single epitope-specific T cell clones. However, when attempting to identify a response to a single (potentially weak) peptide epitope against the entire T cell repertoire, many problems arise which were not and could not have been obvious in earlier research. Some of these problems are examined here, as well as how to overcome them.

3.1 Lessons from B cell epitope mapping

The large scale multi-peptide synthesis methods described here were initially designed to map epitopes recognized by B cells. The detection of B cell recognition

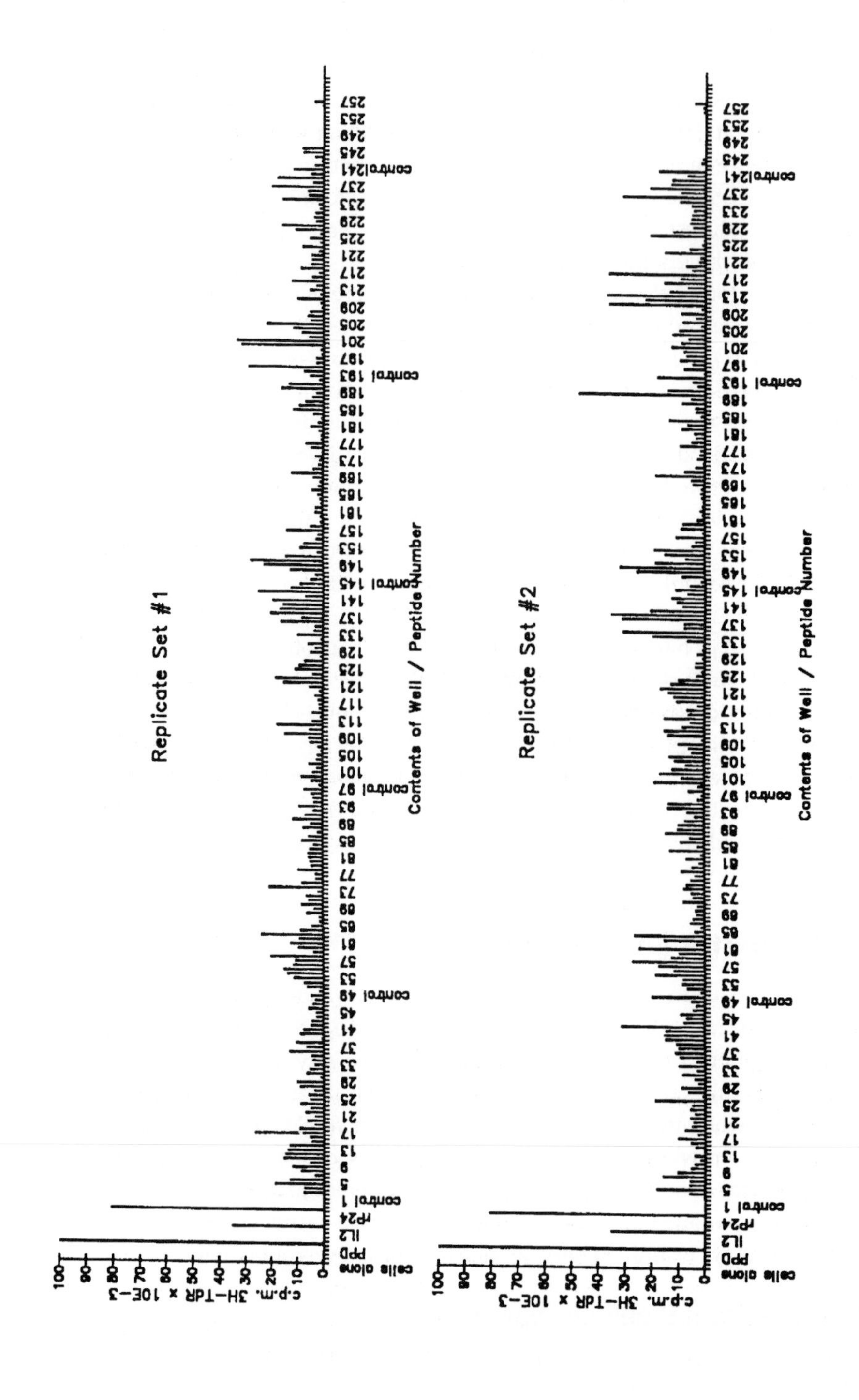
Replicate Set #1
Replicate Set #2
Contents of Well / Peptide Number
c.p.m. 3H-TdR x 10E-3
cells alone
PPD
IL2
rP24
control

Figure 1 A typical large scale T cell response assay using methods similar to those used to map B cell epitopes. The assay shows the PBMC proliferation responses of from a HIV negative donor to HIV-1 gag p24 peptides. PBMC (2×10^5/190 μl U-bottomed well) were cultured with 10 μl of either synthetic test peptide or control reagents: medium for cells alone; PPD at 10 μg/ml; rHIL2 at 10 IU/ml; rHIV1p24 at 10 μg/ml; or synthesis control peptides (marked 'control'), for seven days. The test peptides consisted of a series of overlapping dodecamer peptides beginning at each residue of the HIV-1 (strain BH10) gag p24 sequence. The volume of test peptide used provided a concentration of between 5–10 μg/ml, depending on peptide synthesis yield. Wells were pulsed with 1.0 μCi/well [^{3}H]-test TdR for 6 h before harvesting and scintillation counting. Results of peptide stimulation are expressed as the c.p.m. of individual wells. This assay was performed in duplicate (except for the four initial control tests which were performed in triplicate) and the results of individual replicates shown as either Set 1 or Set 2. The mean of the triplicate replicates of the four initial controls is shown for both data sets.

and stimulation using ELISA is highly specific and is usually amplified. Thus, even a single peptide-binding event can be detected against a cacophony of other events using an ELISPOT assay (see Chapter 2).

Some of the early investigations into large scale peptide mapping of T cell epitopes attempted to mimic B cell assays. For instance, assays that used a single replicate of overlapping 12-mer peptides spanning an entire antigen with an offset of a single residue for consecutive peptides. Additional identical assays were created as replicates. The results of one such assay are shown in *Figure 1* which is representative of many others that were undertaken.

The assay results shown in *Figure 1* demonstrate many of the problems using such a method. Responses to strong stimulants (the controls) are much higher than for any individual peptide. There are little or no strong similarities between the two replicate assays. Although the proliferation of the single negative control well used for these assays is low, multiple replicates of this control would normally resemble the results for the peptide stimulation. Many such assays failed to provide a clear indication of the location of any T cell epitopes within this antigen. Interestingly, the apparent low levels of stimulation from peptides shown in *Figure 1* is the result obtained from a HIV negative donor. Results from HIV positive donors were very similar.

Had this assay been undertaken using B cells with the degree of antibody affinity to the peptides used as the measured outcome, the following result would have been expected. First, the background and negative control levels would have contained no spurious false positives. Secondly, as the series of peptides approached, entered, and departed the epitope, a distinct bell-shaped results curve would be observed, centred about the epitope. Clearly, no such similar results where produced using such strategies with T cells. Even using a single replicate per test, it was expected that the minimal T cell epitope required for stimulation would have occurred in multiple sequential peptides as the series passed through the epitope, providing an indication of the region to be used in further experiments with higher replicates. This was not the case, although several sequence regions which appeared (not statistically) positive more frequently later proved to contain T cell epitopes.

3.2 The nature of the peptide antigen

The current database of known T cell epitopes is rapidly increasing in size, but very few of these has been reduced to minimal length. Clearly, it is important to assess the effects of different end-groups on the ability of an epitope to be fully functional at and near the minimal recognizable length. In most studies much longer peptides have been used (typically 15-mers to 20-mers). Knowledge of the antigen processing mechanisms applicable to peptides added to cell cultures and the effect of end-groups of T cell epitopes may assist in a clearer understanding of the molecular interactions between MHC class II molecules, peptides, and T cells, as well as in the formulation of better peptide vaccines. An important consideration to avoid when selecting potent T cell stimulator peptides for vaccines or immunotherapies, is the potential for a sequence to correspond to, or mimic a natural autoantigen, or other detrimental element (7).

3.3 Protein sequence databases and sequence analysis

Protein sequences may be obtained from a number of possible databases, such as:

(a) Protein Identification Resource (PIR) from the National Biomedical Research Foundation (NBRF, Brookhaven, USA).

(b) GenBank (BBN Laboratories Inc., Cambridge, MA, USA).

(c) On-line service to the Australian National Genomic Information Service (ANGIS) database at the Department of Engineering, University of Sydney.

(d) Swiss-PdbViewer: Glaxo Wellcome Experimental Research http://www.expasy.ch/spdbv/mainpage.htm

4 Peptide synthesis considerations

Peptides incorporated into large scale T cell epitope mapping assays must possess certain basic characteristics. These include low toxicity, minimal residual synthesis materials, solubility, sufficient quantity, and be of the highest possible purity with respect to the desired amino acid sequence. Additional characteristics required of actual peptide antigens for use in large scale mapping assays include:

- easy multiple synthesis capacity
- available in a format which reduces assay set up complexity
- preferably already in a low toxic aqueous form
- almost unrestricted control over peptide length, overlap, and termini during synthesis
- low cost per peptide with minimal excess peptide production

When planning a large scale T cell epitope mapping strategy several peptide-specific aspects need to be considered. These include:

- the length of the sequence to be mapped
- the total number of peptides to be synthesized

- optimal degree of overlap between subsequent peptides to provide the degree of mapping resolution desired should be determined

Both the total number of peptides and the degree of peptide overlap are interrelated with the desired individual peptide length to be used. The choice of which peptide termini to be used must also be considered. Variations in the moieties at either the C- or N-termini can in turn affect T cell proliferation.

4.1 Types of peptides

The use of MultiPin peptide synthesis strategies developed by Geysen *et al.* (1984, 1987), satisfies all of the above criteria (8, 9) (see also ref. 10, and Chapter 2). Resin-based peptide synthesis methods are suitable for the production of large volumes of small numbers of peptides. However, such methods are costly and wasteful when the production of thousands of peptides in small quantities are required for screening. Once epitopes are defined, many synthesis methods including resin methods and scaled up pin methods may be used for further studies.

MultiPin peptides are conveniently cleaved directly into microculture trays and may be synthesized in any arrangement or sequence, allowing the researcher to design their assay plate layout before peptide synthesis and have the peptides produced to match this layout. These peptides are of very high purity and cleaved directly into low toxic, cell compatible buffer solutions allowing direct transfer into cellular assays.

4.2 Features of T cell epitopes

4.2.1 Cores and envelopes of T cell epitopes

There is a wide diversity of peptide sequences and lengths used in the study of T cell epitopes. Within such epitopes there are certain amino acid residues that are:

- critical to an epitope
- less critical but nevertheless required
- neither critical nor required, but the presence of which affects the epitope in some way

T cell epitopes must bind to MHC molecules, and also interact with the T cell receptor. Therefore a T cell epitope core may be defined as residues shared by all immunogenic peptides spanning a T cell epitope region. However, although the core itself may not be immunogenic, it is likely to be essential for immunogenicity and may be defined using multiple, highly overlapping peptides of several (e.g. three) different lengths spanning a T cell immunogenic region of a protein. The essential residues are the smallest possible immunogenic component within all stimulatory peptides spanning this region. In the absence of testing multiple length peptides, the core may be defined as the overlap between all stimulatory overlapping peptides spanning the epitope site. It is noteworthy that different strains of mice may recognize the same core residues of a T cell epitope, but may require different envelope residues flanking the core for an immune response to occur.

The existence of a non-stimulatory core and additionally required envelope residues can be observed in certain T helper cell epitopes. For instance, when examining the epitope ^{593}YSYFPSV^{599} (tetanus toxoid 593–599), is non-stimulatory, but addition of either of the native residues I^{592} or I^{600} confers complete stimulatory capacity to this peptide (11). In addition, the epitope core plus the minimal number of additional envelope residues required for T cell stimulation defines the minimal length epitope of any particular T cell clone or set of T cell clones. There still remains one additional feature common to many reported T cell epitopes that requires definition. Amino acid residues flanking the minimal length epitope which are not essential or required for T cell stimulation, have been observed to affect the degree of T cell stimulation when included in the peptide—such residues may be described merely as 'flanking residues'. Flanking sequences may or may not be significant with respect to the delivery of the antigen for the mode of delivery can also affect the T cell repertoire. Flanking sequences may alter protein topology and thus also affect antigen uptake, processing, and presentation on MHC molecules, indirectly affecting the resultant T cell repertoire and the immune response pattern (12).

4.2.2 *Peptide termini*

Modifications to the amino (N-) or carboxy (C-) terminus of a peptide can alter the stimulatory capacity of a T cell epitope.

Consider the following example. To test the effects of peptide end-groups, peptides based on four different peptide sequences were synthesized, each made in N-terminal acetylated and non-acetylated forms, and each of these made with free acid, methylamide, or DKP C-terminus. The RP9 tetanus toxoid (TT)-specific T cell clone used has been previously shown to respond to certain peptides containing either of the minimal length octapeptides ^{592}IYSYFPSV^{599} or ^{593}YSYFPSVI^{600}, but shorter peptides based on ^{593}YSYFPSV^{599} are non-stimulatory (11, 13). Non-stimulatory peptides based on the 8-mer ^{590}TKIYSYFP^{597}, offset from the determinant region by two residues, were also included as further negative controls.

Figure 2 shows the proliferative response of this clone to two concentrations of each of 24 peptides, representing all four sequences described above, flanked by each of the six combinations of N- and C-terminal groups. Each minimal length octapeptide T cell epitope stimulated the clone as expected, while the heptapeptide and the control octapeptide, offset by two residues, failed to stimulate significantly at the concentrations used. A further experiment using all the above-mentioned peptides at 1.0 μM concentration showed the same pattern of responses. In all cases, peptides containing ^{593}YSYFPSVI^{600}-β-Ala (an β-Alanine spacer) were found to be slightly more effective than those containing ^{592}IYSYFPSV^{599}-β-Ala. Peptides containing ^{593}YSYFPSVI^{600}-β-Ala were as active as control purified peptide p404 (H-^{592}IYSYFPSVI^{600}-$CONH_2$). The activity of

^{593}YSYFPSVI^{600}-β-Ala was generally independent of the carboxyl terminal group added to it. All the non-acetylated ^{592}IYSYFPSV^{599}-β-Ala peptides were of low stimulatory activity in comparison with the acetylated peptides containing the same sequences.

In contrast to the effectiveness of peptides based on ^{593}YSYFPSVI^{600}-β-Ala (*Figure 2*) in stimulating proliferation of the RP9 T cell clone, a peptide with the same sequence but lacking the carboxyl terminal β-Ala group was non-stimulatory at doses up to 5.0 M. The structural motifs proposed which attempt to predict T cell epitopes within antigenic sequences ultimately depend on the propensity of the peptide to form an α-helix (14, 15). An increase in the stimulatory activity of some T cell epitopes has been reported to be associated with N-terminal acetylation (16). This finding was postulated to result from increased stabilization of an α-helical structure within the epitope by removing the unfavourable interaction between the half positive charge dipole of the amino terminal end of the helix and the positively charged N-terminus. This mechanism however, does not adequately account for the findings presented herein (*Figure 2*). First, the proline residue close to the centre of the RP9 T cell epitope precludes the formation of a regular α-helix. Secondly, ^{593}YSYFPSVI^{600}-β-Ala peptides were found to be equally active regardless of N-terminal acetylation.

One possible mechanism to account for the significant positive effect of acetylation is that this modification changes the 'fate' of these peptides during the course of the proliferative assay. Added peptide becomes bound to the MHC class II molecules of the antigen-presenting cell, a kinetically slow step (17), prior to interaction with T cells. At the same time however, peptide is subject to the action of proteolytic enzymes present in the serum-supplemented culture medium and to loss by adsorption to serum proteins. Thus, the dominant effect of acetylation may be to minimize the loss of peptide during culture and thus maximize peptide available for uptake by antigen-presenting cells.

Similarly, the blocking of the C-terminal carboxylic acid group of T cell epitopes has also been reported to enhance stimulatory capacity (18). The proposed mechanism is the cancellation of an unfavourable effect of the carboxyl negative charge with the half charge associated with the carboxyl terminal end of an α-helix within an epitope. In the examples presented here, the absence of any major effect of the carboxyl terminal group on the stimulatory activity of the peptides may be explained by the presence of the β-alanine spacer residue present in each peptide. The conformational freedom of the β-alanine, plus the greater spacing distance provided by the added methylene group of the β-alanine, may have negated any effect of the terminal group on the binding conformation of stimulatory peptides. Alternatively, the uncommon β-alanine residue may discourage degradation of these peptides by proteolytic enzymes that target the carboxyl terminal region of peptides.

The presence of a carboxyl terminal DKP moiety on the RP9 T cell epitope sequence improves its stimulatory capacity, only slightly; at least there is no adverse effect on the capacity of the peptide to be recognised by T cells.

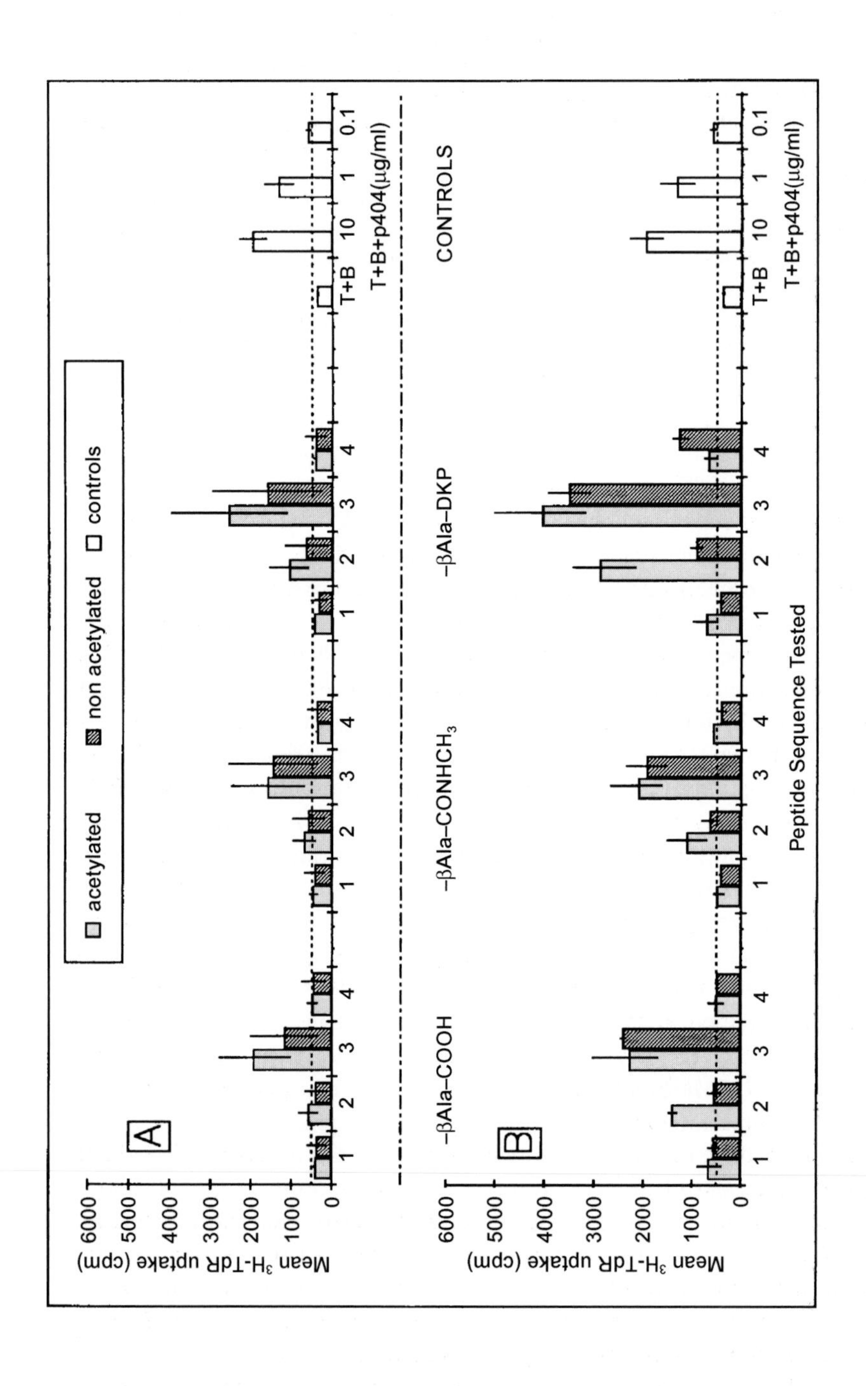

A
B
acetylated
non acetylated
controls
Mean ^{3}H-TdR uptake (cpm)
6000
5000
4000
3000
2000
1000
0
–βAla–COOH
–βAla–CONHCH$_3$
–βAla–DKP
CONTROLS
1
2
3
4
T+B
10
1
0.1
T+B+p404(μg/ml)
Peptide Sequence Tested

Figure 2 Comparison of the responses of the RP9 helper T cell clone to amino and carboxy terminal peptide variants. Cleaved peptides (0.1 or 10 μM) were added to mixtures containing equal numbers of seven day post-stimulation RP9 cloned T cells and log-phase growth irradiated RP B-LCL cells (2×10^4 cells/U-bottom well each) and incubated for 72 h. Wells were pulse labelled with 0.5 μCi/well [^{3}H]-TdR for the final 18 h. DNA harvesting and scintillation counting used standard procedures. Results are expressed as the mean ± SD of triplicate cultures. (A) Test peptides were added at 0.1 μM. (B) Test peptides were added at 10 μM. Test peptide sequences derived from TT were: 1, ^{593}YSYFPSV^{599} (shorter than the minimal epitope); 2, ^{592}IYSYFPSV^{599} (minimal epitope 1); 3, ^{593}YSYFPSVI^{600} (minimal epitope 2); and 4, ^{590}TKIYSYFP^{597} (which does not contain the complete minimal epitope). Control peptide 404 is H-^{592}IYSYFPSVI^{600}-$CONH_2$ which contains both N- and C-terminal residues required by either of the two minimal epitopes for this clone (11, 13). The boxed legend describes the N-terminal peptide group while the C-terminal group is listed between the corresponding results shown in panels A and B (e.g. β-Ala-COOH). T + B (RP9 T cells plus irradiated RP-LCL B cells) is the assay negative control, without peptide. The dotted lines represent the mean + 3SD of this control and therefore the significance level for the assay.

4.2.3 Minimal length of T cell epitopes

The length of peptides used for T cell stimulation varies considerably between different studies. Peptides used as MHC class I restricted T cell epitopes are probably more restricted in variations of length than peptides used as MHC class II restricted epitopes. This is due to the requirement of class I peptides to be constrained at both ends within the peptide binding groove of MHC class I molecules. Nonamer peptides of influenza virus matrix protein have been found to be between 100–1000 times more potent in cytotoxicity assays than corresponding dodecamer peptides (19). Consider the example of alanine-replacement analogues of the influenza A virus matrix cytotoxic T cell epitope ^{57}KGILGFVFTLLV^{68} (M57–68) using episomal plasmids transfected into B-LCLs. HLA-A2 restricted CTL lines specific for this peptide were reported to lyse target cells which endogenously produced an analogue epitope that retained only the five residues (^{60}LGFVF^{64}), flanked by seven alanine residues (AAALGFVFAAAA) with almost equal efficiency to the endogenously produced parent epitope. It is of interest to note that peptide epitope evaluation by functional analysis of alanine-substituted variants, may be used to demonstrate the specific amino acid residues that are contact sites for the T cell receptor. For instance, Gogolak *et al.* (2000) tested the 317–341 region of the human influenza virus (H1N1 subtype) to find that the ^{324}P was the primary contact residue (20).

MHC class II peptides have been reported to comprise a variety of minimal lengths. For instance, the reported minimal length stimulatory peptide from the malarial circumsporozoite (CSP) antigen required only seven residues ^{328}DKHIEQY^{334}, in order to elicit responses from lymph node cells of CSP- or peptide-primed mice (21, 22). In contrast, another T helper cell epitope of ^{953}SFWLRVPKVSASHLE^{967}, described by Panina-Bordignon *et al.* (1989) was reported to be non-stimulatory at lengths less than 14 residues (23, 24).

The length of naturally processed T cell epitope peptides bound to MHC class II molecules is well established to be between 13–25 amino acids, with a majority of peptides found to be 17 amino acids long (25–29). Nonetheless, the length of T cell stimulatory peptides used by different researchers varies considerably, and in many instances efforts have not been undertaken to truncate the peptide in question to determine the precise length of the minimal stimulatory epitope. As a general observation, a T cell epitope peptide comprising 12 amino acids is sufficient to result in T cell stimulation in most experimental circumstances. When Partidos and Kanse (1997) were selecting T helper cell epitopes from a fusion protein of the measles virus using immunized mice, they concluded that the epitopes on chimeric proteins were:

- a function of the amino acid sequence
- dependent where the epitope was located within the chimeric protein
- dependent on the strain of animal used in the experiments (30)

It can not be overemphasized that the length of peptides should be optimized for the system being investigated. For instance, Kammerer *et al.* (1997) who were looking for putative peptides for use in venom immunotherapy concluded that 15-mer peptides (which may be considered comparatively long) were not necessarily sensitive enough to fully define possible T cell epitopes of an allergen. This group recommended the use of longer overlapping peptides to delineate specific antigenic sites (31).

4.2.4 Conformation of T cell epitope peptides

The conformation of short peptides comprising T cell epitopes have been predicted by several groups to comprise amphipathic α-helical structures of 7–11 residues (14, 15, 17, 31, 32). This conclusion has been based on databases of known T cell epitopes and secondary structure prediction algorithms (33, 34). However, more direct methods of structure analysis would refute the suggestion that T cell epitopes have a single dominant secondary structural motif. Studies which have examined the capacity of a large number of synthetic peptides to inhibit T cell responses to known immunogenic peptides have found little correlation between amphipathic tendencies and MHC binding and no correlation between known secondary structure and MHC binding (35).

Conformation is significant, for sequences can 'mimic' one another with respect to shape. Although this is of particular significance with respect to B cell epitopes, recent evidence has confirmed that this also impinges on T cell investigations. In the work of Bach *et al.* (1998), they were trying to identify peptides that mimic sequence motifs of the a known antigen, that is a target for T cell attack and the autoimmune destruction of the pancreas in insulin-dependent (type 1) diabetes mellitus, i.e. the 65 kDa glutamic acid decarboxylase (GAD65). They found that GAD65-specific T cells could be stimulated by a conformational peptide mimic that had little sequence homology with the defined autoantigenic epitope. Their findings have suggested of particularly significance is knowledge of the primary residue contacts with the T cell receptor within the epitope and the MHC class II binding motifs (36).

Nonetheless, secondary structure of peptide antigens are somewhat dependent on the environment in which they are presented, e.g. buffer system, solvent, as conformational analysis using circular dichroism spectroscopy (CD) often demonstrates. Examination of peptide–MHC conformational interactions would suggest that an α-helical peptide would be more difficult to fit into the peptide binding site of a MHC molecule than a peptide in extended conformation. This has been confirmed by X-ray crystallographic studies of MHC class I molecules (37, 38) and of MHC class II molecules (39). Moreover changing the oxidative state of the peptide can also alter its immunoreactivity, e.g. oxidized form containing an intramolecular disulfide bond, but a reduced form contained two thiol groups (40).

The use of sequence databases and algorithms to predict the conformation of short regions within whole protein molecules can be misleading. Kabsch and Sander (1984) analysed identical pentapeptide sequences within proteins of known tertiary structure and found that some identical pentamers were α-helical in some proteins and β-sheets in others. Hence it must be appreciated that antigenicity is also dependent on sequence context, i.e. whether it forms part of a protein or a simple peptide. Those who extol the validity of T cell epitope prediction motifs sometimes overlook this aspect (41).

5 Use of pin peptides in human PBMC proliferation and cell culture assays

Protocol 1

Preparation of peptides and antigens for proliferation assays

Equipment and reagents

- Disposable plastic eight or 12 channel reservoir liners (Costar)
- 96-well microculture trays (Sterilin Ltd.)
- Buffers (see text for details)
- Peptides
- Recall antigens e.g. PPD, Con A mitogen

Method

1. Dilute the peptides cleaved into various buffers (see text for details) into 10 μM (10 × 1 μM) or 10 × final concentration, in complete medium using disposable plastic eight or 12 channel reservoir liners.
2. Add equivalent concentrations of peptide cleavage buffer components such as Hepes and methyl cyanide solvent to complete medium containing either no additional reagents (for negative control wells), or 10 × concentrations of recall antigens, e.g. PPD or Con A mitogen (for positive control wells).
3. Plate 10 × strength reagents plated into appropriate wells of 96-well microculture

Protocol 1 continued

trays at either 10 μl/well for 100 μl assays in V-bottom trays or at 20 μl/well for 200 μl assays in U-bottom trays.

Comparisons of proliferation induced by selected peptides prepared in this manner with freshly prepared identical peptides showed little or no differences.

1. Positions of test and control assay wells on microculture trays were normally designated by a custom-written computer algorithm ('ALLOC', described in Section 6).
2. A typical print-out showing the allocation of wells for an assay is presented in *Figure 5*.
3. Tests of peptide-induced proliferation are usually undertaken in a minimum of 16 replicates.
4. The number of negative control wells distributed over all trays in an assay is generally several times greater.
5. Wells containing positive control antigens are distributed evenly over all trays but usually equalled the same number of replicates used for peptide test wells.

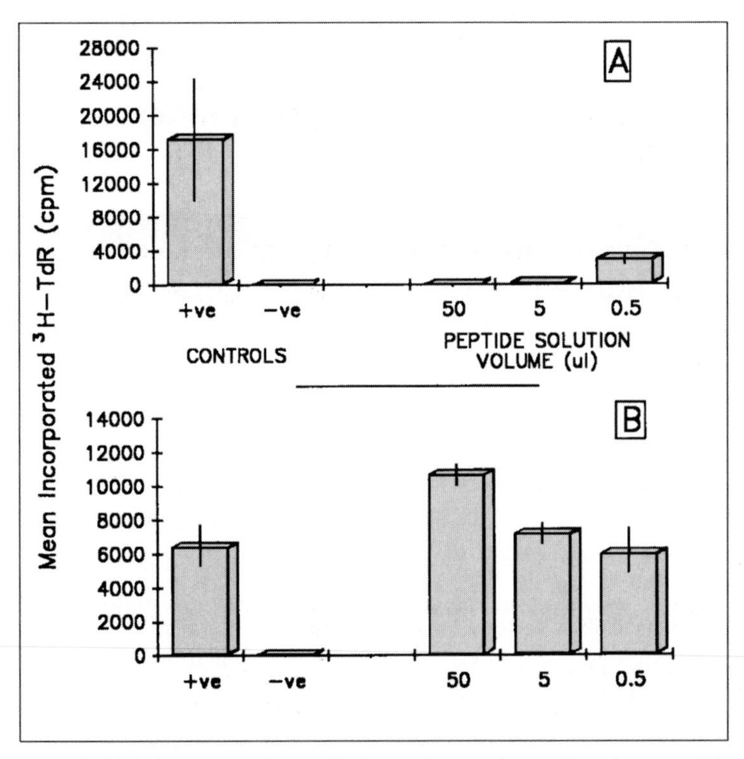

Figure 3 Methylamine toxicity test. Various volumes of a solution phase peptide cleaved using 4% (v/v) methylamine solution were added to PHA-stimulated (2.0 μg/ml PHA) PBMC cultures (5000 cells/200 μl U-bottomed well) and incubated for three days before pulse labelling with 0.5 μCi [^{3}H]-TdR/well and standard DNA harvesting and scintillation counting procedures. Peptide solution was added either (A) without lyophilization treatment to extract the methylamine, or (B) after lyophilization and reconstitution in 0.1 M phosphate buffer pH 7.0.

5.1 Buffer systems

The ability of the cleavage buffer to solubilize peptides as they are cleaved is an important factor in the selection of cleavage buffer components. For instance some buffer systems include organic solvents such as acetonitrile, and water miscible solvating reagents to enhance peptide solubility, e.g. Tween 20, ethanol, methanol, ethanedithiol (EDT), anisol, Teric PE68, dimethyl sulfoxide (DMSO), glycerol, polyvinylpyrrolidone (PVP), and methyl cyanide (acetonitrile). In general it is expedient to test the cellular assay system to find the most effective buffers that provide consistent levels of peptide but do not affect cellular proliferation. Generally < 5% (v/v), acetonitrile has relatively low toxicity within *in vitro* PBMC proliferation assays, even though obviously it would to be toxic *in vivo*. Other reagents perform differently under similar conditions and adversely affected proliferation results. For instance, PVP can enhance mitogen-stimulated proliferation but inhibited antigen-driven proliferation.

A further example of toxicity testing is important to demonstrate as it bears importantly on the results of the peptide end-group moiety studies described in the immediately following subsection. The use of 4% (v/v) aqueous methylamine in the synthesis of C-terminal methylamidated peptides can be highly toxic if not entirely extracted by lyophilization. *Figure 3A* shows the effects of 4% (v/v) methylamide peptide solution prior to lyophilization when added to PHA-stimulated PBMC. After lyophilization and reconstitution of the same solution (*Figure 3B*) no toxicity is observed even when 50 μl of solution are added to the 200 μl cultures.

5.2 Cleavage of peptide and peptide concentration

Buffers used for the cleavage of pin peptides should sterilized by filtration through 0.22 μm filters. Side chain deprotected peptides attached to pins with the DKP-forming moiety may be cleaved in a choice of buffer systems, depending on the multi-pin system used. For example:

- 0.1 M sodium phosphate buffer pH 7
- 0.05 M Hepes (Calbiochem) in 20% (v/v) methyl cyanide (HiPerSolv grade; BDH) at pH 7.6
- 0.1 M Hepes in 40% (v/v) methyl cyanide pH 7.8

The elution buffers of pin-bound peptides may need sonicating after the peptides have been cleaved.

Side chain deprotected peptides attached to pins with a 4-hydroxymethyl benzoic acid (HMB) linker may be cleaved to form either a free acid β-Alanine C-terminus, or a methylamide β-Alanine C-terminal. Where the deprotection process demands extremes of pH, then the solutions must be neutralized before use in proliferation assays. For example: free acid peptides made by soaking in 0.1 M NaOH (150 μl/pin) for 3 h at room temperature are neutralized by the addition of 0.6 M sodium dihydrogen orthophosphate (30 μl/pin).

Methylamide peptides may be generated by soaking the pins in 4% (v/v) methylamide/water for 3 h (150 μl/pin). Unreacted toxic methylamide may be removed by vacuum desiccation over phosphorus pentoxide, but then the peptides need to be reconstituted in the buffer of choice.

High purity with respect to the synthetic peptides is paramount for successful T cell epitope mapping. This can be confirmed from the high performance liquid chromatography and amino acid analysis data. Moreover, the concentration of peptides should also be known in order that precise quantities are utilized in the proliferation assays, and so assays can be repeated. In order to determine the useful concentration range of a solvent or reagent within peptide cleavage buffers, an estimation of its final concentration within *in vitro* assays must be made. This is affected by several factors:

(a) The concentration of the cleaved peptides.

(b) The desired final peptide concentration within *in vitro* proliferation assays determines the volume of cleavage solution/peptide which is added to any particular test.

If each 200 μl well of cleaved peptides contains 50 nmole of peptide (250 μM) and the desired peptide concentration in an assay is 1.0 μM, then a 1:250 dilution of peptide solution is made in the final assay medium. Even if the solvent concentration in the cleaved peptide solution was 50% (v/v), this becomes only 0.2% (v/v) in the final assay. Thus, it is possible to use relatively high concentrations of appropriate solvents in cleaved peptide stock solutions.

5.3 Using human cells

Freshly isolated human cells, and in particular human PBMC, although moderately easy to culture *in vitro*, present unique difficulties for researchers investigating immunological responses to antigenic peptides. Most of these difficulties can be attributed directly to the heterologous population of immune cells present, most of which respond to only a single specific epitope from a potential library of millions. Other difficulties arise as a consequence of the differences between individual donors such as their HLA type (MHC class I and II cell surface receptor types), blood group, and rhesus type. In addition, differences in response levels between cell samples taken from the same donor at different times have often been observed, and may be a result of underlying infections, general health status, or the effects of numerous other factors at the time of blood donation. To minimize variations, optimize response levels, and ensure assay reproducibility to produce meaningful results, researchers must be aware of how many of these difficulties may be overcome.

Fresh, uncloned human T cells require assay conditions different from those used for murine, rodent, or clones. Most assays using T cells from animals (typically mice) utilize concentrated sources of lymphocytes such as lymph nodes or spleen. Although there are ways to extract large numbers of concentrated T cells from humans, these are not always readily available to researchers. The most common source of obtaining human T cells is by phlebotomy. Cell populations obtained using this method contain a wide variety of cells other than T cells that may influence the outcome of stimulation assays. When considering the potential size of the T cell response repertoire, the total number of T cells available from different animal sources, the sample size, and the concentration of T cells within that sample, then blood collections from humans would contain a vastly smaller proportion of the total available repertoire than lymph node collections from mice or rats.

5.3.1 Preparation of PBMC

The preparation of human PBMC for large scale T cell epitope mapping assays is relatively straightforward and uses well known principles. However, to ensure that cells are collected in optimal condition and to assist in the reproducibility of results, several important issues should be observed.

5.3.2 Important issues in the collection of human PBMC

Human PBMC are usually collected using standard phlebotomy procedure. In order to reduce damaged cell numbers and to maximize cell proliferative performance, care must be taken to ensure that the cells are not stressed or damaged. Therefore, collection of whole blood using procedures suited more to those for serum and plasma are not recommended. The use of small (i.e. 10 ml) evacuated blood collection tubes not only requires many manual exchanges but may damage cells, reduce or alter cell surface structures, or critically alter sensitive peptide recognition and response mechanisms. Although often less convenient, collection of blood using traditional 50 ml syringes fitted with butterfly needles and long tubes results in much more reliable and reproducible tests and the highest concentrations of viable cell numbers. Similarly, the use of moderately harsh anticoagulants such as citrate, or rapid changes to the pH of either the whole blood or blood/culture medium mixtures may cause similar problems. Citrated blood was frequently found to contain much lower numbers of viable cells. Heparinized or defibrinated blood is therefore recommended for use in human PBMC T cell epitope mapping studies.

Defibrination using glass beads is an alternative method in preparing blood samples for separating large numbers of healthy viable PBMC. No chemical processes are used and the serum is almost completely cleared of all coagulation factors, resulting in a preparation with very few constraints on its use.

Following collection and immediate anticoagulation, blood is diluted in cell culture medium, e.g. RPMI 1640. However, if the medium is allowed to stand for long periods there is often increases in pH due to CO_2 loss which can easily be observed as a pink or even red colour. For use in PBMC cultures and blood dilution, a pH of 6.9–7.1 was found to be the optimal. Filter sterilized 1 M HCl may be used to acidify RPMI medium until a bright orange colour is obtained. Medium which is somewhat more acidic than this does not appear to damage cells but may become too acidic too quickly in very strongly responding (e.g. mitogenic) cultures. The use of alkaline medium, even pink coloured, frequently resulted in large scale cell death and clumps of cellular debris and DNA during density interface separation procedures to purify the PBMC fractions. Therefore it is highly recommended that slightly acidified RPMI culture medium be used, although several other media including DMEM were also found to be just as good as RPMI.

Protocol 2

Defibrination of whole human blood using glass beads

Equipment and reagents

- Glass beads, 4.5–5.5 mm dia. and 7.5–8.5 mm dia.
- 100 ml glass or polycarbonate screw-capped bottles
- Rotary suspension mixer
- 1 M HCl
- 100 ml fresh whole human blood.

A. Preparation of glass beads

1. Soak the glass beads in 1 M HCl overnight followed by scrubbing and standard tissue culture glassware washing procedures.
2. Place 10–20 g of the glass beads in the 100 ml bottles and sterilize the beads and the bottles by autoclaving.

B. Defibrination of whole blood

1. Immediately after phlebotomy, add up to 100 ml whole human blood aseptically to the bottle of beads and seal tightly.
2. Load and balance the bottle onto the rotary suspension mixer and rotate at 15–20 r.p.m. for 10–15 min. The sound of the beads in the bottle will change from clear to dull when the process is complete.
3. Aspirate the defibrinated whole blood aseptically and separate the cells as described in *Protocol 3*.[a]

[a] The glass beads can be soaked in standard cell culture decontamination solution and then reused repeatedly by following part A above.

5.3.3 Separation of PBMC from whole human blood

Human PBMC are easily separated to high yields using traditional density interface techniques methods (42). Ficoll Paque (ficoll-sodium diatrizoate solution, $\rho = 1.077$, Pharmacia LKB Biotechnology AB) is specifically formulated to optimal osmolarity and density for human PBMC. As mentioned above, the pH of the diluting medium is critical to the success of this process. Minimizing excess centrifugation steps is also advised. Following separation, PBMC may be used directly in assays or stored frozen until use. Frozen storage of cells was not found to affect subsequent assay sensitivity in terms of response levels to mitogens, whole protein antigens, or peptides.

Protocol 3A

Preparation of human PBMC from whole blood

PBMC isolated from defibrinated or anticoagulated human blood using a modified density-interface centrifugation method of Boyum (1964) under aseptic conditions (42).

Equipment and reagents

- Centrifuge tubes (50 ml screw cap)
- 20 ml syringes (sterile)
- Flexible plastic canula (12 inches; sterile)
- Pipette (sterile)
- Human blood
- Ficoll Paque (ficoll-sodium diatrizoate ρ = 1.077 Pharmacia LKB Biotechnology AB)
- Complete medium (freshly made up before use): RPMI 1640 medium supplemented with 2.0 g/litre sodium bicarbonate supplemented with sterile 2 mM L-glutamine, sterile 5 mM Hepes buffer, sterile 50 μg/ml gentamicin antibiotic solution (50 mg/ml stock solution)

Method

1. Prepare a 2:1 mixture of human defibrinated blood with complete medium.
2. Add 30 ml of diluted defibrinated blood to a 50 ml centrifuge tube.
3. Using a sterile flexible plastic canula, gently underlay 13–15 ml Ficoll Paque, then centrifuge at 450 g for 25 min at 20 °C.
4. Using a sterile pipette, collect all but 5 ml of the serum above the PBMC layer.
5. Heat inactive the serum taken from the centrifuge tube by heating at 56 °C for 30–40 min (this is to be used in step 8).
6. Using a sterile pipette, carefully aspirate the PBMC layer with minimal underlying Ficoll Paque into a sterile 50 ml tube. (Multiple layers from a single donor may be added together to a total of 25 ml PBMC per 50 ml tube.)
7. Wash the PBMC by resuspending in an equal volume of complete medium and pelleting the cells by centrifuging at 150 g at 20 °C for 15 min, then resuspend in culture medium. (Note that the PBMC layers contain enough serum within the suspension to negate the need to add serum at this stage.)
8. Resuspend the final pellet in complete medium containing 10% (v/v) heat inactivated autologous serum (as prepared in step 5).

Protocol 3B

Preparation of lymphocytes from whole blood using differential centrifugation on a density gradient

The traditional method with minor modifications remains the best for small-medium sized laboratories without access to expensive blood separation equipment (43).
A number of different methods may be used to separate white blood cells from erythrocytes in whole blood. Here are methods for (A) human cells, and (B) murine cells. Differential

Protocol 3B continued

centrifugation on a density gradient using sterile vessels and solutions is a rapid method and gives high-purity lymphocyte preparations.

Equipment and reagents

- Centrifuge tubes
- Pasteur pipette (sterile)
- Lymphoprep™ (Nycomed Pharm AS)
- Endotoxin-free heparin
- Hanks balanced salt solution (Sigma)
- Culture medium: RPMI 1640 containing supplements (2 mM L-glutamine, 1 mM sodium pyruvate, 100 U/ml penicillin, 100 mg/ml streptomycin, 0.5 mg/ml fungizone, and 10% heat inactivated autologous human serum (see *Protocol 3A*, step 5))

A. Preparation of human lymphocytes

1. Venesect the subject and collect blood into vessel containing endotoxin-free heparin (10 U/ml blood).
2. Make a 1:2 dilution of blood with Hanks balanced salt solution and gently mix together by inversion.
3. Pipette 5 ml Lymphoprep™ into a centrifuge tube, then slowly overlay 10 ml of the diluted heparinized blood.
4. Centrifuge at 800 g for 20 min at 20 °C; lymphocytes are at the interface (middle layer).
5. Using a sterile Pasteur pipette, remove the lymphocytes without disturbing the upper layer.
6. Wash the lymphocytes by resuspending in 10 ml Hanks balanced salt solution, then pellet the cells by centrifuging at 250 g at 20 °C for 15 min.
7. Culture the cells by resuspending in cell culture medium (see below for recommendations on media and condition for epitope mapping).

Equipment and reagents

- Centrifuge tubes
- Ficoll Paque (ficoll-sodium diatrizoate)
- Triosil 75 (available commercially, and is sterile)
- Endotoxin-free heparin in tubes for blood collection
- Cell culture medium (see below for recommended media and conditions)

B. Preparation of murine lymphocytes

1. Collect blood into endotoxin-free heparin, mix blood by inversion.
2. Dilute blood with an equal volume of cell culture medium (minus fetal bovine serum).
3. Make a 9.2% (w/v) Ficoll Paque solution in distilled water, then sterilize by autoclaving.
4. Using aseptic conditions, prepare 50 ml of gradient solution: 43.4 ml of Ficoll solution and 6.6 ml of Triosil 75.

Protocol 3B continued

5. Add 2 ml of gradient solution to a centrifuge tube, then layer 5 ml of the diluted blood and centrifuge at 300 g for 15 min at 4 °C.
6. White blood cells are found at the interface between the upper supernatant and the plasma density gradient.
7. Using a sterile Pasteur pipette carefully remove the supernatant above the interface band of cells.
8. Collect the lymphocyte band, taking care not to collect an excessive amount of gradient solution thereby avoiding contamination with neutrophils.
9. Wash the cells three times by resuspending in 10 ml cell culture medium, and centrifuge at 150 g for 10 min at 4 °C. This will help to remove any residual platelet contamination.

5.4 Cell culture conditions

Cell culture conditions for proliferative assays affect the frequency of spontaneous false positives within both the negative control and test cultures. Therefore it is important to optimize conditions within the laboratory with respect to:

- cell culture medium
- serum supplementation
- antibiotic
- assay vessel
- incubation period
- pulse labelling

5.4.1 Culture medium

Pre-screening culture media and other media components is necessary in most assays, and particularly for large scale T cell epitope mapping using synthetic peptides. This is crucial to reduce all potential sources of background, spontaneous false positive proliferation, as well as provide optimal conditions for peptide-specific cellular expansion. Culture medium components should be screened in a similar fashion to serum using large replicate numbers of both positive and negative controls.

Batches of culture medium from different manufacturers vary in quality, not only in ability to support growth, but also in their tendency to give rise to non-specific proliferation. RPMI 1640 was specifically designed for human lymphocyte culture and gives a consistent performance. High backgrounds created by other media may be caused, at least in part, by components present that have various levels of B cell stimulatory endotoxins, or due to contamination with traces of T cell stimulatory substances, such as bacterial superantigens. Typical results of a standard medium screening assay using 48 replicates per negative control are shown in *Table 1* as an example.

Table 1 Non-stimulated background responses of human PBMC using various brands and batches of medium

Medium[a]	Mean[b]	Range		SD	SI[c]
A	72	41	168	25	356
B	393	144	1497	272	100
C1	427	171	1757	257	98
C2	354	62	2312	338	121
D1	310	98	803	154	111
D2	490	115	4169	565	93
E	613	247	1914	341	51
F(MEM)	1025	109	13747	2631	42
G(DMEM)	659	47	7384	1429	51
H	131	43	555	99	18

[a] Media used - A: ICN-Flow liquid RPMI 1640. B: Gibco liquid RPMI 1640. C: two batches of CSL Liquid RPMI 1640. D: two batches of CSL powdered RPMI 1640. E: Hazelton powdered RPMI 1640. F: CSL liquid MEM. G: CSL liquid DMEM. H: CSL serum-free Monomed liquid medium.

[b] 2×10^5 PBMC in complete medium without antibiotics, with or without antigen (PPD, 10 μg/ml) were incubated in 200 μl wells at 37 °C for five days and pulsed with 0.5 μCi/well [^{3}H]-TdR for 6 h. Results are expressed as the uptake of [^{3}H]-TdR in counts per minute (c.p.m.) of 48 replicates.

[c] Stimulation index = mean of positive antigen response (12 replicates) / mean of background response (48 replicates).

5.4.2 Serum supplementation

Serum as a cell culture supplement is the single most important culture-related factor for producing reliable, reproducible assay results. Unlike murine, rodent, and other cells, human PBMC react in a mitogenic fashion when fetal bovine serum (FBS) is added as a growth supplement, and their antigen-driven responses are suppressed. Thus, all the usual benefits associated with using FBS such as the use of large, pre-tested batches to enhance assay reproducibility, are totally negated.

5.4.2.1 Pooled human serum

For human PBMC proliferation assays, the culture-related factors that affect the frequency of spontaneous false positives within the negative control cultures were the base medium, antibiotic, and serum. The primary culture-related factor which affects the ability of cells to respond strongly to positively stimulating peptides is the serum and to a much lesser extent, the medium.

Therefore large batches of pre-tested sera, e.g. pooled human serum acquired from the National Blood Service, may be utilized. However, it is advisable to use human AB serum to avoid any blood group antibody-driven cross-reaction supplement. Sets of human PBMC cell stocks with known proliferative responses to a range of mitogens, antigens, and peptides are required. The quantity of cells in each set needs to be large enough to screen all individual serum batches for selection as to suitability for inclusion in the pooled human serum batch. For example, four different sets of human PBMCs to evaluate varying responses to

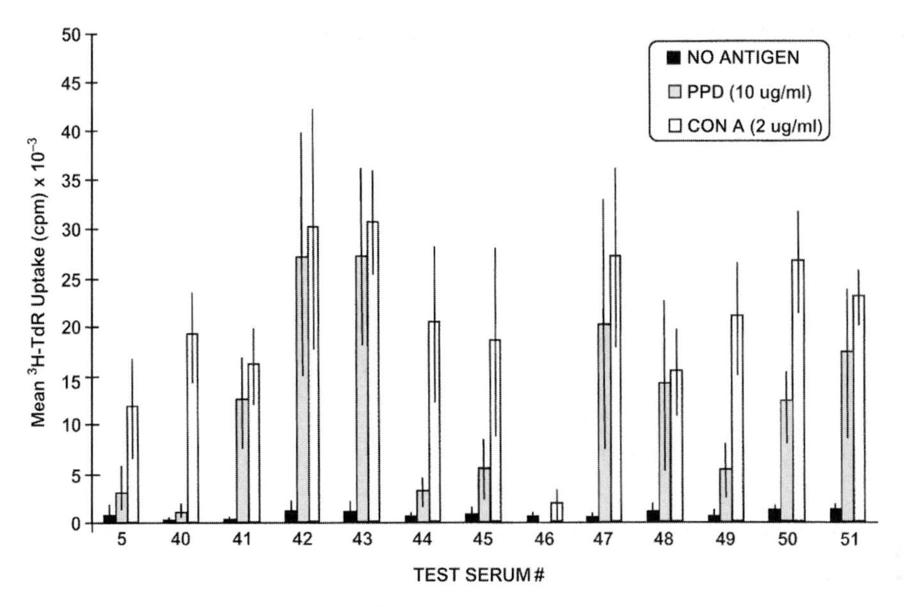

Figure 4 Effects of human serum supplements on human PBMC proliferation. PBMC (2×10^5/well) were cultured in U-bottom microculture wells in complete medium containing 10% (v/v) serum from individual donors. Cultures were without added antigen (no antigen), or had antigen, e.g. purified protein derivative (PPD), or mitogen (Con A) added. Con A was added three days after initiation of the cultures. Cultures were incubated for six days before pulse labelling with 0.5 μCi/well [^{3}H]-TdR for 6 h. Results are expressed as the mean c.p.m. of incorporated [^{3}H]-TdR + SD of eight replicates.

the test stimulators, such as, the mitogen PHA, strong antigens PPD and tetanus toxoid, and a range of T cell epitope peptides, e.g. influenza A virus, ovalbumin. Once suitable cells are defined, donors may donate large volumes of blood (~ 450 ml) from which PBMC are then isolated and stored in frozen aliquots. Each aliquot should be sufficient to screen several dozen individual test sera. Responses to the test stimulators varied from donor to donor even when using the same serum source due to differences in recognition and response to the stimulators provided. Serum screening assays generally used traditional T cell assay methods as the test stimulators usually provided good–excellent proliferation results (*Figure 4*). Sera supporting strong specific proliferation and low backgrounds can then be pooled, aliquoted, and stored at −20 °C or colder.

Protocol 4

Testing pooled serum to use in T cell epitope mapping assays

1 Select suitable human sera by using PBMC of several donors whose responses to known antigens are already well characterized.

Protocol 4 continued

2. Sera supporting strong specific proliferation and low backgrounds can then be pooled, aliquoted, and stored at −20 °C or colder.
3. Screen individual lots of human serum at 10% (v/v) in complete medium for the ability to support the proliferation of antigen- and mitogen-stimulated human PBMC while giving only low background proliferation levels in unstimulated cultures.
4. Assess proliferation as the mean c.p.m. of incorporated tritiated thymidine ([^{3}H]TdR) in four to eight replicate cultures in six day proliferation assays.
5. Use PBMC from three donors with pre-characterized responses to the test antigens, and a standardized 'control' serum batch were used in these assays.
6. Pool serum and sterilize by filtration through sterile 0.22 μm capsule filters (Sartorius GmbH, Germany), dispense into 50 ml or 100 ml aliquots, and store at −20 °C to −70 °C.

The protocol may be modified whereby the variable parameter is the culture medium for quality control purposes.

Table 2 Comparison of PBMC responses to antigens using either autologous serum (auto) or screened pooled human serum (PHS)

Donor	TT(1)[a]		TT(0.1)		PPD(10)		A/Sha(1)		p442(10)		Con A(5)	
	auto	PHS	auto	PHS	auto	PHS	auto	PHS	auto	PHS	auto	PHS
DM[b]	11.4	1.1	9.2	2.8	12.1	3.0	–[c]	–	–	–	9.8	15.5
	(2.4)	(0.5)	(3.3)	(0.8)	(2.7)	(1.0)					(1.5)	(2.8)
DG	25	1.4	–	–	25.1	3.2	–	–	–	–	17	18.7
	(5.0)	(1.2)			(4.5)	(0.6)					(14.5)	(4.5)
RC	–	–	12.5	9.5	15.5	6.6	18.7	2.6	2.3	1.3	28.9	23.2
			(1.5)	(3.8)	(4.9)	(2.2)	(6.1)	(1.0)	(2.6)	(0.9)	(12.1)	(4.8)
IR	27.5	10.7	–	–	40.7	32.8	–	–	1.2	1.3	8.9	21.2
	(10.2)	(2.8)			(5.4)	(3.8)			(0.4)	(0.6)	(1.0)	(3.8)
SP	–	–	34.4	13.5	17.5	5.0	–	–	–	–	13.5	12.8
			(8.1)	(3.1)	(3.6)	(0.9)					(2.5)	(2.3)
PR	2.5	1.9	–	–	7.1	7.9	–	–	–	–	6.3	7.3
	(0.6)	(0.7)			(1.0)	(0.6)					(1.9)	(1.5)
PB	–	–	2.5	3.7	15.4	9.6	–	–	–	–	7.5	22.9
			(1.0)	(0.8)	(3.8)	(1.2)					(1.3)	(1.9)

[a] Antigens tested were: TT, tetanus toxoid at 1.0 and 0.1 Lf/ml; PPD, tuberculin purified protein derivative at 10 μg/ml; A/Sha, A/Shanghai/11/87 whole influenza virus at 1 μg/ml; p442, TT peptide at 10 μg/ml, and Con A at 5 μg/ml. PBMC were incubated for six days in U-bottomed 96-well plates with or without antigen before pulse labelling with [^{3}H]-TdR for 6 h. Antigen tests were performed in eight replicates while cells alone were performed in 56 replicates.

[b] Results are expressed as stimulation indices (SI, mean c.p.m. of stimulated wells/mean c.p.m. of unstimulated wells using same serum). The standard deviation (SD) of the SI is shown in brackets under the SI value.

[c] – Not tested.

5.4.2.2 Autologous serum

An attractive alternative to screened human serum is autologous serum. PBMC and serum can be recovered simultaneously from the same whole blood sample in high yield by defibrination and dilution of blood with RPMI 1640 medium, followed by density interface centrifugation. This process provides more autologous serum than required for a 10% (v/v, final concentration) supplement in culture medium for proliferation assays at the cell densities commonly used. Autologous serum is at least as good as screened human serum (pooled) in supporting cellular proliferation (*Table 2*). Moreover, autologous serum can be heat inactivated at 56 °C for 30 min without loss of growth-supporting qualities. It is noteworthy that to mix 'poor' sera with 'good' sera in an effort to make the supply of serum last longer is not always expedient (*Figure 5*). The economical use of 'good' sera can be made by using them at 5% (v/v) if necessary as there was little observable difference in antigenic proliferation of PBMC at 5% or 10% serum concentrations (*Table 3*).

In most cases media containing autologous serum produced higher stimulation indices than media containing pooled serum. Advantage of using autologous serum:

- no pre-screening is required
- very easily obtained in sufficient quantity for the experiment
- no additional processing to remove anticoagulants or fibrinogen
- may be stored frozen at −70 °C from a previous bleed of the same donor without loss of efficacy

Table 3 Effect of autologous serum concentration in PBMC proliferation assays

		Serum concentration							
Donor	**Antigen**	**5%**		**10%**		**15%**		**20%**	
		mean[a]	**%resp[b]**	**mean**	**%resp**	**mean**	**%resp**	**mean**	**%resp**
C. R.	Bkgd[c]	353(93)	–	357(116)	–	327(107)	–	346(167)	–
	PPD(10)[d]	939(145)	92	1296(232)	100	1620(360)	100	1376(413)	92
	A/Sha(1)	2871(722)	100	4274(1382)	100	5276(1095)	100	6279(1199)	100
	p480(10)	937(316)	46	811(196)	21	1222(591)	67	1273(548)	33
I. H.	Bkgd	1164(289)	–	1229(292)	–	1292(354)	–	969(352)	–
	PPD(10)	4717(1962)	100	5044(560)	100	4729(788)	100	4086(658)	100
	TT(0.1)	3479(956)	100	3813(717)	100	NT	NT	3098(597)	100
	p459(1)	2322(220)	25	2637(236)	42	2640(665)	25	2968(880)	54
R. L.	Bkgd	291(58)	–	363(132)	–	290(89)	–	406(201)	–
	PPD(10)	1016(465)	83	1555(664)	75	1003(326)	100	1302(257)	83
	TT(1)	9074(1768)	100	11712(1084)	100	11373(2725)	100	10349(2035)	92
	A/Sha(1)	5662(795)	100	7553(2137)	100	7129(1839)	100	9323(1556)	100

[a] Mean c.p.m. and SD.

[b] %Responders = % of wells with c.p.m. values above the cut-off determined by the background analysis algorithm (refer to Section 6).

[c] Background = mean c.p.m. of all wells which were below the cut-off.

[d] Antigens tested were: tuberculin purified protein derivative (PPD, μg/ml), tetanus toxin (TT, Lf/ml), A/Shanghai/11/87 whole influenza virus (A/Sha, μg/ml), and peptides p480 and p459 (μg/ml). 12 replicates per group were used.

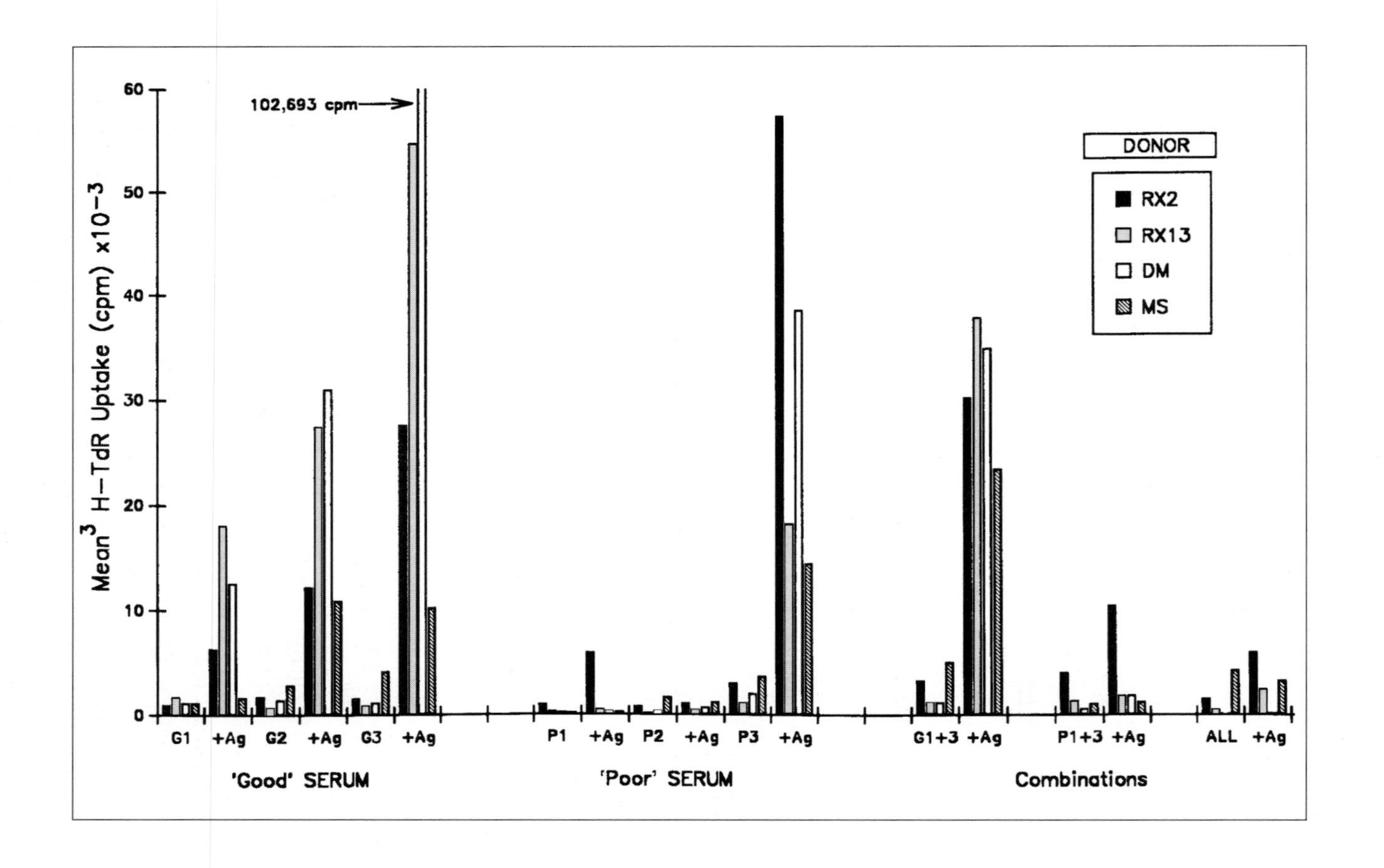
DONOR
RX2
RX13
DM
MS
102,693 cpm
Mean ^{3}H–TdR Uptake (cpm) x10-3
0
10
20
30
40
50
60
G1
+Ag
G2
+Ag
G3
+Ag
'Good' SERUM
P1
+Ag
P2
+Ag
P3
+Ag
'Poor' SERUM
G1+3
+Ag
P1+3
+Ag
ALL
+Ag
Combinations

Figure 5 Pooling of satisfactory ('good') and unsatisfactory ('poor') sera for use in PBMC proliferation assays. Sera which either failed to provide adequate capacity to allow cells to proliferate or that caused high background proliferation was defined as 'poor' sera. Such 'poor' sera may be able to be mixed with sera that provided adequate proliferation support and low background proliferation (defined as 'good' sera) if the 'poor' sera merely lacked essential components that may be oversupplied in 'good' sera. Pooling of such sera would prolong the supply of the serum-supplement for use in proliferation assays. PBMC (2×10^5/well) from four donors (RX2, RX13, DM, and MS) were cultured in complete medium containing 10% (v/v) of the stated serum. 'Good' sera (G1, G2, G3) and 'poor' sera (P1, P2, P3) were tested either individually, as pools of three 'good' (G1 + 3) or 'poor' (P1 + 3) sera, or as a pool of all six sera (ALL). Each lot of PBMC was tested with no added antigen (G1, P1, etc.) or with PPD (10 μg/ml) (+Ag). Incubation and labelling were as for *Figure 4*. Results are expressed as the mean c.p.m. of incorporated [^{3}H]-TdR of eight replicates.

Autologous serum used at a concentration of 5% to 20% (v/v) in cell culture medium is suitable for antigen-driven proliferation. PBMC and serum can be recovered simultaneously from the same whole blood sample in high yield by:

- defibrination of blood using glass beads (see *Protocol 2*)
- dilution of blood with RPMI 1640 medium
- density interface centrifugation (see *Protocol 3*)

5.4.5 Antibiotics

Most cell culture texts advise the use of penicillin and streptomycin (P/S). However the negative control cultures of PBMC from some donors give unexpectedly high frequencies of 'spontaneous' background proliferation in medium containing this antibiotic combination (*Figure 6*). This spontaneous proliferation may be due to penicillin, to which some people are allergic. Gentamicin is free of such problems and does not contribute to spontaneous proliferation. Therefore gentamicin is the antibiotic of choice for the assays described here.

5.4.6 Assay vessels

The effect of culture well shape on antigen-stimulated PBMC proliferation can also affect the success of results in large scale T cell epitope mapping assays. To determine if well shape had an effect on PBMC proliferation assays, we incubated cells in both U-bottom and flat-bottom wells for a range of times with various antigens. The frequency of false positive responses in negative (cells alone) control wells and the frequency of positive responses to strong whole antigens such as tetanus toxoid were found to be largely unaffected by well shape. However, cells responding to known peptides epitopes (e.g. 12-mer epitopes from tetanus toxin) responded more frequently in the U-bottom wells than in flat-bottom wells, at a range of peptide concentrations, regardless of the time of incubation. This may be due to the greater cell–cell contact afforded in U-bottom compared to flat-bottom wells. Thus U-bottom wells are used in the human PBMC assays described here. Other experiments were performed using V-bottom culture wells. Results from these experiments showed that V-bottom culture

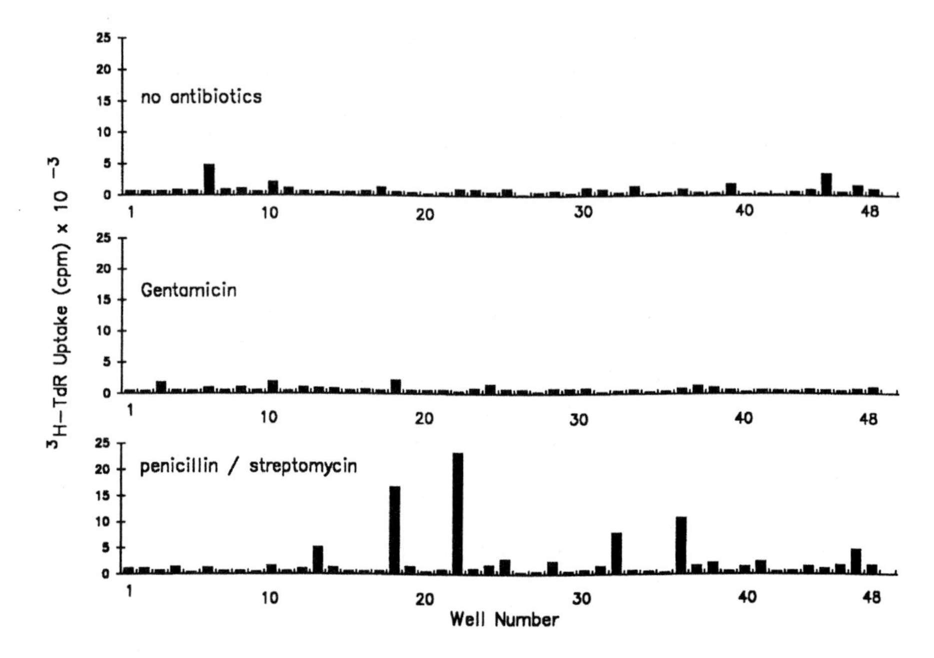

Figure 6 Effect of antibiotics on the background proliferation of human PBMC. PBMC (2×10^5/200 μl well), without added antigen, were cultured in U-bottom microculture wells in complete PHS medium without antibiotics (no antibiotics). In two further groups of wells the medium was supplemented with gentamicin (20 μg/ml), or penicillin (100 IU/ml) and streptomycin (100 μg/ml) (penicillin/streptomycin). After six days, cultures were pulse labelled with 1.0 μCi/well [^{3}H]-TdR for 6 h before harvesting and scintillation counting of the DNA. Results are expressed as the incorporation of [^{3}H]-TdR into DNA in c.p.m. per individual well.

wells could be used successfully if the culture volume was scaled down to 50% and the cell numbers scaled down to 7.5–20% of that used in U-bottom wells (data not shown). Therefore if cultures of 2×10^5 cells in 200 μl of medium were used in a typical U-bottom well assay, only $1.5–4 \times 10^4$ cells in 100 μl medium would be used in V-bottom well cultures under otherwise similar assay conditions to provide comparable uniformity of results.

5.4.7 Incubation period

In proliferation assays, the incubation time and the number of PBMC added per well are interrelated factors that must be adjusted together. It should be noted that high initial cell concentrations and long incubation times might dramatically reduce proliferation. There may be one or more reasons for this:

(a) Exhaustion of nutrients.

(b) Failure of the buffering capacity of the medium to maintain a physiological pH due to an accumulation of acidic metabolic products.

(c) The inhibition of cell growth due to either contact inhibition or decreased nutrient diffusion in dense cell populations.

Use of too few cells per well requires a very large number of test wells for the detection of antigen-specific precursors at biologically significant frequencies (e.g. > 1 per million PBMC). Our studies showed that short incubation times (four days) resulted in the sensitive detection of positive responses due mainly to the low and consistent backgrounds.

5.5 Pulse radiolabelling

Maximum sensitivity of detection of cellular proliferation is vital for the precise calculation of accurate antigen-specific precursor frequencies when undertaking limiting dilution analyses of cells. Optimal experimental conditions for peptide-driven PBMC proliferation should not only avoid non-specific stimulation, but also provide an environment in which specifically stimulated cells can proliferate uninhibited. When evaluating these methods, one should not simply look for the greatest magnitude of thymidine incorporation or stimulation index in such assays, but rather for the highest sensitivity and reliability in detecting and quantifying antigen-specific T helper cells. As a general rule, the optimal time for incubation of PBMC proliferation assays is around four days.

Exogenous thymidine is quickly assimilated into the intracellular pool and used for DNA synthesis during the S phase of the cell cycle (44–46). Therefore, the addition of small quantities of [^{3}H]-TdR (known as 'trace labelling') can be used to measure rates of DNA synthesis, provided the thymidine is of low specific activity (< 2.0 Ci/mmol). However, use of low specific activity thymidine can lead to an excess of total thymidine, changing the conditions to those of 'flood labelling'. Use of high specific activity thymidine (40–80 Ci/mmol) results in cytotoxicity, probably due to radiological damage. We compared the use of multiple small doses of thymidine over several days with a single dose incubated for various periods up to 27 h. Multiple small doses do not seem to significantly enhance the total incorporation of thymidine. Any advantage in total incorporation by antigen-stimulated cultures may be lost due to increased incorporation in the unstimulated controls.

Trace labelling, using a single low dose of high specific activity thymidine, is the preferred choice for use in large scale peptide mapping assays since the cultures are terminated after a short labelling time (6 h) and the intent is primarily to detect significant differences between proliferating and non-proliferating cultures. As thymidine is sold by radioactive content, a small dose of high specific activity is the most economical way to radiolabel. Under these conditions the dose of thymidine is limiting and incorporated c.p.m. are thus proportional to the rate of DNA synthesis at the time of the addition of the radiolabel.

5.5.1 Random T cell background proliferation and response variation

The probable participation of non-specifically stimulated bystander cells within *in vitro* T cell activation assays (often referred to as the 'recruitment' of uncommitted cells) was recognized by Marshall *et al.* (1969). Marshall *et al.* (1969) included such cells in the '3-cell system' of T cell activation, which comprises antigen-presenting cells, antigen-specific T cells, and unprimed T cells that are recruited secondarily during the response (47, 48).

Bystander cells (defined as responsive, non-specifically stimulated T cells and B cells) have been shown to contribute significantly to the magnitude of *in vitro* proliferative responses during antigen-specific proliferation assays. The effects of bystander cells, if allowed to incorporate radiolabel, may be confused with antigen-specific cells. In some limiting dilution assays, irradiated or mitomycin C (MitC)-treated autologous or syngeneic filler cells may be used to maintain a constant cell number in each well (49, 50). However, this method does not reduce the effects of replication-competent bystander cells included in the non-treated, limiting PBMC population being tested. A modified version of the limiting dilution assay utilizes the capacity of primed memory T cells to maintain the ability to proliferate after mild radiation treatment compared to that of unprimed T cells (a component of the bystander cell population) which were unable to proliferate after such treatment. Such a treatment can significantly reduce the numbers of estimated responding cells in limiting dilution assays compared to similar assays using the entire sample PBMC population. However, responses due only to unprimed T cells were not reported and as such, the contribution of unprimed T cells to the overall proliferation of a culture must be considered to be non-antigen specific and similar to the contribution by other bystander cell types.

The recruitment of bystander cells in proliferation assays may be the result of blastogenic factors produced by the specifically stimulated cells. Observations of the effects of bystander cells on PPD-stimulated human PBMC proliferation assays using 5-BrdU and light to treat antigen-specific T cells have shown that T cells treated in such a manner can not replicate without either the presence of bystander cells, or culture fluids from non-treated, PPD-stimulated PBMC. Thus, the proliferative capacity of the PPD-specific cells may be significantly affected by the presence of bystander cells. Moreover, a large number of non-specific T cells may be induced to proliferate as a result of the presentation of PPD to PPD-specific T cells. Thus, population of proliferating PBMC observed *in vitro* after antigen stimulation is composed of a substantial number of non-specific bystander cells and that the activation and proliferation of these cells augments the proliferation of the specifically stimulated population.

6 Data analysis using the ALLOC algorithm

There is a certain degree of inter- and intra-assay variability observed with respect to T cell proliferation. The variety of methods used for reporting the outcome of these tests has led to the development of the ALLOC algorithm to analyse data objectively and empirically, as well as to standardize the method of analysis used for large scale T cell epitope mapping studies.

Analysis of multi-replicate human PBMC proliferation assays for the identification of antigen-specific T cell epitopes may be undertaken using a computer algorithm. The algorithm (ALLOC) designed by Chiron Mimotopes Pty Ltd. estimates the responding cell precursor frequencies by using a novel method of data sampling and Poisson distribution analysis. In total, four different methods were devised for the estimation of the assay background. The majority of the proliferation assay data shown in this chapter used Method 1, described immediately below.

Protocol 5

Method 1 of the ALLOC algorithm

1. Selects random counts per minute (c.p.m.) samples from the total data set (test and control groups of a single assay) assuming that the means of these random samples will be normally distributed (Central Limit Theorem) (51).
2. Sampled results that lie outside the 95% confidence level (t-test) of the determined mean are identified and flagged.
3. Repeat the process 1000 times.
4. Results flagged at least 95% of the time are considered to be non-background and removed from the initial data set.
5. The entire process must subsequently be repeated until no new non-background results are found.
6. Calculate the cut-off c.p.m. value from the mean plus 3SD of all remaining (background) results.
7. Results from the original data set which are above the cut-off are considered to be positive, non-background responses.
8. The estimated mean precursor frequency of responding cells per 10^6 PBMC tested and the 95% confidence limits around the mean of each group (tests and unstimulated negative control) are determined using Poisson statistics, assuming a single-hit model.
9. The lower 95% confidence limit of the precursor frequency mean of each test group is compared for overlap with the upper 95% confidence limit for the negative control group.
10. Non-overlapping confidence limits result in a 99.75% ($P < 0.0025$) confidence level and this test was used to identify statistically significant positive experimental groups.
11. A confidence level of 95% ($P < 0.05$) results when the mean of the precursor frequency of the test group lies above the upper confidence limit of the negative control group.

Protocol 6

Method 2 of the ALLOC algorithm

To determine the background values from proliferation assay.

1. Rank-order the entire data set starting from the low end of the ranked series.

Protocol 6 continued

2. The mean and root-mean-square deviation (SD, standard deviation of a normally distributed population) is determined for the first N values (where N was chosen to be equal to the number of replicates in the negative control group). This number is used as a convenient starting point.
3. Calculate a temporary cut-off value, i.e. the mean + 3SD of these first N values.
4. All the values below the temporary cut-off may used to recalculate a new mean, SD, and cut-off.
5. Repeat the process until no additional values could be added to the background data set.
6. The algorithm permits the number of values to diminish if any of the values within the original set were above the cut-off.
7. Calculate the final cut-off (mean + 3SD of the final background data set). Values with c.p.m. above this cut-off were the number of positives per group and were used to determine the Poisson means and estimates of precursor frequency in an identical operation to that described for Method 1 (*Protocol 5*).

Two additional methods of assay background determination may be used (Methods 3 and 4) when an assay consisted of more positive responses than background. Methods 3 and 4 of the ALLOC algorithm are used for comparison of analysis methods; Method 3 used the actual values of the negative control group and applied the iterative algorithm (described for Method 2, above) to these values only.

Method 4 used the negative control group values only to calculate a mean + 3SD cut-off and did not apply the iterative algorithm to select values which may not fall within a normal distribution. For the determination of assay background, this method is identical to the conventional proliferation assay data analysis method in which the background mean c.p.m. and SD values are calculated using all negative control cultures. However it differs in the use of estimated responding cell precursor frequencies for the quantitative comparison of results.

The ALLOC computer algorithm is menu-driven (*Figure 7*) which allows for the simplified entry of assay parameters during the automatic, optimal 'ALLOC'ation of the positions of test and control assay wells on microculture trays (*Figure 8*). Automatic well allocation by ALLOC uses a set of rules.

Assay design rules included the use of:

(a) A minimum of 16 replicates.

(b) The number of negative control wells to be no fewer than for any test group.

(c) The positioning of assay controls on every tray, the ability to place replicates of a group over multiple trays, the choice of either an 8 or 12 well replication format.

(d) The input of the number of cells used per well in order to estimate precursor frequencies.

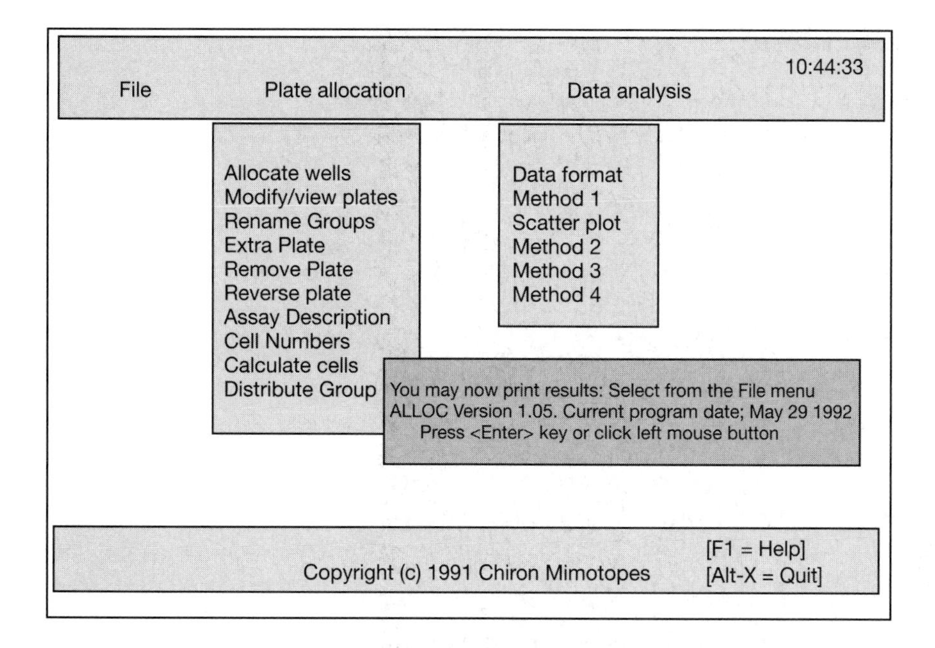

Figure 7 Video screen image of ALLOC, a computer algorithm for the design and analysis of proliferation assays. The algorithm, written at Chiron Mimotopes Pty Ltd. by Mr Paul Bennell and Dr Tom Mason, uses a data analysis method conceived by Dr Mario Geysen and a set of assay layout rules designed by Dr Stuart Rodda and myself. The algorithm is pull-down menu driven, allowing ease of use for the input of various assay parameters. The algorithm designs an assay according to various rules such as minimum numbers of replicates for tests and controls, the requirement for controls on each plate, and the chosen method of cell dispensing. Following the assay design and output of a printed record (shown in *Figure 8*), the results of the assay in the form of raw c.p.m. data from a variety of sources and file types, can be loaded and automatically analysed. The data analysis routines allow several choices as to the method of determining the most appropriate set of cultures comprising the assay background.

Figure 8 Example of computer-assisted assay design. The automatic placement of test wells for a typical multi-replicate human PBMC proliferation assay in a 96-well microculture tray, designated by the 'ALLOC' computer algorithm, is shown. Only five of the nine trays of this assay are depicted. Trays are labelled sequentially Plate 1, Plate 2, etc., while the positions of individual replicates of each test group are defined using the same (group) number. Generally, the group name 'Group *n*' was renamed by the operator during the set-up routine of the algorithm to a more descriptive title, such as those used for groups 45–49. The algorithm used a set of rules which assist in the design of reliable and consistent assays. Two examples of these rules are (i) the use of various control wells on every tray and (ii) the total number of negative control wells cannot be less than the mean number of replicates in each test group. The algorithm used a menu-type questionnaire system which allows the operator to define many assay parameters such as the total number of test groups, the desired numbers of replicates per test group, whether the preferred layout is column or row orientated, the anticipated numbers of cells per well, and the types of controls required. After parameter definition, the algorithm then designed a suitable assay layout which could be modified later, if required, in the event of changes made during the actual assay set-up. A summary of the requirements for the assay was also printed (shown at the bottom of the figure). The layout was then stored and later recalled for the automatic analysis of c.p.m. data on completion of the assay. *See overleaf*→

```
T-Cell assay: n8envpol                                Page 1
          Plate   1                         Plate   2
       H  G  F  E  D  C  B  A         H  G  F  E  D  C  B  A  Group
Names
  1    1  1  1  1  1  1  1  1         1  1  1  1  1  1  1  1    1 Group 1
                                                                2 Group 2
  2    2  2  2  2  2  2  2  2         2  2  2  2  2  2  2  2    3 Group 3
                                                                4 Group 4
  3    3  3  3  3  3  3  3  3         3  3  3  3  3  3  3  3    5 Group 5
                                                                6 Group 6
  4    4  4  4  4  4  4  4  4         4  4  4  4  4  4  4  4    7 Group 7
                                                                8 Group 8
  5    5  5  5  5  5  5  5  5         5  5  5  5  5  5  5  5    9 Group 9
                                                               10 Group
10
  6    6  6  6  6  6  6  6  6         6  6  6  6  6  6  6  6   41
GP120(0.5)
                                                               42
P24(0.5)
  7    7  7  7  7  7  7  7  7         7  7  7  7  7  7  7  7    N Cells
alone
                                                               P1 Tet Tox
(1)
  8    8  8  8  8  8  8  8  8         8  8  8  8  8  8  8  8   P2 Con A
                                                               11 Group
11
  9    9  9  9  9  9  9  9  9         9  9  9  9  9  9  9  9   12 Group
12
                                                               13 Group
13
 10   10 10 10 10 10 10 10 10        10 10 10 10 10 10 10 10   14 Group
14
                                                               15 Group
15
 11    N  N  N  N  N  N  N  N         N  N  N  N  N  N  N  N   16 Group
16
                                                               17 Group
17
 12   P1 P1 41 41 42 42 P2 P2        P1 P1 41 41 42 42 P2 P2   18 Group
18
                                                               19 Group
19
          Plate   3                         Plate   4
       H  G  F  E  D  C  B  A         H  G  F  E  D  C  B  A
  1   11 11 11 11 11 11 11 11        11 11 11 11 11 11 11 11   20 Group
20

  2   12 12 12 12 12 12 12 12        12 12 12 12 12 12 12 12
```

```
 2  12 12 12 12 12 12 12 12     12 12 12 12 12 12 12 12

 3  13 13 13 13 13 13 13 13     13 13 13 13 13 13 13 13

 4  14 14 14 14 14 14 14 14     14 14 14 14 14 14 14 14

 5  15 15 15 15 15 15 15 15     15 15 15 15 15 15 15 15

 6  16 16 16 16 16 16 16 16     16 16 16 16 16 16 16 16

 7  17 17 17 17 17 17 17 17     17 17 17 17 17 17 17 17

 8  18 18 18 18 18 18 18 18     18 18 18 18 18 18 18 18

 9  19 19 19 19 19 19 19 19     19 19 19 19 19 19 19 19

10  20 20 20 20 20 20 20 20     20 20 20 20 20 20 20 20

11   N  N  N  N  N  N  N  N      N  N  N  N  N  N  N  N

12  P1 P1 41 41 42 42 P2 P2     P1 P1 41 41 42 42 P2 P2
           Plate 9
    H  G  F  E  D  C  B  A  Group Names
 1  45 45 45 45 45 45 45 45  45 gp120(10)
                             46 gp120(5)
 2  46 46 46 46 46 46 46 46  47 p24(10)
                             48 p24(5)
 3  47 47 47 47 47 47 47 47  49 p24(2)
                             N Cells alone
 4  48 48 48 48 48 48 48 48  0 Empty
                             0 Empty
 5  49 49 49 49 49 49 49 49  0 Empty
                             0 Empty
 6  N  N  N  N  N  N  N  N


 7  N  N  N  N  0  0  0  0


 8  0  0  0  0  0  0  0  0    Cell requirements for assay 4n8envpol
(9 plates)

 9  0  0  0  0  0  0  0  0    Total number of wells in assay: 820
                              Total number of cells required:
164,000,000
10  0  0  0  0  0  0  0  0    Total negative controls       : 76
                              Total controls - positive #1  : 16
11  0  0  0  0  0  0  0  0    Total controls - positive #2  : 16


12 0  0  0  0  0  0  0  0
```

Poor estimations of assay backgrounds to determine cut-off points above which lie the positive values can lead to wide variations in the interpretation of a positive response. The algorithms used in ALLOC Methods 1 and 2 initially examine the complete assay data set for background values without consideration of the groups from which the values arise. This represents a more objective means to establish the population of background values. Both Methods 1 and 2 remove values, deemed to be part of the background, from subsequent calculations of the mean and SD values of positively responding cultures in test and control groups, providing a much more logical method of comparing the results from these groups as described below.

7 Example of assay conditions for large scale mapping of T cell epitopes

7.1 Individual steps in the T cell epitope mapping protocol

(a) Peptides cleaved into 0.05 M Hepes/20% (v/v) acetonitrile, depending on peptide yield.

(b) Culture medium comprising screened Hepes-buffered RPMI 1640 with 10% autologous serum or screened, pooled human serum, plus gentamicin.

(c) Culture 2×10^5 PBMC/well in multiple wells (eight or more) of U-bottom trays or 2–4 $\times 10^4$ PBMC/well in V-bottom trays for four days, depending on the availability of cells.

(d) Trace label with 0.25 μCi/well of high specific activity (40–80 Ci/mmol) [^{3}H]-TdR for the final 6 h of the incubation period.

(e) Harvest the cells and count the incorporated label (*Protocol 7*).

(f) Analyse the results using the ALLOC algorithm Method 1 that takes into account any positive cultures in the negative control group and any non-responding cultures in test groups.

Data Analysis Strategies Development of an algorithm to estimate the responding cell precursor frequency within proliferating PBMC cultures.

Many variations to this basic procedure may be made with respect to:

- cell concentrations
- well shape and volume
- incubation times
- pulse labelling parameters
- antigen concentrations
- number of replicates

Protocol 7

Assay set up procedure

Equipment and reagents

- U-bottom trays or V-bottom trays
- Glass fibre filter mats (LKB-Wallac type-A)
- Semi-automated harvester (model 1295, LKB-Wallac Oy)
- Piston type, oil-less vacuum pump (Thomas Industries Inc.)
- PBMC
- [^{3}H]-TdR: purchased as sterile aqueous solution, usually at a high specific activity of 40–60 Ci/mmol and in 5.0 mCi (185 MBq) quantities (ICN Biomedicals)

Method

1. Add PBMC in either CPHSM[a] or CAM[b] at 1.25–2.5 × 10^5 cells per 180 μl in U-bottom trays, or at 1.5–4 × 10^4 cells per 90 μl medium in V-bottom trays.
2. Incubate the trays at 37 °C for four days.
3. Pulse label with [^{3}H]-TdR for the final 6 h of this period.
4. Harvest the cells onto glass fibre filter mats, then perform scintillation count.

Pulse labelling with radionucleotide and scintillation counting

1. Use tritiated thymidine ([methyl-^{3}H]-thymidine, [^{3}H]-TdR) for pulse labelling procedures, purchased as sterile aqueous solution, usually at 40–60 Ci/mmol and in 5.0 mCi (185 MBq) quantities.
2. Dilute 5 ml (5.0 mCi) of reagent in complete medium so that 10 μl of solution provides 0.25 or 0.5 μCi [^{3}H]-TdR respectively. (Undiluted and diluted radiolabel solutions may be stored at 4 °C for up to two months.)

Pulse labelling is used as a measure of cellular proliferation, using the direct relationship between the increment of [^{3}H]-TdR incorporation into daughter DNA and cellular replication.

3. Generally, 100 μl or 200 μl cultures in 96-well cell culture trays are pulse labelled by the addition of 10 μl or 20 μl (0.25 or 0.5 μCi) of diluted [^{3}H]-TdR respectively, to provide a dose of 2.5 μCi [^{3}H]-TdR per ml of culture.
4. Return the pulse labelled trays to the incubator for 6 h.
5. Harvest the cells with a semi-automated harvester (model 1295) attached to a piston type, oil-less vacuum pump.

(Cells may be stored at 4 °C overnight, or at −20 °C for periods of up to one week. The latter two procedures were used when immediate harvesting was not possible.)

6. Lyse cells by osmotic pressure to release and remove internalized, non-incorporated [^{3}H]-TdR.
7. Collect cellular DNA onto pre-printed glass fibre filter mats.
8. Dry filter mats in a microwave oven operating at 500 watts for 5 min, or overnight in a fume cupboard.
9. Count beta emissions in a scintillation counter.

[a] Complete Pooled Human Serum Medium.
[b] Complete Autologous Serum Medium

7.2 Example of results obtained

7.2.1 Standard mean ± SD analysis of the example data set

(a) Results expressed as the mean c.p.m. ± SD responses for the 48 replicate tests used in the pooled peptide example proliferation assay are shown in *Figure 9A*.

(b) Positive responses observed to peptide pools 1 and 2.

(c) Peptide pools 1 and 2 have been 'decoded' to examine the responses to individual peptides within these pools.

(d) The mean ± SD c.p.m. results of 12 replicate tests of these individual peptides are shown in *Figure 9B*.

7.2.2 [^{3}H]-TdR uptake frequency analysis of the example data set

(a) Frequency distribution analysis of the 48 individual c.p.m. datum within each pool of the assay, shown in *Figure 9*, was undertaken.

(b) The results, expressed in Log_{10} c.p.m. distribution packets, are shown in *Figure 10*.

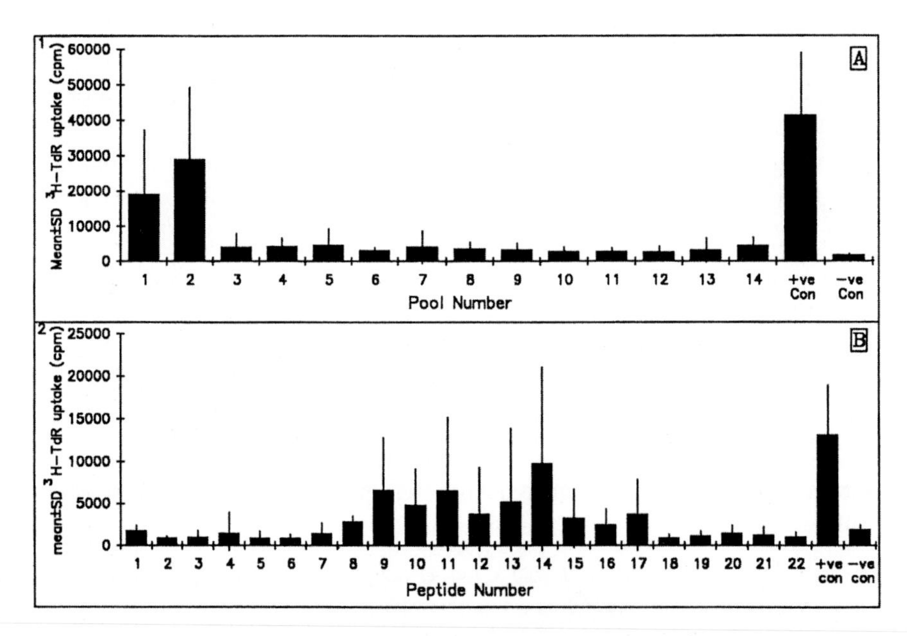

Figure 9 Peptide pooling: an example experiment. Human PBMC (2 × 10^{5}cells/200 μl U-bottom well) were cultured with fully overlapping 12-mer peptides spanning the entire sequence of the MPB-70 antigen (a component of PPD from *Mycobacterium tuberculosis*) for seven days and pulse labelled with 1.0 μCi/well [^{3}H]-TdR for 6 h followed by standard harvesting and scintillation counting procedures. (A) Peptides were pooled into 14 pools each containing 1.0 μl of each of eleven 12-mer peptides and cultures set up in 48 replicates. (B) Individual peptides from pools 1 and 2 in (A) were tested at 1.0 μl per well in replicates of 12 wells each. The positive antigen was 10 μg/ml whole MPB-70 antigen and negative control wells contained no antigen. Results are expressed as the mean ± SD of replicate cultures.

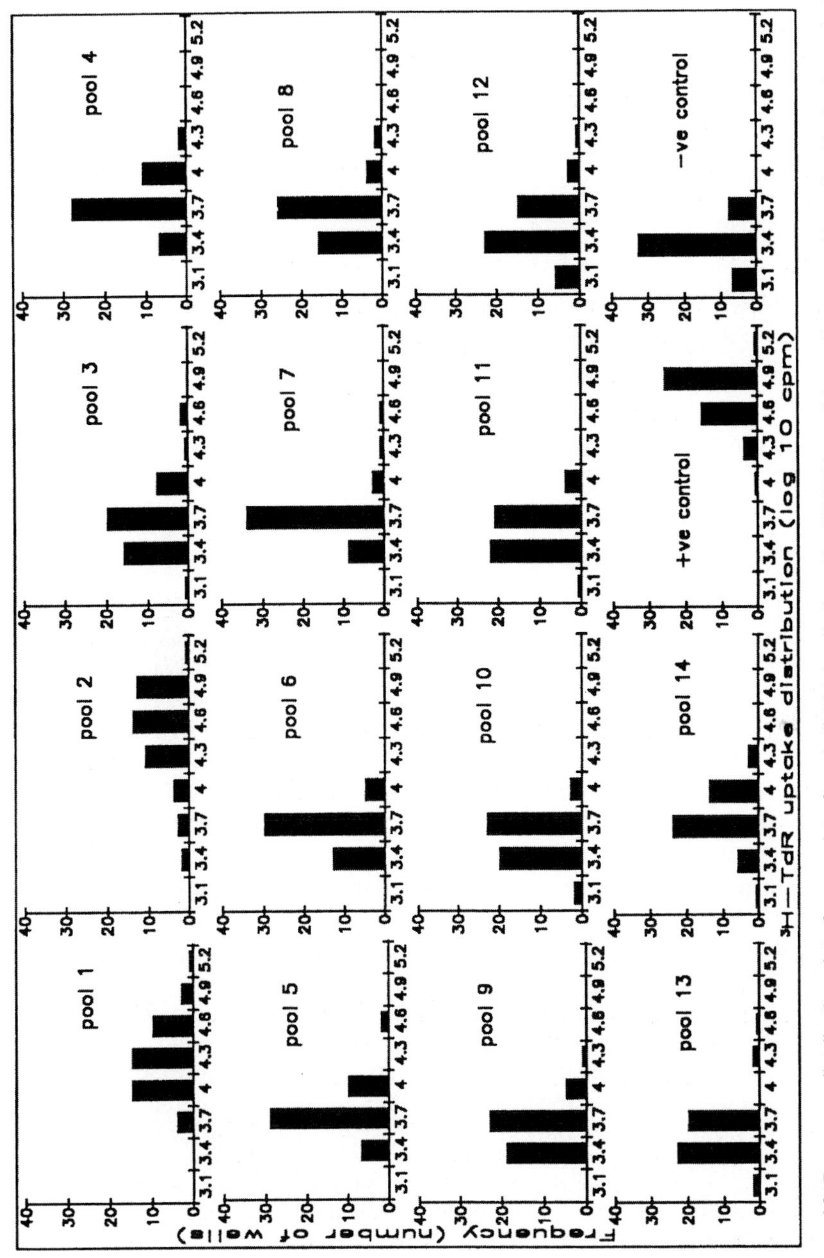

Figure 10 Frequency distribution plot of c.p.m. data from individual wells from the MPB-70 peptide pooling assay (described in the legend of *Figure 9*). The c.p.m. values of each culture were Log_{10} transformed. Results of each pool are shown and expressed in terms of the frequency of cultures within each class interval (size = 0.3, with the first class beginning at 3.1).

group	mean	SD
1	19156.458	18402.949
2	29198.981	20381.155
3	4230.7354	4064.051
4	4377.5208	2662.4241
5	4829.9396	4867.5581
6	3257.65	1132.0887
7	4262.1354	4794.0862
8	3712.3646	2115.62
9	3474.2938	2216.7211
10	3019.6938	1516.785
11	2974.9396	1302.4184
12	2765.4958	2000.4531
13	3379.9063	3683.2948
14	4687.9167	2635.5979
pos	41524.444	17930.839
neg	1902.8458	686.55928
neg	8 + 3SD=	3962.5237

Figure 11 Percentage of positively responding cultures to pooled MPB-70 peptides. This data is recalculated from the experiment described in the legend for *Figure 9*. Results are expressed as the percentage of cultures (out of 48 replicates) with c.p.m. values greater than the mean + 3SD of the negative control.

(c) Most of the responses of individual replicate cultures within pools 1 and 2 showed enhanced levels of [^{3}H]-TdR uptake. The responses of most cultures within these pools could be clearly distinguished from responses within the negative control group (and most other groups) at a cut-off level of about 10 000 c.p.m.

(d) Rearrangement of this same data, expressed in terms of the number of wells per pool with c.p.m. values exceeding the mean + 3SD of the negative control group, is shown in *Figure 11*.

(e) In this assay the SD of the negative control group are relatively low (36% of the mean) compared to some PBMC proliferation assays encountered. The strength of the response to recognized peptides are relatively high (96%, 94%, and 100% of wells tested to pools 1, 2, and the positive control MPB-70 whole antigen respectively) indicating a relatively high precursor frequency of specific responding T cells in the PBMC sample.

This example assay identifies several clear-cut, strong positive responses but also indicates that other regions of the sequence may also be immunogenic but not immunodominant.

References

1. Schellekens, P. T. A. and Eijvoogel, V. P. (1968). *Clin. Exp. Immunol.*, **3**, 571.
2. Du Bois, M. J. G. J., Huismans, D. R., Schellekens, P. T. A., and Eijsvoogel, V. P. (1973). *Tissue Antigens*, **3**, 402.
3. Van Oers, M. G. J., Pinkster, J., and Zeijlemaker, W. P. (1978). *Eur. J. Immunol.*, **8**, 477.
4. Sohnle, P. G. and Collins-Lech, C. (1981). *J. Immunol.*, **27**(2), 612.
5. Knight, S. C. (1982). *J. Immunol. Methods*, **50**, R51.
6. Hensen, E. J. and Elferink, D. (1984). *Hum. Immunol.*, **10**, 95.
7. Linnemann, T., Wiesmuller, K. H., Gellrich, S., Kaltoft, K., Sterry, W., and Walden, P. (2000). *Ann. Oncol.*, **11** (Supp 1), 95.
8. Geysen, H. M., Meloen, R. H., and Barteling, S. J. (1984). *Proc. Natl. Acad. Sci. USA*, **81**, 3998.
9. Geysen, H. M., Rodda, S. J., Mason, T. J., Tribbick, G., and Schoofs, P. G. (1987). *J. Immunol. Methods*, **102**, 259.
10. Shukla, D. D., Tribbick, G., Mason, T. J., Hewish, D. R., Geysen, H. M., and Ward, C. W. (1989). *Proc. Natl. Acad. Sci. USA*, **86**(21), 8192.
11. Ho, P. C.-L., Mutch, D. A., Winkel, K. D., Saul, A. J., Jones, G. L., Doran, T. J., *et al.* (1990). *Eur. J. Immunol.*, **20**, 477.
12. De Berardinis, P., Guardiola, J., and Manca, F. (1997). *Hum. Immunol.*, **54**(2), 189.
13. Suhrbier, A., Rodda, S. J., Ho, P. C., Csurhes, P., Dunckley, H., Saul, A., *et al.* (1991). *J. Immunol.*, **147**, 2507.
14. Margalit, H., Spouge, J. L., Cornette, J. L., Cease, K. B., Delisi, C., and Berzofsky, J. A. (1987). *J. Immunol.*, **138**(7), 2213.
15. Rothbard, J. B. and Taylor, W. R. (1988). *EMBO J.*, **7**(1), 93.
16. Sette, A., Lamont, A., Buus, S., Colon, S. M., Miles, C., and Grey, H. M. (1989). *J. Immunol.*, **143**(4), 1268.
17. Rothbard, J. B., Busch, R., Howland, K., Bal, V., Fenton, C., Taylor, W. R., *et al.* (1989). *Int. Immunol.*, **1**(5), 479.
18. Allen, P. M., Matsueda, G. R., Adams, S., Freeman, J., Roof, R. W., Lambert, L., *et al.* (1989). *Int. Immunol.*, **1**(21), 141.
19. Bednarek, M. A., Samir, S. Y., Gammon, M. C., Porter, G., Tamhankar, S., Williamson, A. R., *et al.* (1991). *J. Immunol.*, **147**(12), 4047.
20. Gogolak, P., Simon, A., Horvath, A., Rethi, B., Simon, I., Berkies, K., *et al.* (2000). *Biochem. Biophys. Res. Commun.*, **270**(1), 190.
21. Good, M. F., Pombo, D., Maloy, W. L., De La Cruz, V. F., Miller, L. H., and Berzofsky, J. A. (1988). *J. Immunol.*, **140**(5), 1645.
22. Good, M. F., Pombo, D., Quakyi, I. A., Riley, E. R., Houghten, R. A., Menon, A., *et al.* (1988). *Proc. Natl. Acad. Sci. USA*, **85**, 1199.
23. Panina-Bordignon, P., Tan, A., Termijtelen, A., Demotz, S., Corradin, G., and Lanzavecchia, A. (1989). *Eur. J. Immunol.*, **19**, 2237.
24. Panina-Bordignon, P., Demotz, S., Corradin, G., and Lanzavecchia, A. (1989). *Cold Spring Harbor Symp. Quant. Biol.*, **54**, 445.
25. Demotz, S., Grey, H. M., Appella, E., and Sette, A. (1989). *Nature*, **342**, 682.
26. Rudensky, A. Y., Preston-Hurlburt, P., Hong, S. C., Barlow, A., and Janeway, C. A. (1991). *Nature*, **353**, 622.
27. Chicz, R. M., Urban, R. G., Lane, W. S., Gorga, J. C., Stern, L. J., Vignali, D. A., *et al.* (1992). *Nature*, **358**(6389), 764.
28. Hunt, D. F., Michel, H., Dickinson, T. A., Shabanowitz, J., Cox, A. L., Sakaguchi, K., *et al.* (1992). *Science*, **256**, 1817.
29. Newcomb, J. R. and Cresswell, P. (1993). *J. Immunol.*, **150**(2), 499.

30. Partidos, C. D. and Kanse, C. (1997). *Mol. Immunol.*, **34**(16–17), 1105.
31. Kammerer, R., Kettner, A., Chvatchko, Y., Dufour, N., Tiercy, J. M., Corradin, G., *et al.* (1997). *Clin. Exp. Allergy*, **27**(9), 1016.
32. DeLisi, C., Cornette, J., Margalit, H., Cease, K., Spouge, J., and Berzofsky, J. A. (1987). In *Immunogenicity of protein antigens: repertoire and regulation* (ed. E. E. Sercarz and J. A. Berzofsky), Vol. 1, Chap. 2A, pp. 35–42. CRC Press, Inc., Boca Raton, Florida.
33. Hopp, T. P. and Woods, K. R. (1981). *Proc. Natl. Acad. Sci. USA*, **78**(6), 3824.
34. Fauchere, J. L. and Pliska, V. (1983). *Eur. J. Med. Chem.*, **18**, 369.
35. Smolenski, L. A., Kaumaya, P., Atassi, M. Z., and Pierce, S. K. (1990). *Eur. J. Immunol.*, **20**, 953.
36. Bach, J. M., Otto, H., Jung, G., Cohen, H., Boitard, C., Bach, J. F., *et al.* (1998). *Eur. J. Immunol.*, **28**(6), 1902.
37. Guo, H. C., Jardetzky, T. S., Garrett, T. P. J., Lane, W. S., Strominger, J. L., and Wiley, D. C. (1992). *Nature*, **360**, 364.
38. Silver, M. L., Guo, H. C., Strominger, J. L., and Wiley, D. C. (1992). *Nature*, **360**, 367.
39. Brown, J. H., Jardetzky, T. S., Gorga, J. C., Stern, L. J., Urban, R. G., Strominger, J. L., *et al.* (1993). *Nature*, **364**, 33.
40. Oldstone, M. B. A., Tishon, A., Lewicki, H., Dyson, H. J., Feher, V. A., Assa-Munt, N., *et al.* (1991). *J. Virol.*, **65**(4), 1727.
41. Kabsch, W. and Sander, C. (1984). *Proc. Natl. Acad. Sci. USA*, **81**, 1075.
42. Boyum, A. (1964). *Nature*, **204**, 793.
43. Hay, F. C. and Westwood, O. M. R. (2001). *Practical immunology* (4th edn). Blackwell Science, Oxford (in press).
44. Cleaver, J. E. (1967). *Radiat. Res.*, **30**(4), 795.
45. O'Leary, J. J., Mehta, C., Hall, D. J., and Rosenberg, A. (1980). *Cell Tissue Kinet.*, **13**, 21.
46. O'Leary, J. J., Hanrahan, L. R., Mehta, C., and Rosenberg, A. (1980). *Cell Tissue Kinet.*, **13**, 41.
47. Marshall, W. H., Valentine, F. T., and Lawrence, H. S. (1969). *J. Exp. Med.*, **130**, 327.
48. Tse, H. Y., Schwartz, R. H., and Paul, W. E. (1980). *J. Immunol.*, **125**, 491.
49. Bishop, D. K., Ferguson, R. M., and Orosz, C. G. (1990). *J. Immunol.*, **144**(4), 1153.
50. Mullins, R. J., Roche, P., Adams, E., Jones, P., Chen, S., Theuvenet, W., *et al.* (1992). *Immunol. Cell Biol.*, **70**, 277.
51. Diem, K. and Seldrup, J. (1986). In *Geigy scientific tables* (ed. C. Lentner), p. 202. Ciba-Geigy Corporation: Basel, Switzerland.

Published methods to detect cellular antigenic stimulation

1. Nowell, P. C. (1960). Phytohemagglutinin: An indicator of mitosis in cultures of normal human leukocytes. *Cancer Res.*, **20**, 462.
2. Jasinka, J. and Michalowski, A. (1962). The effect of radiation and chemotherapy on the incorporation of thymidine into Ehrlich carcinoma cells and rat bone marrow *in vitro. Nowotwory*, **12**, 321.
3. Hirschhorn, K., Kolodny, R. L., Haskem, N., and Bach, F. (1963). Mitogenic action of phytohemagglutinin. *Lancet*, **ii**, 305.
4. Bain, B. and Lowenstein, L. (1964). Genetic studies on the mixed leukocyte reaction. *Science*, **145**, 1315.
5. Schellekens, P. T. A. and Eijvoogel, V. P. (1968). Lymphocyte transformation *in vitro.* I. Tissue culture conditions and quantitative measurements. *Clin. Exp. Immunol.*, **3**, 571.
6. Rosenthal, A. S., Davie, J. M., Rosenstreich, D. L., and Blake, J. T. (1972). Depletion of antibody-forming cells and their precursors from complex lymphoid cell populations. *J. Immunol.*, **108**(1), 279.
7. Du Bois, M. J. G. J., Huismans, D. R., Schellekens, P. T. A., and Eijsvoogel, V. P. (1973).

Investigation and standardization of the conditions for micro-lymphocyte cultures. *Tissue Antigens*, **3**, 402.

8. Shevach, E. M. and Rosenthal, A. S. (1973). Function of macrophages in antigen recognition by guinea pig T lymphocytes. II. Role of the macrophage in the regulation of genetic control of the immune response. *J. Exp. Med.*, **138**, 1213.
9. Knight, S. C. and Farrant, J. (1978). Comparing stimulation of lymphocytes in different samples: Separate effects of numbers of responding cells and their capacity to respond. *J. Immunol. Methods*, **22**, 63.
10. Rella, W. (1978). The mixed lymphocyte response in whole blood: Technical aspects. *J. Immunol. Methods*, **21**, 237.
11. Van Dam, R. H., Van Kooten, P. J. S., and Van Der Donk, J. A. (1978). *In vitro* stimulation of goat peripheral blood lymphocytes: Optimization and kinetics of the response to mitogens and to allogeneic lymphocytes. *J. Immunol. Methods*, **21**, 217.
12. Van der Zeijst, B. A. M., Stewart, C. C., and Schlesinger, S. (1978). Proliferative capacity of mouse peritoneal macrophages *in vitro*. *J. Exp. Med.*, **147**, 1253.
13. Chandler, P., Matsunaga, T., Benjamin, D., and Simpson, E. (1979). Use and functional properties of peripheral blood lymphocytes in mice. *J. Immunol. Methods*, **31**, 341.
14. Corradin, G. and Chiller, J. M. (1979). Lymphocyte specificity to protein antigens. II. Fine specificity of T-cell activation with cytochrome c and derived peptides as antigenic probes. *J. Exp. Med.*, **149**, 436.
15. Corrigan, A., O'Kennedy, R., and Smyth, H. (1979). Lymphocyte membrane alterations caused by nylon wool column separation. *J. Immunol. Methods*, **31**, 177.
16. Hsia, S., Wilkinson, R. S., and Amos, D. B. (1979). Mixed lymphocyte reactions in serum-free medium. *J. Immunol. Methods*, **27**, 383.
17. Milthorp, P. and Richter, M. (1979). The cells involved in cell-mediated and transplantation immunity in the rabbit. XII. The establishment of the optimum conditions for the demonstration of a consistent response for the circulating white blood cells in the mixed lymphocyte reaction. *J. Immunol. Methods*, **27**, 339.
18. O'Brien, J., Knight, S., Quick, N. A., Moore, E. H., and Platt, A. S. (1979). A simple technique for harvesting lymphocytes cultured in terasaki plates. *J. Immunol. Methods*, **27**, 219.
19. De Jong, B., Anders, G. J. P. A., Zijlstra, J., and Van Der Meer, I. H. (1980). Chinese hamster lymphocyte cultures. Relationship between lymphocyte proliferation, cell concentration, culture time and culture area. *J. Immunol. Methods*, **34**, 295.
20. Farrant, J., Clark, J. C., Lee, H., Knight, S. C., and O'Brien, J. (1980). Conditions for measuring DNA synthesis in PHA stimulated human lymphcytes in 20 μl hanging drops with various cell concentrations and periods of culture. *J. Immunol. Methods*, **33**, 301.
21. Kagan, J. and Ben-Sasson, S. Z. (1980). Antigen-induced proliferation of murine T-lymphocytes *in vitro*. I. Characterization of the lymphocyte culture system. *J. Immunol. Methods*, **37**, 15.
22. Ben-Sasson, S. Z. and Kagan, J. (1981). Antigen-induced proliferation of murine T-lymphocytes *in vitro*. II. The effect of different macrophage populations on the antigen-induced proliferative response. *J. Immunol. Methods*, **41**, 321.
23. Corradin, G. and Chiller, J. M. (1981). Lymphocyte specificity to protein antigens. III. Capacity of low responder mice to beef cytochrome c to respond to a peptide fragment of the molecule. *Eur. J. Immunol.*, **11**, 115.
24. Holt, P. G., Leivers, S., and Warner, L. A. (1981). Optimal culture conditions for *in-vitro* antigen-induced proliferation of rat lymph node cells. *J. Immunol. Methods*, **44**, 205.
25. Needleman, B. W. and Weiler, J. M. (1981). Human lymphocyte transformation induced by mitogens and antigens in a serum-free tissue culture system. *J. Immunol. Methods*, **44**, 3.
26. Arnold, E. A., Katsnelson, I., and Hoffman, G. J. (1982). Proliferation and differentiation

of hematopoietic stem cells in long-term cultures of adult hamster spleen. *J. Exp. Med.*, **155**, 1370.

27. Felsberg, P. J., Serra, D. A., Mandato, V. N., and Jezyk, P. F. (1983). Potentiation of the canine lymphocyte blastogenic response by indomethacin. *Vet. Immunol. Immunopathol.*, **4**, 533.
28. Bernhard, M. I., Pace, R. C., Unger, S. W., and Wanebo, H. J. (1986). The influence of incubation time and mitogen concentration on lymphocyte blastogenic response: Determination of conditions that maximize population differences. *J. Immunol.*, **124**(2), 964.
29. Knight, S. C. (1987). Lymphocyte proliferation assays. In *Lymphocytes: a practical approach* (ed. G. G. B. Klaus). IRL Press, Oxford.
30. Lundin, K. E. A., Bosnes, V., and Gaudernack, G. (1989). Human T lymphocyte clones: Influence of culture conditions and optimization of proliferative assays. *Scand. J. Immunol.*, **30**, 83.

Chapter 4
Combined B cell and T cell epitopes

Sowsan F. Atabani
Department of Pediatrics, Molecular Immunology Division, Children's Hospital Research Foundation, University of Cincinnati, Cincinnati, OH 45229-3039

1 Introduction

B and T cell epitopes can either be overlapping or contiguous within a single sequence. Indeed, numerous studies have determined that up to, and within, a synthetic peptide sequence as small as ten amino acids, the B and T cell recognition sites may be distinct (1). Surface immunoglobulins are capable of binding to both linear sequences of amino acid residues and conformational epitopes which represent non-adjacent amino acid sequences brought into close proximity within a native protein, whereas the T cell receptor recognizes linear processed peptide epitopes bound to major histocompatibility complexes on molecules. B cells are among antigen-presenting cells capable of presenting peptide epitopes to $CD4^{+}$ T cells. T cells produce cytokines and other immunomodulators that provide help for antibody production and induction of memory B cells. A single B and T cell epitope comprises the minimal built-in subunit able to trigger T and B cell co-operation *in vivo* (2). Moreover, studies have shown that overlapping B and T cell epitopes within a synthetic peptide sequence does not impair its B cell immunogenicity (3). The ability of a free synthetic peptide representing a single amino acid sequence, which varies minimally between strains, to act as a functional immunogen able to induce both long-lasting cellular and antibody responses has major implications for the design of future vaccines. Indeed, there exists numerous examples of peptides which are immunogenic in the free, uncoupled state. These results suggest that a satisfactory single sequence can readily induce full protection of all vaccinated animals against a severe challenge with the virulent parent virus (4, 5). Therefore, the topographical relationship between T and B cell epitopes within a native protein sequence is of great interest. This chapter describes experimental methods by which to identify combined and overlapping B and T cell epitopes within the same amino acid sequence with the aid of synthetic peptides.

2 Molecular mapping of antigenic and immunogenic epitopes

2.1 Identification of antigenic epitopes *in vitro*

The identification of amino acid sequences which comprise antigenic determinants from protein antigens is the initial stage in determining sequences which are important for the induction of an effective, protective immune response. Antigenic determinants fall into two major categories:

(a) Linear or continuous epitopes, which are made up of linear sequences of amino acid residues.

(b) Conformational or discontinuous epitopes which represent non-adjacent amino acid sequences that are brought into close proximity by the folding of the protein in its native conformation.

The only method capable of precise identification of the contact amino acid residues of paratopes and epitopes is X-ray crystallography of antibody:antigen complexes. Other procedures applied in an attempt to identify antigenic determinants on proteins include predictive algorithms based on amino acid sequence data (see also Chapter 3). Further methods for identification of antigenic sites include analysis of particular substitutions and/or deletions in the sequence of various proteins by site-directed mutagenesis (see Chapter 10), or induced under immunological pressure, and the use of monoclonal antibodies raised to the native protein (see Chapter 7) to probe enzymatically digested protein fragments or peptide sequences derived from the original protein, all as described in detail elsewhere within this manual.

2.1.1 Identification of linear antigenic epitopes using solid phase peptide synthesis with either monoclonal antibodies or polyclonal antisera

A common experimental approach utilized for the identification of linear antigenic regions *in vitro* has been the synthesis of sets of overlapping peptides corresponding to the known amino acid sequence of a protein. Solid phase synthesis techniques allow the synthesis of a large number of peptides on a solid phase support. In the pin method (6), the peptide is synthesized on a polystyrene pin embedded in a matrix in such a way that each pin fits into a single well of a 96-well microtitre plate. Otherwise, overlapping peptides may be used directly following synthesis using resin-based methods which ultimately result in free, soluble peptides in large (mg) amounts, although this may be time-consuming in the case of a large number of peptides. The synthesized peptides are then tested for reactivity with neutralizing antibodies raised to the native protein, in an appropriate assay system (see also Chapter 2). A disadvantage to this method is that it is mostly linear antigenic determinants which are recognized, although in some cases antibodies may bind to peptides which are separate on the linear sequence, suggesting that they may represent components of conformational epitopes on the native protein. A large number of human monoclonal antibodies may be used to identify principal linear neutralizing B cell epitopes.

In the case of polyclonal antisera, interpretation of epitope mapping data may be more difficult as human sera may exhibit high background binding to short peptide sequences and the presence of low affinity antibodies with relatively weak binding to antigenic epitopes, which may result in a decrease in the signal-to-noise ratio. Therefore, it is advisable to use a large panel of polyclonal sera for epitope mapping to allow for the identification of dominant epitopes.

Protocol 1

Solid phase indirect enzyme-linked immunosorbent assay (ELISA) for peptide antigenicity by simple adsorption of peptides on microplates

Equipment and reagents

- Polyvinyl chloride 96-well microtitre plates
- ELISA plate reader
- Synthetic peptides at a concentration of 0.5–5.0 μg/ml in peptide buffer
- Peptide buffer: 0.1 M carbonate/bicarbonate buffer pH 9.6
- Washing buffer: PBS, 0.1% Tween 20 pH 7.2
- Blocking buffer: PBS, 1% BSA
- Diluent: PBS, 1% BSA
- Appropriate horseradish peroxidase-conjuated IgG antibody
- Phosphate/citrate buffer: 0.2 M sodium orthophosphate, 0.1 M citric acid pH 5.05
- Substrate: freshly made solution of 0.004% hydrogen peroxide and 0.5 mg/ml *o*-phenylenediamine (Sigma) in phosphate/citrate buffer
- Stopping solution: 2 M sulphuric acid

Method

1. Coat the wells of a 96-well microtitre plate with the appropriate synthetic peptides at an optimal concentration of 0.5–5.0 μg/ml in coating buffer and add 50 μl/well. Uncoated wells serve as control wells.
2. Incubate overnight at 4 °C.
3. Wash the plate thoroughly with washing buffer at 250 μl/well approx. four times.
4. Add 150 μl/well of blocking buffer to block excess adsorption sites.
5. Incubate plate for 2 h at 37 °C.
6. Blot the plate dry.
7. Add the appropriately diluted test antiserum. The dilutions to be tested will depend on the sample in which anti-peptide antibodies are to be tested. In the case of human antibodies raised to the whole protein, an appropriate starting range may be 10^{-1} to 10^{-3}. Add 50 μl/well of diluted sample and incubate at 37 °C for 45 min.
8. Wash and blot plate as described above.
9. Add the appropriate peroxidase-conjugated IgG antiserum at the predetermined dilution for a final volume of 50 μl/well and incubate at 37 °C for a further 45 min.

Protocol 1 continued

10 Wash and blot plate (as above).

11 Enzymatic detection is carried out by the addition of 50 μl/well of substrate and incubate at room temperature for up to 30 min in the dark.

12 The reaction is stopped by the addition of 50 μl/well of stopping solution.

13 Read plates immediately on an ELISA plate reader at 490 nm.

14 The reactivity of serum with each peptide is expressed for each sample as the corrected optical density (OD) value at A_{490}, which is obtained for each sample by subtraction of optical density in control wells from the optical density obtained in peptide antigen wells.

2.2 Identification of T cell epitopes *in vitro*

Epitopes recognized by T cells may be determined by the ability of isolated peptides to stimulate protein-specific T lymphocytes (6). The most straightforward method for determination of the presence of a helper T cell epitope within an identified sequence is by the *in vitro* stimulation of primed lymphocytes, from immunized or infected individuals, with the antigenic peptide in specific lymphoproliferative assays. The identification of T cell epitopes is less problematic since these epitopes are naturally recognized as peptide fragments presented by antigen-presenting cells in association with their major histocompatibility complex (MHC) molecules on the cell surface. Therefore, the overlapping peptides are used free in solution. The use of soluble peptides in a tritiated thymidine uptake assay is a common and efficient method for screening of T cell epitopes. However, the functional recognition of T cell epitopes only in context of MHC molecules, which are polymorphic within an outbred population, remains the major obstacle, although some peptides appear to 'promiscuously' bind to a range of MHC molecules (7, 8). The major disadvantages to using multiple synthesis of soluble peptides for identification of T cell epitopes are shared with those described for B cell epitope mapping, in particular that each peptide may not be present in equal amounts as it is difficult to monitor every reaction during the process of synthesis.

Protocol 2

Identification of immunogenic T cell epitopes *in vitro*

Equipment and reagents

- Sterile 96-well round-bottom tissue culture plates
- Liquid betaplate scintillation counter
- Ficoll-Hypaque solution (Amersham Pharmacia Biotech)
- MLR medium: RPMI 1640 medium (Gibco), 10% autologous serum (see Chapter 3), 0.1 M Hepes buffer (Gibco), 50 μM 2-mercaptoethanol, 2 mM glutamine, 100 U/ml penicillin, and 100 μg/ml streptomycin

Protocol 2 continued

- Heparinized venous blood from individuals sensitized to the native protein
- HBSS (Gibco)[a]
- Mitogen, e.g. PHA
- Peptide antigens at concentrations of 1–50 μg/ml dissolved in MLR medium
- Tritiated thymidine [^{3}H]-thymidine

Method

1. Test the primed cells from a large number of individuals with different MHC class II molecules in order to determine the ability of the peptide to bind 'promiscuously'.
2. Collect approx. 35–50 ml of blood by venesection and separate mononuclear cells as described previously (see Chapter 3).
3. Dispense 20 μl/well of antigenic peptides (e.g. at a concentration of 30 μg/well) and mitogen into the microtitre plate.
4. Perform all concentrations in triplicate.
5. Resuspend mononuclear cells in complete RPMI 1640 and adjust cells to 1.1×10^6/ml.
6. Add 180 μl of cells (in total 1×10^5) to each well and in the control wells.
7. Cover the plates with lid and seal with micropore tape.
8. Incubate in humidified 5% CO_2 incubator for 72–96 h.
9. Pulse cultures by the addition of 0.5–1.0 μCi/well of [^{3}H]-thymidine in complete RPMI medium into each well and incubate for a further 16 h.
10. Harvest cells onto glass filter fibres.
11. Measure [^{3}H]-thymidine incorporation by liquid scintillation spectroscopy using a betaplate scintillation counter.
12. The results are expressed as the mean of triplicates (c.p.m.). A stimulation index (SI) is expressed for each triplicate by dividing mean radioactivity (c.p.m.) of stimulated cells by that of unstimulated cells. In most studies, T cell proliferation is considered positive if the stimulation index of the test sample exceeds an arbitrary cut-off point in the range of two or three times background levels, defined as the uptake of [^{3}H]-thymidine by T cells in the presence of medium alone or with an irrelevant peptide.

[a] Medium used to transport or manipulate tissues/cells e.g. spleen.

3 *In vivo* analysis of immunogenicity and antigenicity

The potential of these *in vitro* determined epitopes to induce T and B cell immunogenic responses *in vivo* remains to be elucidated. Once the appropriate epitope has been identified, their potential immunogenicity *in vivo* should be tested with the use of experimental animals. This epitope should induce T cell proliferation that recognizes the native protein as well as induces antibodies which cross-react with the native protein *in vivo*. The ability of a free peptide to stimulate lymphocyte proliferation *in vitro* and generate significant cross-reactive antibodies, which in the case of viral peptides induce virus neutralization following immunization, indicates the presence of both a B cell and T cell epitope within its sequence.

The peptides may also be used to test for immunogenicity. Before use for immunization procedures, crude peptides are initially purified with the use of a reverse-phase high performance liquid chromotography (HPLC, Bio-Rad) and the homogeneity of the purified peptide determined by reverse-phase HPLC and laser-desorption time-of-flight mass spectrometry (LaserMat, FinniganMat, UK). For induction of polyclonal anti-peptide antibodies, the free peptide is emulsified with an adjuvant and administered either intraperitoneally, intramuscularly, or subcutaneously into the experimental animal. For peptide immunogens, a booster immunization is almost always necessary for activation of memory B cells with the production of high titres of specific IgG antibodies of a high affinity for the immunizing peptide. Additional booster immunizations may be administered to maintain high titres of anti-peptide antibodies.

Protocol 3

Immunization using synthetic peptides representing identified epitopes

Reagents

- Appropriate adjuvant, e.g. Freund's complete adjuvant (FCA), incomplete Freund's adjuvant (IFA), or aluminium hydroxide
- Purified synthetic peptides
- Experimental animals, e.g. mice, guinea-pigs, rabbits, and monkeys

Method

1. Prepare 10–100 μg/ml of free peptide emulsified 1:1 (v/v) in IFA or mixed with 25% aluminium hydroxide.
2. Immunize a sufficient number of laboratory animals intraperitoneally or subcutaneously with 0.5–1.0 ml of this vaccine.
3. Immunize control animals with an equal volume of PBS or adjuvant alone.
4. Boost animals with an equal immunizing dose of peptide as before, two to four weeks following primary immunization, up to a total of four subsequent booster immunizations.
5. Collect blood samples prior to priming, before each boost, and at regular intervals thereafter.
6. To examine the genetically determined cellular immune responsiveness to a linear peptide sequence, various strains of animals representing different major histocompatibility complex haplotypes may be used for immunization procedures (9).

3.1 Investigation of the core sequence of B cell and T cell epitopes

The fine-specificities of the B and T cell epitopic sites are then determined as the minimum number of amino acid residues which are recognizable by both the anti-peptide and anti-protein antibodies and also which induce significant uptake of thymidine by activated T cells. To further characterize and delineate an epitope, a nested series of shorter overlapping, as well as truncated synthetic peptides derived from the original region, based upon the amino acid sequences previously published, may be utilized. Another strategy to define minimal peptide sequences recognized by antibodies is based on 'chimeric' peptides in which small peptides derived from the original sequence are synthesized within the sequence of other peptides known to form a secondary structure conformation, so as to maintain the structure of the minimal epitope (10). However, in the case of anti-peptide antibodies raised to a synthetic peptide immunogen, it is essential to bear in mind that the polyclonal population will contain antibodies which are capable of recognizing the peptide immunogen in various conformations.

Using a solid phase ELISA assay, it is only possible to detect antibody above a certain threshold of affinity for each peptide coating density. Therefore, it is difficult to make an accurate quantitative analysis of anti-peptide antibodies following detection using a solid phase ELISA with direct adsorption of peptides on microtitre plates (as described in *Protocol 1*).

Protocol 4

Detection of *in vivo* anti-peptide antibody production using indirect ELISA

Equipment and reagents

- See *Protocol 1*

Method

1 Coat ELISA plates with homologous peptide and incubate at 4 °C overnight, as described in Protocol 1.

2 Following repeated washes, add 50 µl/well of anti-peptide serum starting at an initial high dilution if serum is obtained from a hyperimmunized animal, e.g.10^{-3}, with serial twofold dilutions in diluent thereafter. Incubate plate at 37 °C for 1 h.

3 Wash the plate thoroughly with washing buffer.

4 Add the appropriate enzyme conjugate at the suggested dilution in diluent for a final volume of 50 µl/well and incubate at 37 °C for a further hour.

5 Continue the ELISA as described (*Protocol 1*).

6 Calculate the anti-peptide antibody titre as $\log_{10}$ of the reciprocal of the antibody dilution reacting with the homologous peptide to give an OD value greater than 2 standard deviation (SD) above the mean absorbance of the reactivity of the normal, pre-immunization serum, at the lowest dilution, with the peptide.

Table 1[a] Fine mapping of an antigenic peptide

Assay	Sequence	No. samples positive (%)[b]	Mean relative affinity[c] (M^{-1}) ± SD
A[d]			
P32	A N C A S I L C K C Y T T G T	100	$(2.0 \pm 0.9) \times 10^{7}$ *
P32-1	A N C A S I L C	11.1	$(2.6 \pm 0.1) \times 10^{5}$
P32-2	C A S I L C K C	33.3	$(4.3 \pm 1.5) \times 10^{5}$
P32-3	S I L C K C Y T	11.1	$(1.2 \pm 5.0) \times 10^{5}$
P32-4	L C K C Y T T G T	16.7	$(4.5 \pm 1.4) \times 10^{5}$
B[e]			
	A N C A S I L C K C Y T T G T	100	$(2.0 \pm 0.9) \times 10^{7}$
	A N S A S I L S K S Y T T G T	0	–
	A N S A S I L C K S Y T T G T	0	–

[a] Modified from a table given in ref. 4.

[b] Peptides were used as solid phase antigens in a direct ELISA. The cut-off for positive reactivity of each serum sample with the peptide is expressed as the mean optical density at A_{490} + 2SD for the negative control samples, at a serum dilution of 1/200.

[c] Assessed by a solid phase inhibition ELISA, using peptides as fluid phase inhibitors; (* $p < 0.001$).

[d] Percentage positivity and mean relative affinity of 18 human polyclonal antisera for the peptide P32 as a 15-mer and as overlapping 8-mers.

[e] Recognition of 18 human antisera with two further 15-mer peptides in which two or three of the cysteine residues were replaced by serines.

To map the epitope responsible for binding to the antibody, a set of overlapping and truncated peptides modelled on the original peptide sequence may be synthesized and adsorbed directly onto the microtitre plate and tested for the ability of the resulting peptides to bind to antibody directed against the parent peptide (see *Table 1*).

3.2 Inhibition ELISA for the measurement of antibody affinity as a method to determine the B cell epitope

The binding of antibody to peptide is a function of antibody concentration as well as antibody affinity for the epitope. Measurement of protein-specific antibody affinity to synthetic peptides has previously been used for the identification of antigenic determinants on surface proteins. These studies have shown that antibodies bind with the greatest affinity to the peptide that mostly resembles, in length and conformation, the B cell epitope within the native protein (11, 12).

An inhibition ELISA is selective for antibodies which bind well with the parent protein in solution, therefore, an inhibition ELISA using shortened peptide analogues with truncations of both the amino and carboxy terminus may be useful for mapping the exact antibody binding site (see *Table 1*). In this case, a constant working dilution of antibody is incubated with the solid phase original peptide length and an increasing concentration of test peptides in the fluid phase. However, the results of an inhibition ELISA should always be considered as relative affinity rather than absolute values and thus preferably used for comparative purposes.

Protocol 5

Inhibition ELISA for mapping of the antibody binding site using chimeric peptides and shortened peptides with truncations of the amino or carboxy terminus for mapping of the binding site

Equipment and reagents

- See *Protocol 1*

Method

1. In an inhibition ELISA, the antibody dilutions are initially titrated by ELISA on plates coated with the peptide under study, as described in *Protocol 3*.
2. From this titration ELISA, determine the antibody dilution which gives an absorbance of 0.5, which is to be used for the affinity assay.
3. This dilution is then made up in diluent and incubated for 1 h with 0.5 $\log_{10}$ dilutions of the peptides under study as fluid phase inhibitors.
4. Transfer these preparations to a microtitre plate pre-coated with the parent peptide.
5. For each antibody sample, three adjacent wells received no inhibitor to represent 100% binding.
6. Incubate the microtitre plate for a further hour at 37 °C.
7. Complete the ELISA as described above (*Protocol 1*).
8. Antibody affinity is determined as the reciprocal of the molar concentration which gives 50% inhibition of binding and these values represent estimated average antibody affinity.

3.3 Determination of precise helper T cell epitopes by proliferative and cytokine responses

As with determination of B cell epitopes, to determine the exact residues which act as the helper T cell epitope, overlapping and progressively shorter, truncated versions of the peptides, as well as analogues with single amino acid substitutions within the original sequence for determination of the contribution of individual amino acids, are synthesized and used as *in vitro* stimulants for the lymphocytes obtained from animals following immunization with the original parent synthetic peptide. Stimulation can be assessed by the incorporation of tritiated thymidine as a measure of lymphocyte proliferation or by measurement of the production of cytokines such as IL-2.

For smaller experimental animals, e.g. mouse or rat, the spleen or draining lymph nodes, e.g. popliteal, are used as source of lymphocytes whereas peripheral blood mononuclear cells (PBMC) are utilized when immunogenicity studies are carried out on larger laboratory animals, e.g. monkey.

Mononuclear cells are harvested and tested for proliferation upon re-stimulation with both the native protein as well as the homologous peptide and its corresponding truncated peptides, as described above used for antibody-binding activity, *in vitro*. $CD4^+$ T cell lines or clones may be generated, however this procedure is tedious and time-consuming, whereas unexpanded cells from spleen or lymph nodes have been shown to produce successful results.

Protocol 6

Identification of precise T cell epitopes by using a T cell proliferative assay

Equipment and reagents

- See *Protocol 2*

Method (see *Protocol 2*)

1. One week following immunization (*Protocol 3*) the spleen or relevant lymph nodes are carefully removed and gently pressed into single cell suspensions.
2. Resuspend in MLR medium.
3. Incubate the mononuclear cells (10^5/well) in the presence of irradiated autologous spleen cells ($1–3 \times 10^3$/well) and 20 μg/well of appropriate synthetic peptides for 96 h.
4. Analyse the peptide-specific T cell proliferative responses after the addition of 0.5 μCi/well of [^{3}H]-thymidine and determination by standard liquid scintillation counting.
5. Following the determination of stimulation indices, the shortest peptide which elicits a significant uptake of [^{3}H]-thymidine with significant proliferative responses may be designated the helper T cell epitope.

The correlation of lymphocyte proliferation with IL-2 production suggests that the responding lymphocytes are helper T cells. IL-2 production and activity should correlate with the proliferative responses and indicate that T helper cells respond to peptide stimulation.

Protocol 7

Cytokine-specific ELISA

Equipment and reagents

- Flat-bottomed 96-well microtitre plates (Nunc)
- Monoclonal anti-IL-2 capture antibody (Cambridge BioScience)
- ELISA plate reader
- Secondary anti-IL-2 antibody (Cambridge BioScience)
- Coating buffer: carbonate/bicarbonate buffer pH 9.6
- Washing buffer (see *Protocol 1*)

Protocol 7 continued

- Blocking buffer: PBS in 2% BSA
- Diluent (see *Protocol 1*)
- TMB buffer[a]
- Stopping solution
- Recombinant human IL-2

Method

The supernatants from splenic/lymph node cell cultures described above may be collected at 24–48 h and stored at −20 °C until assayed.

1. Coat microtitre plates with 50 μl/well of monoclonal anti-IL-2 capture antibody at 4 μg/ml in bicarbonate buffer.
2. Incubate overnight at 4 °C.
3. Wash plates and blot dry (see *Protocol 1*).
4. Block plates with 100 μl/well of blocking buffer.
5. Leave at room temperature from 30 min up to 2 h.
6. Add different dilutions of culture supernatants in diluent at 50 μl/well. Add all samples to be tested in duplicate.
7. Generate a linear standard curve using recombinant human IL-2 at 4 ng/ml in serial doubling dilutions in RPMI 1640 medium (without FCS), plus a zero.
8. Incubate at room temperature for 1 h.
9. Follow with a thorough washing (see *Protocol 1*).
10. Add 50 μl of the secondary anti-IL-2 antibody per well at a concentration of 2 ng/ml, in diluent.
11. Incubate at room temperature for an extra hour.
12. Repeat wash step 9.
13. Add streptavidin–peroxidase at 100 μl/well of a 1:8000 working dilution diluted in blocking buffer.
14. Incubate for 30 min at room temperature.
15. Repeat wash step 9.
16. Develop reaction with substrate solution (100 μl of TMB in each 10.5 ml of substrate buffer). Add 100 μl/well and incubate at room temperature for up to 30 min, depending on how quickly colour develops.
17. Stop reaction with 50 μl/well of 2 M sulphuric acid.
18. Read plate immediately at absorbance of 450 nm.

[a] A standard ELISA buffer (commercially available)

4 General summary

The ability to determine a single relevant antigenic and immunogenic site within a target protein is thus the basis for the development of synthetic vaccines for the prevention of and protection against disease. The combination of a B cell and helper T cell epitopes within a sequence is essential for the induction of both cellular and mature antibody responses. By testing the reactivity of anti-peptide antibody and mononuclear cells with sets of synthetic peptides, the smallest fragments which retain antibody-binding activity and T cell stimulating activity may be identified.

Ultimately, the results of these experiments following *in vivo* priming should be in good agreement with the *in vitro* analysis of antigenicity and immunogenicity to validate the assignment and designation of T and B cell epitopes within a single core peptide sequence. A single amino acid sequence that elicits both cellular and antibody responses among a large outbred population and which are cross-reactive with the native protein would be ideal for inclusion in a subunit vaccine for the future.

References

1. Lehner, T., Walker, P., Smerdon, R., Childerstone, A., Bergmeier, L. A., and Haron, J. (1990). *Arch. Oral. Biol.*, **35**, 39S.
2. Brumeanu, T. D., Casares, S., Bot, A., and Bona, C. A. (1997). *J. Virol.*, **71**, 5473.
3. Harris, D. P., Vordermeier, H. M., Arya, A., Bogdan, K., Morens, C., and Ivanyi, J. (1996). *Immunology*, **88**, 348.
4. Atabani, S. F., Obeid, O. E., Chargelegue, D., Aaby, P., Whittle, H. C., and Steward, M. W. (1997). *J. Virol.*, **71**, 7240.
5. DiMarchi, R., Brooke, G., Gale, C., Cracknell, V., Doel, T., and Mowat, N. (1986). *Nature*, **329**, 629.
6. Geysen, H. M., Meleon, R. H., and Barteling, S. J. (1984). *Proc. Natl. Acad. Sci. USA*, **81**, 3998.
7. Sinigaglia, F., Guttiriger, M., Kilgus, J., Doran, D. M., Matile, H., Etlinger, H., *et al.* (1988). *Nature*, **336**, 778.
8. Francis, M. J., Hastings, G. Z., Campbell, R. O., Rowlands, D. J., Brown, F., and Pent, N. T. (1989). In *Vaccines '89* (ed. R. A. Lerner, H. Ginsberg, R. M. Chanock, and F. Brown), p. 437. Cold Spring Harbour Laboratory, New York.
9. Collen, T., DiMarchi, R., and Doel, T. R. (1991). *J. Immunol.*, **146**, 749.
10. Nicholas, J. A., Mitchell, M. A., Levely, M. E., Rubino, K. L., Kinner, J. H., Harn, N. K., *et al.* (1988). *J. Virol.*, **62**, 4465.
11. Brown, S. E., Howard, C. R., Zuckerman, A. J., and Steward, M. W. (1984). *J. Immunol. Methods*, **72**, 41.
12. Brown, S. E., Zuckerman, A. J., Howard, C. R., and Steward, M. W. (1984). *Lancet*, **2**, 184.
13. Relf, W. A., Cooper, J., Brandt, E. R., Hayman, W. A., Anders, R. F., Pruksakorn, S., *et al.* (1996). *Peptide Res.*, **9**, 12.

Chapter 5
CTL epitopes

Tim Elliott
Cancer Sciences Division, University of Southampton School of Medicine, Level F, Centre Block, Southampton General Hospital, Tremona Road, Southampton SO16 6YD, UK.

John S. Haurum
Institute of Cancer Biology, The Danish Cancer Society, Strandboulevarden 49, 2100 Copenhagen, Denmark.

1 Introduction

Cytotoxic T lymphocytes (CTL) are primarily responsible for the clearance of acute virus infections. Thus, for example in mice, recovery from influenza A infection correlates with the induction of CTL before a neutralizing antibody response emerges (1). In humans, a rise in the frequency of influenza-specific CTL correlates with a reduction in virus titres in nasal washes (2, 3), and similar observations have been made for RSV (4), HSV (5), measles virus (6), and hepatitis A virus (7). CTL may also be important in controlling chronic viral infections such as EBV (8), CMV (9), and HIV (10). Some intracellular bacterial infections such as *Listeria monocytogenes* (11) as well as parasitic infections such as *Theileria parva* (12) and *Plasmodium yoelii* (13, 14) also appear to be susceptible to attack by CTL. The protective effect of CTL has been amply demonstrated in several cases by adoptive transfer experiments (3–5, 12, 15–19). In addition to providing protection against infectious agents, CTL are thought to provide some degree of protection against spontaneous tumours, by virtue of their ability to detect subtle quantitative and qualitative antigenic differences in transformed cells (reviewed in ref. 20).

1.1 T cell receptor (TcR) recognizes a complex between MHC and antigenic peptide

The highly variable antigen receptor of CTL recognizes a complex ligand comprising an antigenic peptide bound to a class I MHC molecule on the surface of the target cell (reviewed in ref. 21). Different class I MHC alleles bind a broad yet limited spectrum of peptide ligands depending on the chemical nature of their peptide binding site. The molecular basis for this specificity is outlined in *Figure 1a* and *Appendix 1* (at the end of this chapter).

The variable domain of the TcR makes contact with both the class I MHC

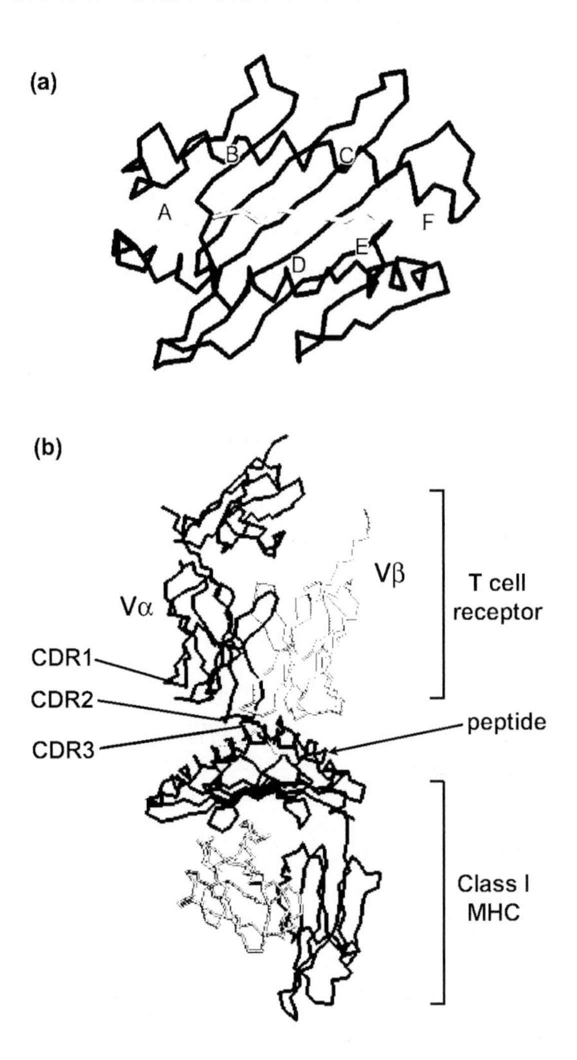

Figure 1 (a) Top view of the class I MHC molecule, showing the peptide binding groove formed by two antiparallel alpha helices above a platform of beta sheet. The backbone of a bound peptide is also shown. The peptide binding groove is defined at both ends by non-polymorphic pockets (A and F) into which fit the N- (A) and C- (F) termini of a bound peptide. Four other pockets or depressions (labelled B–E) are lined with polymorphic amino acids and have evolved to accommodate particular side chains of the bound peptide. (b) Side view of the class I MHC–peptide complex and its recognition by the T cell receptor. Only the Vα and Vβ domains of the receptor are shown. Only the complementarity determining regions (CDRs) of Vα are indicated, but those of Vβ are also visible.

molecule and the bound peptide ligand as shown in *Figure 1b*. From the two first TcR–MHC–peptide structures solved by X-ray crystallography (22, 23), it appears that the TcR alpha chain makes most of the contact with the N-terminal end of the peptide and the 'left-hand' end of the MHC molecule while the beta chain shadows the C-terminal end of the peptide and the 'right-hand' end of the MHC molecule. In the structure of the highest resolution (23), most contacts with the

MHC molecule itself are via the CDR1, 2, and 3 loops of Vα. Primary contacts with the peptide ligand are made by CDR1 of Vα (which recognizes the N-terminal end of the peptide) and CDR3 of Vβ (which recognizes the middle and C-terminal end of the peptide), with secondary contributions from CDR3 of Vα and CDR1 of Vβ. The CDR1 and 2 hypervariable loops are encoded within the V gene segments and the CDR3 hypervariable loops encoded by the VDJ junctional region (summarized in ref. 24).

The CTL co-receptor molecule CD8 also makes contact with the MHC, and along with the CD3–TcR complex forms a large multimolecular assembly which can deliver an intracellular activation signal to resting CTL when the TcR binds to its cognate ligand (reviewed in ref. 25).

1.2 Antigen processing

The MHC–peptide complex is formed inside the endoplasmic reticulum (ER) and is the product of a complex series of biochemical events known as antigen processing. These events are summarized in *Figure 2*.

The processing of class I MHC-restricted CTL epitopes begins in the cytosol. Proteins can become exposed to this compartment in a variety of ways. For example:

(a) They can be synthesized there on free ribosomes before being targeted to other intracellular compartments (like nucleus, mitochondria, or even the ER).

(b) They can be dislocated there after having been co-translationally transported into the lumen of the ER.

(c) They can be artificially delivered there by osmotic lysis of pinosomes or by conjugation to a fusogen.

(d) They can be delivered there from macropinosomes. This latter route may be important in priming a CTL response since it appears to be particularly active in dendritic cells.

Among the proteases in the cytosol which are involved in class I MHC-restricted antigen processing, the proteasome is the best studied and cleaves peptide bonds on the carboxyl side of basic, hydrophobic/aromatic, and acidic residues. Ubiquitinated proteins are particularly effective substrates. The activity of the proteasome is in part regulated by two subunits (LMP2 and LMP7) encoded in the MHC class II region, which are interferon-gamma (IFN–γ) inducible. Some CTL epitopes have been shown to be sensitive to the presence of LMP2 and/or LMP7 whereas others are insensitive. Some CTL epitopes appear to be generated in the cytoplasm by a proteasome-independent pathway which suggests the presence of other as yet uncharacterized proteases in this compartment.

The peptide products are translocated across the ER membrane by a specific transporter called the transporter associated with antigen processing (TAP, reviewed in ref. 26). TAP is polymorphic although the functional consequence of this has only been documented in the rat. TAP can transport peptides with of up to 40 amino acids, but with an optimum of around 13 amino acids. Human TAP display little sequence selectivity (although peptides with proline at the first three positions or at the C-terminus are strongly disfavoured) whereas murine TAP is more selective and in addition disfavour charged residues at the last position. TAP is tightly bound to a molecule called tapasin (which has been described as the third TAP chain) whose function is still un-clear. It is essential for the antigen presenting function of some but not all alleles and needs to be present for class I molecules to bind to TAP. It also binds to peptides *in vitro*, although it is not yet clear whether this is relevant to its function *in vivo*.

Once in the lumen of the ER, long precursors of CTL epitopes can be trimmed by as yet unidentified proteases to yield peptides of a length which are optimal for binding to class I MHC molecules. Peptide binding and class I MHC assembly are linked phenomena in that the incorporation of a peptide ligand into the peptide binding groove of the newly synthesized heavy chain (HC) is an essential step in the formation of a stable HC-β_2-m heterodimer. *In vivo*, this assembly involves a number of cofactors, including the ER-resident, calcium-binding chaperones calnexin and calreticulin. Newly synthesized class I HC-β_2-m heterodimers bind simultaneously to calreticulin, tapasin, and TAP in the ER, where they are held until they encounter a stabilizing peptide ligand. Only then are they released into the secretory pathway. The mechanism of this retention/release is unknown, but is thought to centre around a peptide-induced conformational change in the class I MHC molecule which can be detected by a cofactor responsible for the retention of incompletely assembled class I molecules.

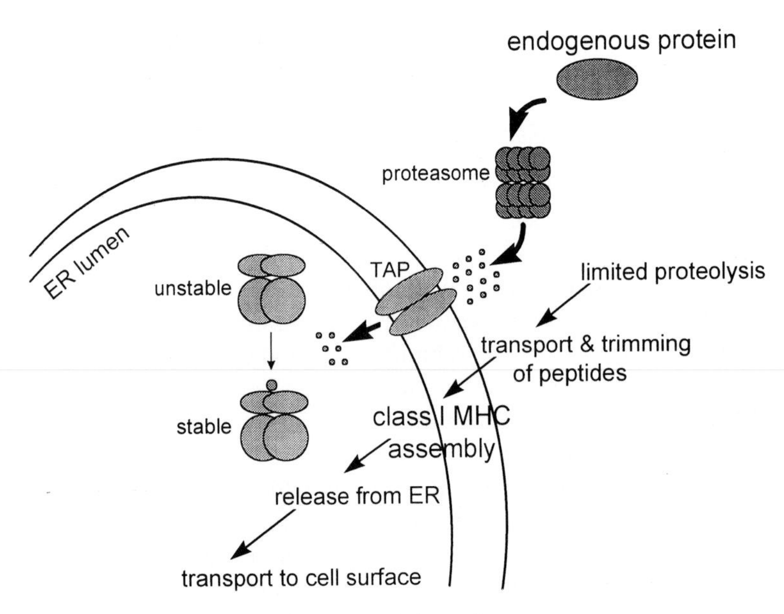

Figure 2 The class I MHC antigen presentation pathway.

1.3 Immunodominance

The CTL response to a pathogen or tumour is characterized by the phenomenon of immunodominance. This term was first used by Zinkernagel and Doherty (27) to describe the existence of high, low, and non-responder alleles to different viruses. In more recent years it has become clear that the CTL response to a particular pathogen is usually limited to the recognition of a single peptide epitope, and that the choice of this epitope is dependent on the class I MHC alleles that are expressed by the host (30–32). The same is true for CTL responses to minor transplantation antigens (28) and tumour antigens (29).

An example of the scale of immunodominance is shown in Figure 3 where it can be seen that of the 200 000 or so peptides of between 5 and 50 residues that can be generated from the ten proteins encoded by the influenza A genome (total number of amino acids around 4500), 4000 are nonamers. Only 26 of these have sequences which predispose them to bind to the H-2K^{d} molecule expressed by BALB/c (see *Figure 1* and *Appendix 1* at the end of this chapter), and of these, only two are normally recognized following experimental infection. Thus only one in 100 000 peptides (of between 5 and 50 residues) elicit a CTL response.

Nowak *et al.* have recently established a mathematical model of immunodominance in terms of competition between immune responses to different epitopes (33). Persistent stimulation will lead to an equilibrium state where the virus load (or any other antigenic challenge) is reduced to a point where the rate of CTL stimulation is the same as the rate of CTL decay. Thus, in cases where multiple CTL responses have equal decay rates, such an equilibrium is only possible for the 'most sensitive' CTL response and this will therefore dominate CTL responses *in vivo*. Therefore, natural CTL responses are often characterized by a hierarchy of specificities with subdominant (or immunorecessive) epitopes becoming apparent only when a response to the immunodominant epitope has been eliminated, for example by exhaustion of CTL to the immunodominant

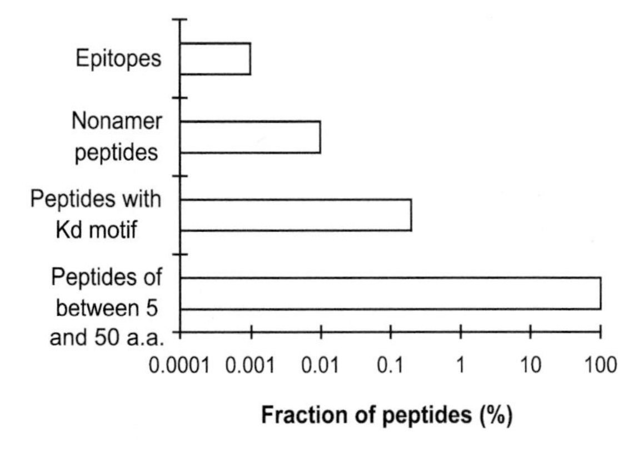

Figure 3 An illustration of immunodominance from data shown in ref. 60.

Table 1 Factors affecting immunodominance

Intracellular concentration of protein antigen
Recruitment of protein to proteolytic enzymes, e.g. by ubiquitination
Specificity of enzymes
Rate of peptide generation
Rate of peptide degradation in the cytosol and ER
Specificity of TAP-mediated transport
Trimming of peptide epitopes in the ER
Assembly of peptide with MHC heavy chain and β_2-m
Stability of MHC–peptide complex at the cell surface / number of complexes
Hole in the T cell repertoire
Induction of anergy versus stimulation of CTL proliferation
CTL competition and immunosuppression

epitope, or as a result of mutations in or around the immunodominant epitope (29, 34, 35).

Many factors could affect the rate of CTL stimulation and decay, leading to the establishment of an immunodominant CTL response to a complex antigenic challenge, and some of these are outlined in *Table 1*.

What follows is a guide to the available strategies for identifying CTL epitopes with an emphasis on how available information about the factors which influence immunodominance can be used to predict the specificity of a CTL response to a complex antigenic challenge. The ability to do this successfully would represent an important step forward in the design of vaccines.

2 Indirect methods of CTL epitope identification using synthetic peptides

Identification of target epitope(s) has been successfully achieved in the past by screening CTL, generated in response to a viral infection, for their ability to recognize a panel of overlapping synthetic peptides. Thus, without any knowledge of the peptide binding motif of the H-2D^b molecule, the immunodominant CTL epitope generated in C57Bl/6 mice infected with influenza A was shown to be contained within a 14-amino acid peptide (36). This approach is only feasible in cases where either the genome of the infective agent is small or where the target antigenic protein has been identified. Target proteins have been identified by introducing cloned candidate genes or gene fragments into target cells by transfection or delivery using a recombinant vaccinia virus, before determining whether the target cell can be recognized by CTL generated to the whole pathogenic organism (36, 37). Techniques for assaying CTL recognition are summarized below.

The assessment of antigen-specific T cell frequencies *in vivo* may be performed by the limiting dilution technique. However this is very time-consuming, and the result is influenced by the clonogenicity of the T cells *in vitro*. Recently, the

ELISPOT technique (39) has been adapted by several groups to allow the quantification of epitope-specific CTL precursor frequencies. This assay is based on the detection of peptide-induced release of the cytokines IFN-γ or tumour necrosis factor-α (TNF-α) by single CTL from, e.g. peripheral blood upon stimulation with peptide antigen (40–43).

Protocol 1

Methods for identifying CTL: assays for lytic activity

Equipment and reagents

- Round-bottomed 96-well plates, U-bottomed 96-well plates
- Microplate centrifuge carrier
- CO_2 incubator
- [^{51}Cr]: supplied as sodium chromate (Nycomed Amersham)
- PBS, RPMI, 1640 FCS
- R10: RPMI 1640 with 10% FCS
- 1% Triton X-100
- CytoTox96™ kit (Promega)
- ELISA-based kit (Boehringer Mannheim)

A [^{51}Cr] release

1. To label target cells, place 1×10^6 washed targets in a well of a round-bottomed 96-well plate.
2. Spin using a microplate centrifuge carrier (1000 r.p.m. for 5 min) and aspirate the medium. Resuspend the cell pellet in approx. 20–100 μl [^{51}Cr] (supplied as sodium chromate). Add an equal volume of peptide solution (10 μM) in PBS (or virus in RPMI 1640) and incubate for 1 h at 37 °C in a CO_2 incubator.
3. Pellet the cells and remove the supernatant (radioactive waste). Wash the cells four times in the 96-well plate using R10. All washes are radioactive. Dilute the cells in R10 to $0.3–1 \times 10^5$/ml and dispense 100 μl/well in a U-bottomed 96-well plate. For sensitization with viruses it may be necessary to culture the cells for 1–4 h to allow sufficient levels of antigen expression.
4. Wash the CTL twice in R10 and resuspend at an appropriate concentration for the desired effector : target ratio (E:T ratio) and dispense in 100 μl/well. Perform each experimental point in triplicate. For each target type include:
   (a) Spontaneous [^{51}Cr] release (targets + 100 μl R10).
   (b) Total release (targets + 100 μl of 1% Triton X-100).
   (c) Non-specific CTL-mediated lysis (targets not pulsed with peptide or virus + 100 μl CTL).
5. Co-culture targets and CTL for 4 h at 37 °C in a 5% CO_2 incubator. Pellet cells and harvest, e.g. 50 μl of the supernatant for gamma counting to assess [^{51}Cr] release as a result of target cell lysis. Calculate the specific lysis (%) = ((mean c.p.m. experimental release − mean c.p.m. spontaneous release) / (mean c.p.m. total release - mean c.p.m. spontaneous release)) × 100%.

Protocol 1 continued

B BLT esterase release

1. CTL degranulate in response to antigenic stimulation, releasing perforin and several serine esterases.
2. These enzyme activities can be detected with a chromogenic substrate, for example the BLT esterase, as described in ref. 38.

C LDH release

1. A kit for the detection of the cytosolic enzyme lactate dehydrogenase released from lysed target cells is available from Promega (CytoTox96™).
2. This assay has the advantage that radioactive labelling is not required. We have found it to be as sensitive as the [^{51}Cr] release assay in a direct comparison.

D Release of BrdU-labelled low molecular weight DNA release

1. An ELISA-based kit is available from Boehringer Mannheim in which target cells are grown overnight in BrdU to label DNA before being exposed to CTL.
2. Apoptotic cell death results in the release of low molecular weight DNA into the culture medium which is then sampled and assayed using an anti-DNA capture antibody and an enzyme-conjugated anti-BrdU second antibody.
3. Again, we found this assay to be as sensitive as the [^{51}Cr] release assay.

Protocol 2

Methods for identifying CTL: assays for lymphokine secretion

Equipment and reagents

- Nitrocellulose-bottomed 96-well plates (Millipore, MAIP N45)
- Dissection microscope
- Anti-IFN-γ antibody (e.g. 1-D1K from Mabtech)
- See *Protocol 1*
- Biotinylated secondary antibody (e.g. mAb 7-B6-l-biotin from Mabtech)
- Peripheral blood mononuclear cells (freshly isolated using standard Lymphoprep protocols)
- Tween 20
- Avidin–enzyme conjugate (AP-avidin, Calbiochem—Novabiochem)
- Enzyme substrate

Method

1. Coat nitrocellulose-bottomed 96-well plates overnight at 4 °C with 7.5 μg/ml of anti-IFN-γ antibody (e.g. 1-D1K from Mabtech) in 75 μl sterile PBS. Wash the wells with PBS (× 6) before blocking with R10 (see *Protocol 1*) for 2 h at 37 °C (200 μl/well).

Protocol 2 continued

2 Add peripheral blood mononuclear cells to each well in triplicates using a range of cell counts, depending on the expected CTLp frequencies for your antigen. Initially, useful cell counts are 1.25×10^4, 2.5×10^4, 5×10^4, 1×10^5, 2×10^5, and 4×10^5 cells per well in 100 μl R10.

3 Add to each well 1–2 μM peptide in an additional 20 μl of R10 and incubate overnight (14–16 h) at 37 °C with 5% CO_2. On the following morning wash the wells with PBS containing 0.05% Tween 20 (× 6).

4 Add biotinylated secondary antibody at 0.5 μg/ml in 75 μl PBS containing 1% BSA (PBS/BSA) with 0.02% NaN_3. Incubate 3 h at room temperature (RT). Wash the wells (× 6) with PBS containing 0.05% Tween 20.

5 Add to each well 75 μl of the avidin–enzyme conjugate (diluted 1:1000 in PBS/BSA) with 0.02% NaN_3. Incubate for 1 h (RT). Wash with PBS (× 6) containing 0.05% Tween 20.

6 Mix the enzyme substrate: 1/10 vol. nitro-blue-tetrazolium chloride, 1/20 vol. of bromo-chloro-indolyl phosphate (both from 1 mg/ml stock solutions in 0.1 M ethanolamine pH 9), 1/500 vol. of $MgCl_2$ (2 M stock), and ethanolamine (0.1 M, pH 9.0) to the final volume. Add 75 μl substrate per well and leave for 20–60 min. Stop the reaction with tap-water upon the emergence of dark purple spots.

7 Count the spots using a dissection microscope and express the CTLp frequency as a fraction of the total number of mononuclear cells added per well.

Briefly, nitrocellulose-bottomed microtitre wells are coated with monoclonal antibodies against the cytokine of choice, fresh PBMC are added to the wells together with the test antigen, and after an overnight incubation, the plates are developed with a suitable biotinylated secondary cytokine-specific antibody, followed by AP-avidin, and the development with an appropriate substrate. The number of spot-forming cells can then be counted under a dissection microscope, and the CTLp frequency expressed as a fraction of the total number of cells added. A convenient anti-IFN-γ ELISPOT kit is available from Mabtech, Uppsala, Sweden, but other antibodies may work equally well.

Since the ability of a peptide to bind with high affinity to a class I MHC molecule is probably the single most important factor in determining which epitopes dominate a CTL response, this criterion is often used to reduce the number of candidate peptides entered into such a screen. In practice, candidate epitopes are frequently selected using allele-specific peptide binding motifs as a guide, based on the assumption that peptides containing this motif will be preferentially selected for presentation to CTL and that those with the highest binding affinity will dominate the response. While, in the absence of any other information regarding the CTL response, this is a reasonable assumption, it is important to know how useful it is in practice—given the other factors which could influence immunodominance, as outlined in *Table 1*.

2.1 Peptide binding to class I MHC molecules

Many assays have been developed to measure the binding of peptides to class I MHC molecules. Due to the complex nature of class I MHC assembly with peptides, and the fact that peptide binding and class I assembly are linked phenomena (44), few of the assays actually yield true thermodynamic or kinetic binding constants, although they are often reported as such. In general, binding is best represented as the half-maximal concentration of peptide required to effect a specific experimental read-out (C_{50}), for example saturation of detectable class I molecules or inhibition of index-peptide binding and so on. It is generally not possible to compare binding 'affinities' measured in different binding assays. Nevertheless, surprising correspondence is seen between assays, and in some cases (for example the assembly assay) C_{50} correlates well with the actual equilibrium constants (K_A). Peptide binding assays employ either soluble or cell-associated class I MHC molecules and are outlined in *Table 2*. Methods for the most useful and widely employed binding assays are given below.

2.1.1 Techniques employed to measure peptide binding to class I MHC molecules

There are almost as many different ways of measuring peptide binding to class I molecules as there are laboratories investigating the phenomenon. It is import-

Table 2 Peptide binding assays

Assay	Reagents	References
Assembly assay	Soluble peptide + [^{35}S]-methionine-labelled lysate of TAP-negative cell	45–47
Radiolabelled peptide	[^{125}I]-peptide + detergent solubilized class I MHC molecule	48–50
β_2-m dissociation	Soluble peptide + refolded, recombinant class I MHC molecule	51
Surface plasmon resonance	Immobilized peptide + detergent solubilized class I MHC molecule	35
Gel filtration	Soluble peptide + detergent solubilized class I MHC molecule	52
Thermal stability	Soluble peptide + recombinant class I MHC molecule	53
Light scattering	Soluble peptide + soluble class I MHC	54
Radiolabelled peptide	[^{125}I]-peptide + live cells	55–59
Stabilization of 26 °C-induced class I MHC—FACS	Soluble peptide + live cells	58, 60
Stabilization of 26 °C-induced class I MHC—CPEIA	Soluble peptide + live cells	61
Binding to acid-stripped class I MHC molecules	Soluble peptide + live cells	62–64
CTL competition	Soluble peptide + live ^{51}Cr-labelled target cells + CTL	65, 66

ant to note that different assays measure different aspects of the class I MHC–peptide interaction and it is therefore not always possible to compare data generated using different techniques. We have attempted to group techniques according to their general approach. The most accessible assays are given in detail and other entries are intended as a guide. References are generally to original articles in which the particular technique is described in sufficient detail to allow the newcomer to perform the experiment.

Protocol 3

Peptide binding to class I MHC: Techniques using soluble class I MHC: assembly assay

Equipment and reagents

- CO_2 incubator
- Eppendorf tubes
- TAP-deficient cell line
- Test peptide
- [^{35}S]-methionine
- MetCys-free R10: methionine/cysteine-free RPMI with 10% dialysed FCS and 10 mM Hepes
- PBS
- [^{35}S]-labelled 'Promix' (Nycomed Amersham)
- TBS: Tris–HCl-buffered saline pH 7.4
- NP-40, PMSF, pepstatin, leupeptin, and iodoacetamide
- Conformation-specific monoclonal antibody e.g. W6/32
- 10% suspension of protein A–Sepharose (Sigma)
- SDS–PAGE or IEF PAGE
- *Staphylococcus aureus* cells (Pansorbia)
- Sealed humid chamber containing activated charcoal

Method

1. In a TAP-deficient cell line (see Appendix 2 at the end of this chapter), metabolically label a cohort of class I MHC molecules with [^{35}S]-methionine. Harvest 5×10^6 cells per experimental point, wash once in PBS (37 °C). Resuspend at 10^7 cells/ml in MetCys-free R10 and culture for 40 min in a humid CO_2 incubator at 37 °C. Pellet the cells and resuspend the pellet in MetCys-free R10 (10^8 cells/ml) before adding 25–50 μCi [^{35}S]-labelled 'Promix'. Return the resuspended pellet to the incubator for 20–30 min. The radiolabels tend to be crude preparations and could include radioactive volatiles. We advise the use of a sealed humid chamber containing a little activated charcoal in which to contain the culture vessels. Wash the cells once in ice-cold PBS, split into Eppendorf tubes according to sample number, and place on ice.
2. Lyse cells in TBS pH 7.4 with 0.5% NP-40 (lysis buffer) and 0.5% Mega-9 (Sigma), 5 mM EDTA, 2 mM PMSF, 2 μg/ml pepstatin, 2 μg/ml leupeptin, and 5 mM iodoacetamide at a concentration of 5×10^6 cells/ml, and place 1 ml aliquots into individual Eppendorf tubes containing dilutions of test peptide. Incubate on ice for 30 min then spin out the nuclei at 13 000 r.p.m. for 5 min. Place supernatants in new Eppendorf tubes containing 50 μl freshly washed *Staphylococcus aureus* cells (Pansorbin) and incubate

Protocol 3 continued

overnight at 4 °C with rotation, in order to pre-clear and to allow unstable class I MHC molecules to dissociate. For alleles which are relatively stable in their peptide non-ligated form overnight at 4 °C we have introduced a brief heating step after lysing the cells in the presence of peptide (47). This is 60 °C, 5 min for HLA-B*2705; 55 °C, 2 min for H-2K^k; 55 °C, 5 min for HLA-A11; and 45 °C, 5 min for HLA-A3.

3. Precipitate stable MHC–peptide complexes with a conformation-specific monoclonal antibody (e.g. W6/32, see Appendix 3 at the end of this chapter). Spin out the Pansorbin (15 500 r.p.m., 15 min) and place the supernatant in fresh tubes containing sufficient mAb to give a final concentration of 5–10 μg/ml. Incubate for 90 min at 4 °C then add 50 μl of a 10% suspension of protein A–Sepharose and continue to incubate at 4 °C with rotation for 60 min. Wash the pellet four times in lysis buffer (without addition of Mega-9 and protease inhibitors). Fractionate the immunoprecipitate electrophoretically using SDS–PAGE wherever an allele-specific mAb is used (45), or IEF PAGE when only a monomorphic determinant is possible (46). Quantify the amount of immunoprecipitated class I by autoradiography and densitometry or phosphorimaging.

(a) Special reagents: conformation-specific monoclonal antibodies, TAP-deficient cell expressing the allele of interest.

(b) Advantages: rapid screening of large numbers of peptides over a wide concentration range. No requirement for an allele-specific conformation-specific antibody when IEF is used. Measurements approximate well to K_A for binding. Broad range, and sensitive (K_A between 10 μM and 1 nM can be measured).

Other techniques have been described which measure the binding of peptides to soluble class I MHC molecules but these require either a chemically modified peptide ligand or highly specialized expertise. They are therefore not ideally suited for the screening of large numbers of potential peptide epitopes. They are included here for the sake of completeness and appropriate references are given for those wishing to pursue the assays in more detail.

i. Radiolabelled peptide binding

(a) **To class I molecules prior to purification**. Increasing concentrations of ^{125}I-labelled (48–50, 59) peptide is added to a lysate of a TAP-deficient cell line expressing the appropriate class I MHC molecule in the presence of excess β_2-m (1 μM) (48). The class I MHC–peptide complexes are then immunoprecipitated with a conformation-specific antibody, as described above, and counted.

(b) **To purified class I molecules** (49, 50, 54, 67). Around 10% of class I molecules in a B cell line are peptide-receptive and this fraction is much higher for some class I molecules expressed in CHO cells (for example D^b (68), K^d (53), B27 (A. McMichael, personal communication). These can be immunopurified from 1% NP-40 lysates by immunoaffinity purification (49, 50, 67). Genetically engineered, secreted forms of several class I alleles have been produced in CHO cells (53), L cells (35), and SP6 cells (69) and can be isolated from cell culture supernatant by immunoaffinity chromatography (35), preparative isoelectric focusing, and gel filtration (68). Labelled peptide is added as above, again in the presence of excess of β_2-m (1 μM). Separation of bound from free ligand is achieved by gel filtration. These assays can be adapted to a competition assay in which increasing concentrations of cold peptide can be used to inhibit the binding of a sub-saturating dose of radiolabelled index peptide. Binding affinity is then given as the half-maximal dose required to completely inhibit binding.

Dissociation rate constants can be measured by assaying the rate of dissociation of labelled peptide from loaded class I molecules in the presence of 10- to 100-fold molar excess of cold peptide (48, 50). Association rates have also been measured this way (48, 50).

ii. Rate of dissociation of β_2-m from refolded class I MHC–peptide complexes

Recombinant class I HC in the form of purified inclusion bodies from *E. coli* is denatured in the presence of 8.8 M urea and then added to a solution of test peptide in a small volume. This is incubated for 1 h before adding ^{125}I-labelled β_2-m, diluting out the denaturant, and incubating for a further 14 h at 4°C. Class I MHC–peptide complexes are then isolated by gel filtration, an excess of cold β_2-m added, and the dissociation rate of labelled β_2-m measured (51).

iii. Surface plasmon resonance

Peptide analogues of known class I binding peptides containing cysteine residues at positions which are unimportant with respect to class I MHC binding are immobilized on the gold-dextran surface of a biosensor chip. Soluble, purified, peptide-receptive class I molecules at various concentrations are then exposed to the peptide-modified surfaces where binding results in a change in the surface plasmon resonance of the chip (35).

This technique can measure both thermodynamic and kinetic constants since the signal is recorded in real time. As an indirect assay (i.e. using test peptide to inhibit the binding of class I to immobilized index peptide) it would be suited to screening and has the advantage of using very little peptide and class I protein with high sample throughput.

iv. Fluorescence spectroscopy

Two fluorescence-based techniques have been described to measure both the thermodynamic and kinetic binding constants to peptide-receptive, soluble class I MHC molecules, prepared as described in ref. 53. The first of these measures the energy transfer from the tryptophan residues of the class I MHC molecule to dansyl-labelled peptides (70); while the second measures the perturbation of class I HC-encoded tryptophan fluorescence that occurs upon peptide binding (68). This latter technique is the most accurate available for measuring K_A, k_a, and k_d for peptide binding to pre-assembled, peptide-receptive class I MHC molecules. These are, however analytical approaches and not suited to screening.

Peptide-receptive, thermolabile class I molecules accumulate on the surface of TAP-negative cells when they are cultured at 26 °C (58). These can be stabilized against denaturation at 37 °C by peptide binding. This observation has been exploited by several groups to measure the binding of peptides to 'temperature-induced' class I molecules (35).

Protocol 4

Peptide binding to class I MHC:Techniques using live cells: surface stabilization of temperature-induced class I MHC molecules

Equipment and reagents

- U-bottomed 96-well plates
- CO_2 incubator
- TAP-negative cells
- RPMI 1640 containing 5% FCS or serum-free medium e.g. OptiMEM
- Peptide
- Allele-specific mAb
- PBS containing 0.1% (w/v) (PBS/BSA)
- FITC-conjugated second antibody

Method

1. TAP-negative cells expressing the appropriate class I allele are cultured overnight at 26 °C at around 1×10^6 cells/ml (3×10^5 cells per experimental point) in RPMI 1640 containing 5% FCS or in serum-free medium. Cells are then pelleted and resuspended in medium (1.5×10^6 cells/ml) at 26 °C and dispensed into the wells of a U-bottomed 96-well plate (0.2 ml/well).
2. Increasing concentrations of peptide are then added and the plates are incubated at room temperature for 1–2 h before increasing the temperature of incubation to 37 °C (by placing the plate in a humid CO_2 incubator) for 2–4 h. Spin the plates at 1500 r.p.m. for 4 min and aspirate the supernatant.
3. Wash the pellet twice in ice-cold PBS then resuspend in 100 μl of an allele-specific mAb at a final concentration of 5–10 μg/ml. Incubate for 1 h at 4 °C then wash three times in PBS/BSA.
4. Add FITC-conjugated second antibody (e.g. 1/100–1/1000 dilution of goat anti-mouse IgG) and incubate for a further 30 min. Wash the cells three times in PBS/BSA then fix in PBS containing 1% paraformaldehyde for 5 min at 4 °C. Analyse the cells by flow cytometry.

(a) **Modifications**: in order to prevent the appearance at the cell surface of new class I molecules, brefeldin A (5 μg/ml), anisomycin (12.5 μg/ml), and cyclohexamide (5 μg/ml) can be included during the 37 °C incubation step (60). This addition also allows the rate of disappearance of class I MHC–peptide complexes from the cell surface to be measured at 37 °C (35, 71) if samples are removed for staining at increasing time intervals during the incubation.

(b) In place of FITC-second antibody staining, one group has measured the extent of allele-specific antibody binding by cell panning enzyme immunoassay (72). Round-bottomed 96-well plates are coated with anti-mouse antibody (e.g. a 1/1000 dilution of purified goat anti-mouse antiserum) to capture the appropriate allele-specific monoclonal antibody. Peptide loaded cells are then transferred to the plates (2×10^5 cells/well) and incubated for 45 min at room temperature before extensive washing. The cells are then lysed in 1% Triton X-100 in PBS and a chromatic substrate for the released cellular enzyme LDH is added and allowed to develop. Absorbance is then measured with a microplate reader.

(c) **Special reagents**: appropriate TAP-negative cell line and allele-specific antibody.

(d) **Advantages**: straightforward and simple making it well suited to screening. It also facilitates the measurement of the rate of disappearance of the complex from the cell surface of living cells which could be of major importance in determining the immunosuperiority of a particular CTL epitope.

(e) **Disadvantages**: this assay is somewhat less sensitive than the assembly assay. Requires an allele-specific antibody.

i. Binding to acid-stripped class I molecules (59, 62)

In cases where a TAP-negative cell line expressing the desired class I MHC allele is unavailable, cell surface class I molecules expressed on normal cells can be stripped of their bound peptide without denaturing them. These cells can then be treated like the TAP-negative cells described above.

Wash cells (usually lymphoblastoid cell lines, 3×10^5 per experimental point) expressing the allele of interest in cold PBS. Quickly resuspend the pellet in 131 mM citric acid, 66 mM Na_2HPO_4 with pH according to allele. The pH optimum for stripping A1, A24, and B7 is 3.1, for A2.1 the pH optimum is 3.2, and for A3 optimum is 2.9. After 60–90 sec, neutralize by adding a large excess of serum-free IMDM, pellet, and resuspend cells at 1×10^6 cells/ml in the same medium containing test peptide, 1.5 μg/ml β_2-m, and 5 μg/ml of the allele-specific antibody.

Incubate for 4 h at room temperature, then wash and stain cells with a fluorescent-labelled second antibody as above. Analyse by flow cytometry.

Modifications: one group (59) has used acid-stripped cells to develop a competition assay in which the binding of a FITC-labelled index peptide is competed out by unlabelled test peptide. This has the advantage of observing peptide binding directly and not relying on the binding of monoclonal antibodies to peptide–MHC complex for detection of binding. The disadvantage is that suitable index peptides for FITC modification must be found and the conjugate prepared.

ii. Radiolabelled peptide

The binding of radioiodinated peptides to cell surface class I MHC has been measured (55). Here, cells and various concentrations of radioiodinated peptide are allowed to come to equilibrium for around 7 h at 23 °C or overnight at 4 °C. Bound and free peptide are separated by a combination of two 20 sec washes followed by spinning the cells through oil (e.g. 84% silicon oil, 16% paraffin oil) and quantitated by gamma-counting. Non-specific binding is measured in the presence of a 500-fold molar excess of an unlabelled peptide known to bind to the same MHC protein. Equilibrium constants are determined from Scatchard plots. Having established an appropriate labelled index peptide, the assay adapts well to a competition assay to measure binding of unlabelled peptides. In this form, the assay would be suitable for peptide screening. Kinetic constants (k_a and k_d) have also been measured this way (55).

2.2 Correspondence between motifs and binding efficacy

Many studies have been carried out to test what proportion of peptides that have a suitable peptide binding motif actually bind to a given class I MHC allele (35, 51, 60, 61, 73–75).

In general, it has been found that only around 30% of peptides chosen on the basis of their primary motif (see *Figure 1* and *Appendix 1* at the end of this chapter for more information on motifs) actually bind using one of the assays shown in *Table 2*. Thus, for example, of the 60 *Plasmodium falciparum* peptides chosen for their HLA-B*5301 binding motif, eight bound to that allele at a level which was comparable to the positive control peptide (76). In a comprehensive analysis of 91 synthetic peptides derived from HBV carrying the primary HLA-A*0201 motif (50, 74) it was found that binding varied over several orders of magnitude (5000 nM $< K_A <$ 0.5 nM) with high affinity binders (i.e. $K_A <$ 50 nM) representing 25% of the total number. Our own experience, using the assembly assay is that this proportion is highly allele-dependent, with a fairly good correspondence between predicted binding and actual binding for, e.g. HLA-A2, -B7, and -B35, whereas HLA-B27, and -B51 exhibits a poor correlation between predictions and experimental results. There are various reasons for the high number of 'false positives' when primary motifs are used to predict peptide ligands for individual alleles. The most important reason is the contribution of other 'secondary' or 'subsidiary' anchors to binding, as well as important negative (inhibitory) contributions from residues outside the anchor positions (44, 61, 77). In addition, some properties of synthetic peptides in solution could have a negative effect on binding in the assays.

The importance of including information about secondary anchor residues, when attempting to predict peptide binding to class I MHC, was illustrated in a study of 240 nonamers chosen from papilloma virus type 16 E6 and E7 proteins with an HLA-A*0201 binding motif. The primary motif only predicted 27% of the high affinity binding peptides, whereas the use of an expanded motif incorporating information about the contribution of auxiliary anchors, raised this figure to 73% (67).

Prediction of CTL epitopes based on the allele-specific primary motif may therefore be of limited value. More sophisticated algorithms which incorporate information about the contribution of all peptide side chains to the overall binding of a particular peptide have now been developed in an attempt to improve the prediction of CTL epitopes (75, 78, 79). The most useful of these algorithms are outlined in *Protocol 5*.

Protocol 5

Methods for predicting class I MHC binding sequences

The best predictive algorithms are those that take into account as much experimental information as possible. Two basic approaches have proved to be successful, which assigns a binding coefficient to each amino acid as it appears in the context of an octamer or nonamer. Thus, primary anchors score high (positive), secondary anchors lower, and residues which are known to be deleterious to binding score low (negative). Three such algorithms are described.

A. http://bimas.dcrt.nih.gov:80/molbio/hla_bind

1. This is a very user-friendly web site developed by Dr Ken Parker (75, 80). Protein or nucleotide sequences obtained from electronic databases can be entered directly and are handled rapidly.
2. The binding coefficient files for 40 human and mouse (and one cattle) class I molecules are available with 8-mer, 9-mer, and 10-mer options for ligand predictions.
3. The user can elect to see the results for the top 250 predicted binders, which are ranked according to their predicted binding constant which is an estimate of the half-time of dissociation.
4. All the relevant background information is clearly given at the web site.

B. Data from pool sequence analysis

1. This method was originally devised by Davenport *et al.* (79) for the prediction of peptide sequences binding to class II MHC molecules. The peptide scoring matrix (derived from pool sequence analysis of peptide mixtures eluted from MHC molecules) is however extremely useful for applying to class I MHC molecules.
2. Copies of the computer program which simply scans an unknown protein sequence iteratively, one amino acid at a time (calculating an allele-dependent score for each amino acid at each position) are available by contacting Dr I. HoShon by e-mail: MX % 'ivan_ho_shon@sydpcug.org.au'.

C. Data from binding analysis of random sets of peptides

1. A similar technique has been described for HLA-A*0201 by D'Amaro *et al.* using a detailed motif for peptide binding derived from large random sets of peptides using a cellular binding assay has been employed (81, 82).
2. The computer program is available from Dr J. D'Amaro at the Department of Immunohematology and Blood Bank, University Hospital Leiden, PO Box 9600, 2300 RC Leiden, The Netherlands.

2.3 Correspondence between peptide binding and immunogenicity / immunodominance

The ultimate value of being able to predict peptide binding based on a sequence motif is in the prediction of T cell epitopes. Implicit in this approach is the assumption that efficiently binding peptides will be the dominant target epitopes during a complex antigenic challenge. Where motifs have been used to predict CTL epitopes successfully (76, 83, 84) the immunodominant peptides bind to their presenting class I MHC molecule with high affinity. Given the fact that there are many factors which influence immunodominance (see *Table 1*), it is important to know how useful binding data might be in forecasting whether a peptide will be a dominant target for CTL. Several groups have attempted to correlate peptide binding efficacy with immunogenicity or immunodominance in a CTL response to whole protein (35, 55, 60, 64, 74, 85–88). In simple experimental systems, the correspondence is found to be good. Thus, for example the CTL response to ovalbumin in C57Bl/6 mice is focused on one dominant and one subdominant epitope restricted by H-2K^b. CTL to the subdominant epitope (OVA_{55-62}) which has a K_A 30 times lower than the immunodominant epitope $OVA_{257-264}$ only become apparent during high-dose immunization (35). In a more complex system, Deng *et al.* (60) found that five out of the top seven influenza A-derived peptides chosen for their H-2K^d binding motif (26 in all) were recognized by CTL generated in BALB/c mice primed with influenza A virus. There was no correlation, however between the magnitude of response and peptide binding and the immunodominant epitope is ranked sixth among the binders (60). A similar weak correlation was found for the large panel of HLA-A*A0201-binding HBV peptides described in Section 2.2 (74). Here, although the peptides that were most frequently recognized by acute hepatitis patients (50% or greater) all had a K_A less than 50 nM, responses to lower affinity peptides (K_A between 50 and 700 nM) were also fairly frequent. One out of three patients also recognized a peptide with a K_A of > 25 000 nM which would have scored negative in most binding assays. Conversely, two peptides with very high affinities (K_A = 14 and 19 nM) were infrequently recognized (15% and 19%, respectively). Vitiello *et al.* (89) have used 26 peptides with predicted HLA-A*0201 binding motifs derived from HBV polymerase which bind with a wide range of affinities (8 nM to > 5 μM) to immunize HLA-A*0201 transgenic mice. Nine out of 26 peptides were immunogenic and all of these were among the top 50% of binders (all had K_A < 250 nM). However, it is worth noting that three peptides with relatively high K_A (50–80 nM) were not immunogenic. The same group went on to find that only four out of the nine epitopes were generated as a result of natural processing of the antigen in target cells, and although the best binder was among these four and the worst binder was not, there was no correlation between binding affinity and the ability to be recognized by CTL. Other groups have noted that peptides which are poor binders can nevertheless act as immunodominant epitopes (55, 87, 88).

One study has shown that the correlation between binding and immunogenicity can be assay-dependent (67). Using a competition assay on live cells to derive binding affinities based on the concentration of peptide required to inhibit index-peptide binding by 50%, only around half of the immunogenic peptides were among the highest binders ($IC_{50} < 5$ mM), and over 90% among the high and intermediate binders ($IC_{50} < 15$ mM). However, when the rate of persistence of the peptide–MHC complex was measured at the surface of live cells at 37 °C, it was found that a half-life of > 3 h predicted 90% of immunogenic peptides. Importantly, the use of 'affinity' as a predictor gave rise to a significant number (5/17) of false positives, whereas no false positives were found with the second assay. In this study, there were notable examples of peptides which apparently had a high affinity but which did not persist at the cell surface. This has also been noted by Levitsky for an HLA-A*1101-binding peptide from EBV EBNA 4 (71).

Two systematic studies have been carried out (60, 74) in which the binding and immunogenicity or immunodominance of a set of predicted CTL epitopes have been compared. It was found that actual epitopes fall into the top 20 experimentally determined binders. Only 60–70% of the predicted epitopes are amongst the top 20 binders (using the algorithms described in *Protocol 5*). This analysis shows that as expected, experimental binding is a better forecaster of epitopes than predictive algorithms, which means that it is always worth checking the binding of a large number of sequences suggested by an algorithm. From a practical point of view, the prediction of false positives is of little consequence, and one way of avoiding these would be to measure actual half-times for the persistence of peptide–MHC complexes at the cell surface. However, prediction of false negatives is a more serious problem and has been observed in several instances (67). Several epitopes have been described which show low or no detectable binding to their presenting allele in various assays.

We would suggest using an algorithm to select around 50 peptides per 5000 amino acids and screen their binding using any of the above assays. This might also include measuring the rate of disappearance of the complex from the cell surface where this can be measured.

2.3.1 Other factors which influence immunodominance

It is worth taking note of the many other factors (also listed in *Table 1*), in addition to the strength and longevity of the peptide–MHC interaction which have been shown to contribute to immunodominance in specific cases.

i. Level of expression of antigenic protein

The extraction of complex mixtures of peptide ligands from class I molecules and the subsequent purification and identification of the most abundant peptides in this mixture has shown that they are usually derived from abundant cellular proteins. This strongly suggests that the level of expression of a protein can determine the level of expression of a particular peptide ligand (see below).

ii. Processing

Both the overall rate and the specificity of the enzymes involved in the initial processing steps in the cytosol have been shown to influence the efficacy of antigen presentation. Townsend first showed that a block in antigen presentation of influenza virus nucleoprotein (NP) to NP-specific CTL could be relieved by promoting the rapid ubiquitin-dependent degradation of NP in the cytosol (90). More recently it has been shown using the same technique to regulate the intracellular turnover, that a rapidly degraded form of recombinant HIV-1 nef is four to five times more immunogenic than a slowly degraded form (91). Slow degradation by the proteasome also appears to be responsible for the crypticity of CTL epitopes encoded within EBNA-1 of EBV (92). In the case of the OVA epitopes discussed above, Niedermann *et al.* (93) found that the immunodominance of $OVA_{257-264}$ over OVA_{55-62} could be due to a highly proteasome-sensitive cleavage site in the centre of the latter epitope, illustrating that epitope cleavage may be highly dependent on its flanking sequence. Once in the ER, proteases may continue to trim precursor epitopes (94), and this may explain why some epitopes are destroyed when they overlap with others (95).

iii. Post-translational modification

It is now clear that T cells can specifically recognize small glycan modifications on MHC-bound glycopeptides (96–100) and that glycopeptide-specific recognition can dominate the T cell response to a glycoprotein antigen (97). Such epitopes will not be identified using screening strategies with synthetic (unmodified) peptides. In addition, some predicted binders may be glycosylated in such a way as to destroy binding (101, 102). Glycosylation of target antigens can also affect their ability to be processed under certain conditions.

iv. Transport

TAP shows some substrate specificity (26). There is some evidence both for (103) and against (60) the involvement of TAP-mediated peptide transport in influencing immunogenicity.

v. TcR repertoire

A peptide which binds well to a class I molecule may mimic a self-peptide which has caused the deletion of potentially reactive CTL during thymic development. Restricting the T cell repertoire can have a profound effect on immunodominance. This has been demonstrated by comparing the immunodominant CTL response to influenza A virus in wild-type C57Bl/6 mice with B6 mice transgenic for a TcR beta chain which is not overly represented following influenza infection. In the former case, the immunodominant CTL response was to NP, whereas in the transgenic animals, the response was to NS1 and MP (104).

vi. CTL sensitivity

Between 1 and 20 000 peptide–MHC complexes are required to stimulate a CTL (55, 56, 105–107). Thus, good binders may still be weakly immunogenic if no high affinity TcR exists to recognize it. al-Ramadi *et al.* (108) have pointed out that while the correlation between TcR avidity and the biological function of CTL (lysis and IL-3 release) is generally good, there are examples of CTL expressing TcR with very low avidity for their cognate ligand which nevertheless are stimulated to give strong biological responses. In another example, the dominance of one CTL specificity over another correlated with higher avidity, and was shown to operate at the level of competition for the antigen-presenting cell (APC) (28). The exact mechanism for this is unknown but could be as simple as competition for space on the APC surface, or for APC-derived stimulatory factors.

vii. Immunosuppression

Immunosuppression of an immunorecessive epitope by a dominant one located in the same protein has been shown for influenza A NP epitopes presented in BALB/c mice (60) and has also been described for SV40-T (109, 110), LCMV (111, 112), and HIV (34). The mechanism for this suppression is not well understood. Other examples have been described in which dominance of one specificity over another operates at the level of competition for antigen-presenting cells expressing both epitopes (113).

viii. Cytokines

At least one example has been documented in which immunodominance within the CTL response could be governed by the intensity of an accompanying (class II restricted) TH1 response to the same antigen (114).

3 Direct identification of CTL epitopes

Where a CTL response has been identified but there is no information regarding the potential source of the epitope, for example in the case of infection with a pathogen for which there is no or limited information about its genome, or where the target antigen could be any self-protein, a different approach must be taken. Examples of such cases might be the identification of target antigens recognized by protective CTL which have been identified in parasitic diseases, or in the identification of tumour antigens. Two approaches are currently available.

3.1 Direct identification of CTL epitopes by immuno-isolation and sequencing

If a CTL clone has been generated against a tumour cell expressing an unknown antigen, the peptide presented may be identified by immunoaffinity isolation of MHC–peptide complexes from tumour cell lysates, followed by peptide separation, and identification of the fraction which reconstitutes recognition of autologous antigen negative target cells in a standard [^{51}Cr] release assay. The recognized peptide contained within this fraction may then be attempted to be sequenced by the use of Edman degradation or tandem mass spectrometry as developed by Hunt and colleagues (115–118).

After having established the CTL clone and determined the restriction element presenting the unknown antigen, a sufficient amount of a cell line expressing the antigen needs to be cultured in roller bottles or cell factories *in vitro*. When using a starting amount of 10^{10} cells expressing an individual peptide antigen at an estimated average level of ten copies per cell, the peptide starting material equals a total amount of 10^{11} copies or just below 200 fmol peptide. Losses incurred during affinity purification procedures and subsequent peptide extraction and chromatography steps may amount to as much as 99%, which would then leave a final yield in the range of 2 fmol of individual peptide ligand. This just exceeds the threshold for mass spectrometry-based peptide sequencing in the best facilities operating today. However, large variations may be observed between experimental settings.

B-LCL express around 10^6 MHC molecules per cell. Thus, assuming that 10^{10} cells equals 1.7×10^{-8} mol MHC heavy chain (10^{16} MHC molecules, 6×10^{23} molecules per mol), and using 45 kDa as the molecular weight of the MHC heavy chain, this roughly equals 700 μg MHC heavy chain (45×10^9 μg/mol $\times 1.7 \times 10^{-8}$ mol MHC heavy chain). Thus, as a rough guide use 10 mg of murine monoclonal antibody to purify 700 μg of class I MHC heavy chain from 10^{10} B-LCL. If the sequencing is found to require more peptide, increase the amount of cells and antibody accordingly.

3.1.1 Purification of peptide antigens presented by class I MHC molecules

Protocol 6

Preparation of immunoaffinity columns

Equipment and reagents

- Column
- TBS: 50 mM Tris–HCl, 150 mM NaCl pH 8.0 (prepared and used at 4 °C)
- PAS (protein A Sepharose)
- MHC-specific monoclonal antibody

Method

1. Determine experimentally the amount of protein A Sepharose (PAS) needed to bind the relevant immunoglobulin. The use of small columns reduces procedure time and non-specific loss of peptide. Pour the appropriate volume of PAS into an empty column and allow to settle without flow. Wash with cold TBS.
2. Pass a solution of PAS-purified MHC-specific monoclonal antibody over the column twice in TBS (0.5 mg/ml). The total amount of antibody used should exceed the capacity of the column by about 50% to ensure saturation. Measure the absorption of the antibody solution at 280 nm before and after passage through the column to determine the amount of antibody bound to the column. Wash the column with TBS and store at 4 °C.
3. Small pre-packed 1 or 5 ml columns (e.g. HiTrap, Pharmacia) are a convenient, but somewhat more expensive alternative which allows the safe automation of procedures on an FPLC system.

Protocol 6 continued

4 Other protocols for the preparation of immunoaffinity columns may be used, including periodate oxidation of immunoglobulin carbohydrate or cyanogen bromide activated beads.

Protocol 7

Lysis of cells

Equipment and reagents

- Dounce homogenizer
- Centrifuge
- 0.2 μm low protein binding filter (e.g. Acrodisc, Pall Gelman Laboratories)
- PBS
- Lysis buffer: 150 mM NaCl, 20 mM Tris-HCl, 1% CHAPS (Boehringer Mannheim) pH 8.0
- Protease inhibitors (see below)

Method

1 Harvest the cultured cells and wash twice in ice-cold PBS. The cell pellets may be frozen at −80 °C at this point. All subsequent steps should be conducted at 4 °C.

2 Resuspend the frozen cell pellet at 10^8 cells/ml of cold lysis buffer containing freshly added protease inhibitors (see below), and leave to lyse for 60 min at 4 °C.

3 Dounce homogenize and centrifuge the lysate at 100 000 g for 1 h (Ti45 rotor at 35 500 r.p.m.). Carefully decant the supernatant and filter through a 0.2 μm low protein binding filter changing the filter every 5–15 ml as it becomes clogged.

Protease inhibitors (should be added immediately prior to use):

	Stock	Ratio added (v/v)
100 μM iodoacetamide	18.5 mg/ml in lysis buffer[a]	1:1000
5 μg/ml aprotinin	5 mg/ml in lysis buffer[a]	1:1000
10 μg/ml leupeptin	10 mg/ml in lysis buffer[a]	1:1000
10 μg/ml pepstatin	10 mg/ml in methanol[a]	1:1000
5 mM EDTA	0.5 M[b]	1:100
0.05% Na azide	20% in water[a]	1:500
1 mM PMSF	17.4 mg/ml in isopropanol[c]	1:100

[a] Store at −20 °C.

[b] Store at room temperature.

[c] Store at −20 °C, dissolves easily as it warms.

Protocol 8

Immunoaffinity isolation of class I MHC–peptide complexes

Equipment and reagents

- Immunoaffinity columns
- SDS–PAGE
- Eppendorf tubes
- Lysis buffer (see *Protocol 7*)
- 20 mM Tris pH 8.0, 150 mM NaCl
- 20 mM Tris pH 8.0, 1 M NaCl
- 0.2 M acetic acid
- TBS (see *Protocol 6*)
- PBS, azide

Method

1. Connect the immunoaffinity columns using short tubing. Use a Sepharose pre-column. The second should contain an irrelevant antibody bound to PAS to derive a negative control extract. Subsequently, place the columns containing allele-specific antibodies, and finally place a column with a pan-class I MHC-specific antibody (e.g. W6/32).
2. Equilibrate all columns with 5 vol. of lysis buffer at 1 ml/min. Subsequently, run the filtered lysate through the columns (0.5–1 ml/min). Note that if the lysate has been frozen then re-centrifugation and re-filtration are usually necessary due to aggregation.
3. Wash the columns with the following buffers (during this wash the ultrafiltration units, as described below):
   (a) 2 column volumes of lysis buffer.
   (b) 20 column volumes of 20 mM Tris pH 8.0, 150 mM NaCl.
   (c) 20 column volumes of 20 mM Tris pH 8.0, 1 M NaCl.
   (d) 20 column volumes of 20 mM Tris pH 8.0.
4. Allow excess buffer to drain from the column. Add one column volume of 20 mM Tris pH 8.0 and resuspend gently. Remove 50 μl of the suspension for quantitation of bound class I MHC molecules by SDS–PAGE. Allow excess liquid to drain.
5. Elute each column with 0.2 M acetic acid whilst collecting fractions in Eppendorf tubes. Monitor the pH with pH paper. Most of the peptide will elute with the acid front. Elute with three additional column volumes of 0.2 M acetic acid.
6. Regenerate columns by washing with two further column volumes of 0.2 M acetic acid followed by 20 column volumes of TBS. Finally equilibrate columns with PBS containing 0.05% azide and store at 4 °C.
7. To re-use columns, return to step 2, *Protocol 6*.

Protocol 9

Peptide extraction

Equipment and reagents

- Eppendorf tubes
- 23-gauge needle
- SDS–PAGE
- Filter unit reservoirs: 5000 Dalton cut-off units (Millipore Corporation, UFC4LCC25)
- Glacial acetic acid

Method

1. Add glacial acetic acid to the eluate to a final concentration of 10% acetic acid. Perforate the cap of each Eppendorf with a 23-gauge needle, place in a boiling water-bath for 5 min, and chill on ice. Centrifuge to pull splash off the sides of the tube.
2. Transfer the acid eluates to the filter unit reservoirs. It is critical to use 5000 Dalton cut-off units to exclude β_2-m. Prior to use, the units should be pre-wet with 1 ml of 10% acetic acid, and spun for 1 h with a further 2 ml of 10% acetic acid to wash the membrane and reservoirs. Centrifuge the acid eluate at 3500 g for 5 h at 4 °C. Check the units after 2 h to verify the integrity of the filters. Save both the filtrate and retentate.
3. Transfer the filtrate to Eppendorf tubes, concentrate by vacuum centrifugation without heating to a final volume of 250 μl, and store at −80 °C.
4. Estimate the yield of class I MHC heavy chain by SDS–PAGE analysis of the beads removed above. Remove excess liquid from beads, add 50 μl of 2 × sample buffer, mix, and boil for 5 min. Pellet the beads and analyse 30 and 15 μl of the supernatant, as well as 1–2 μl of the high molecular weight retentate from the Millipore filter on a 12% SDS–PAGE gel stained with Coomassie blue. In parallel lanes run reference amounts of ovalbumin in the range of 0.5–5 μg and compare with the intensity of the class I MHC heavy chain bands to estimate the total amount of class I MHC complexes purified.

Protocol 10

Reversed-phase HPLC (RP-HPLC) separation of extracted peptide material

In order to minimize sample loss, use PEEK tubing and loops which come in contact with peptide. Also, use a HPLC system of high specification, built around appropriate pumps, mixer, tubing, and fraction collector, in order to allow a low flow rate of 200 μl/min (narrow-bore chromatography). Use HPLC grade reagents throughout.

Protocol 10 continued

Equipment and reagents

- Narrow-bore C18 column (2.1 mm × 3 cm, 5–7 μm particles, 300 Å pore size)
- Polypropylene screw-cap tubes
- Solvent A: 0.1% TFA in water
- Solvent B: 0.085% TFA in 60% acetonitrile, 40% water

Method

1. Load the peptides onto a narrow-bore C18 column at a flow rate of 200 μl/min of 5% solvent B in solvent A.
2. Elute the column with a gradient of 5–15% (5 min), 15–60% (50 min), 60–100% (10 min) of solvent B in solvent A.
3. Collect fractions in polypropylene screw-cap tubes and freeze on dry ice.
4. Further separation of the peptides in each fraction is usually necessary, and second dimension chromatography can be performed by substituting TFA with heptafluorobutyric acid (HFBA) or hexafluoroacetone (HFA).

Protocol 11

Identification of fractions containing peptide antigen

1. From each RP-HPLC fraction collect aliquots equivalent to 10^9 cells and add to 2500 autologous, ^{51}Cr-labelled, antigen negative target cells in 100 μl of RPMI supplemented with 10 mM Hepes and 5% FCS.
2. Incubate for 2 h at 37 °C before the addition of CTL (E:T ratio 30:1) and continue as a conventional chromium release assay (see *Protocol 1A*). Perform in duplicate and include duplicate wells containing only target cells with eluate to control for a toxic effect of the eluate.
3. Perform second dimension chromatography of any recognized fraction(s) and repeat the identification of fraction(s) containing recognized peptide as above.
4. Hopefully, having isolated a peptide species, it is now possible to identify the peptide, e.g. by tandem mass spectrometry!

3.2 Direct identification of CTL epitopes by expression cloning

The technique used to express cloned genes encoding antigens recognized by tumour-specific CTL clones was developed by Thierry Boon and colleagues and has been used to identify genes encoding novel antigens recognized by melanoma-specific CTL (119, 120). The method involves the transfection of a cell line such as COS-7 cells or HeLa cells which does not express the antigen in question with a cDNA library prepared from tumour cells. First it is necessary to identify the MHC-restriction element of the tumour-specific CTL clone by testing for MHC allele-specific monoclonal antibody-mediated inhibition of chromium release or TNF-release CTL assays. Alternatively, assuming that the recognized antigen is expressed by other tumour cell lines, the restriction element may be identified using allogeneic tumour target cell lines, sharing only one class I MHC molecule with the autologous cell line. Finally, it is possible to transiently transfect the relevant class I MHC molecule into target cells for use in TNF release assays.

Protocol 12

Preparation of RNA and cDNA

Equipment and reagents

- Oligo(dT)-cellulose columns (Amersham Pharmacia Biotech)
- Size exclusion chromatography (using Sephacryl S-500 HR)
- Guanidinium thiocyanate
- cDNA synthesis kit (SuperScript) (Gibco BRL)
- RNase H negative reverse transcriptase (e.g. Gibco BRL)
- Radiolabelled nucleotides
- *Sal*I adapter
- Restriction enzymes: *Not*I and *Sal*I
- Plasmid (e.g. pcDNAI/Amp)

Method

1. Lyse 10^8 tumour cells in guanidinium thiocyanate, homogenize, and separate on a CsCl gradient.
2. Collect total RNA and purify poly(A) mRNA using oligo(dT)-cellulose columns.
3. Poly(A) purified mRNA is used as a template for reverse transcriptase using an oligo(dT) primer containing a *Not*I site and 18 Ts from the cDNA synthesis kit (SuperScript) with RNase H negative reverse transcriptase.
4. The reaction is performed with radiolabelled nucleotides, so that the material can easily be purified by size exclusion chromatography (using Sephacryl S-500 HR) after synthesis of the second strand, by collecting radioactive fractions for further procedures.
5. The longer fragments have less chance to ligate, especially when a lot of smaller fragments are present. Thus, perform four separate ligation reactions using the earliest fraction, which contains the longest cDNA, and perform six ligations with each of the following two fractions. A *Sal*I adapter is added to the 5′ end, and all fragments are cut with *Not*I and *Sal*I, before cloning into the digested plasmid (e.g. pcDNAI/Amp).
6. After the ligation, precipitate and pool all reactions and electroporate as described below.

Preparation of bacteria. Recombinant plasmids are electroporated into *E. coli* such as JM101 or DH5αF′IQ (BRL) to be selected with ampicillin. Electroporate the complete material in 30 separate bacterial suspensions and plate these on 15 large plates, using less than 20 ng of ligation product per 10^9 bacteria. On separate small agar plates, seed 20 and 100 μl bacterial suspension and count on the following day to evaluate the number of colonies obtained. Every bacterial colony represents a unique cDNA, usually 4–800 000 in total. Take the large plates, mix all colonies, and store at −80 °C in larger stocks and small aliquots.

Protocol 13

Transfection of COS-7 or HeLa cells

Equipment and reagents

- Flat-bottomed 96-well plates, V-bottomed microtitre plates
- DMEM with 10% FCS
- LB medium
- Isopropanol
- TE, RNase
- DMEM-NS: DMEM supplemented with 10% de-complemented, DNase-free NUSerum
- DEAE–dextran
- Chloroquine
- PBS, DMSO

Method

1. On the day before transfection seed 15 000 recipient cells per flat-bottomed 96-well in 100 μl DMEM with 10% FCS.
2. Thaw one aliquot of bacteria and titrate. Then plate out bacteria to obtain 10 000 colonies and divide these into around 100 pools and amplify in LB medium. Keep an aliquot of each pool stored at −80 °C for later recovery of a recognized gene.
3. Precipitate plasmids using isopropanol and resuspend in 50 μl of TE with 20 μg/ml RNase. One pool of 100–200 colonies should contain approximately 100 ng DNA/μl.
4. Prepare DEAE–dextran mixtures in V-bottomed microtitre wells by the sequential addition of:
   (a) 1–3 μl cDNA library plasmid DNA.
   (b) 5 μl of 10 mM Tris, 1 mM EDTA pH 7.4 containing 200 ng of the vector with the MHC gene to be co-transfected.
   (c) 5 μl DMEM-NS.
   (d) 35 μl DMEM-NS containing 0.8 mg/ml DEAE–dextran and 200 μM chloroquine. For HeLa cells use instead 0.4 mg/ml DEAE–dextran.
5. Remove the media from the recipient cells and add 30 μl of cDNA/DEAE–dextran mixture to each well in duplicates. Incubate at 37 °C for 4 h.
6. Gently flick microtitre plates and add 50 μl of PBS with 10% DMSO. Incubate for 2 min at room temperature.
7. Remove supernatant and add 200 μl DMEM supplemented with 10% FCS.
8. Incubate for 48 h at 37 °C, discard medium, and add 1000–2000 CTL in IMDM containing 10% human serum and 25 U/ml IL-2. After 24 h at 37 °C measure release of TNF into the supernatant.
9. Alternatively, some laboratories use assays for GM-CSF release, or perform T–T cell fusion, generating T cell hybridomas, which produce LacZ after triggering.

Recovery of a positive cDNA clone: if a transfectant is recognized by the CTL clone, the pool of cDNA can be recovered from the corresponding pool of frozen bacteria and recloned. The positive bacteria are then isolated by testing smaller pools of bacteria, and eventually single colonies, until you have a single cDNA clone, which can be sequenced and characterized.

References

1. Yap, K. L. and Ada, G. L. (1978). *Scand. J. Immunol.*, **7**, 73.
2. McMichael, A. J., Gotch, F. M., Noble, G. R., and Beare, A. S. (1983). *N. Engl. J. Med.*, **309**, 13.
3. Lin, Y. L. and Askonas, B. A. (1981). *J. Exp. Med.*, **154**, 225.
4. Cannon, M. J., Openshaw, P. J., and Askonas, B. A. (1988). *J. Exp. Med.*, **168**, 1163.
5. Bonneau, R. H. and Jennings, S. R. (1990). *Virology*, **174**, 599.
6. Sissons, J. G., Colby, S. D., Harrison, W. O., and Oldstone, M. B. (1985). *Clin. Immunol. Immunopathol.*, **34**, 60.
7. Vallbracht, A., Maier, K., Stierhof, Y. D., Wiedmann, K. H., Flehmig, B., and Fleischer, B. (1989). *J. Infect. Dis.*, **160**, 209.
8. Moss, D. J., Rickinson, A. B., and Pope, J. H. (1978). *Int. J. Cancer*, **22**, 662.
9. Borysiewicz, L. K., Morris, S., Page, J. D., and Sissons, J. G. (1983). *Eur. J. Immunol.*, **13**, 804.
10. McMichael, A. J. and Walker, B. D. (1994). *AIDS*, **8** (suppl 1), s155.
11. Zinkernagel, R. M., Blanden, R. V., and Langman, R. E. (1974). *J. Immunol.*, **112**, 496.
12. Goddeeris, B. M., Morrison, W. I., Teale, A. J., Bensaid, A., and Baldwin, C. L. (1986). *Proc. Natl. Acad. Sci. USA*, **83**, 5238.
13. Weiss, W. R., Mellouk, S., Houghten, R. A., Sedegah, M., Kumar, S., Good, M. F., *et al.* (1990). *J. Exp. Med.*, **171**, 763.
14. Rodrigues, M., Nussenzweig, R. S., Romero, P., and Zavala, F. (1992). *J. Exp. Med.*, **175**, 895.
15. Sing, A. P., Ambinder, R. F., Hong, D. J., Jensen, M., Batten, W., Petersdorf, E., *et al.* (1997). *Blood*, **89**, 1978.
16. Heslop, H. E., Ng, C. Y., Li, C., Smith, C. A., Loftin, S. K., Krance, R. A., *et al.* (1996). *Nature Med.*, **2**, 551.
17. Harty, J. T. and Bevan, M. J. (1992). *J. Exp. Med.*, **175**, 1531.
18. Walter, E. A., Greenberg, P. D., Gilbert, M. J., Finch, R. J., Watanabe, K. S., Thomas, E. D., *et al.* (1995). *N. Engl. J. Med.*, **333**, 1038.
19. Riddell, S. R., Watanabe, K. S., Goodrich, J. M., Li, C. R., Agha, M. E., and Greenberg, P. D. (1992). *Science*, **257**, 238.
20. McMichael, A. J. and Bodmer, W. F. (ed.) (1992). *A new look at tumour immunology*. Cold Spring Harbor Laboratory Press, New York.
21. Bell, J. I., Owen, M. J., and Simpson, E. (ed.) (1995). *T cell receptors*. Oxford University Press, Oxford.
22. Garcia, K. C., Degano, M., Stanfield, R. L., Brunmark, A., Jackson, M. R., Peterson, P. A., *et al.* (1996). *Science*, **274**, 209.
23. Garboczi, D. N., Ghosh, P., Utz, U., Fan, Q. R., Biddison, W. E., and Wiley, D. C. (1996). *Nature*, **384**, 134.
24. Bjorkman, P. J. (1997). *Cell*, **89**, 167.
25. Weiss, A. and Littman, D. R. (1994). *Cell*, **76**, 263.
26. Elliott, T. (1997). *Adv. Immunol.*, **65**, 47.
27. Zinkernagel, R. M. and Doherty, P. C. (1979). *Adv. Immunol.*, **27**, 51.

28. Pion, S., Fontaine, P., Desaulniers, M., Jutras, J., Filep, J. G., and Perreault, C. (1997). *Eur. J. Immunol.*, **27**, 421.
29. Dudley, M. E. and Roopenian, D. C. (1996). *J. Exp. Med.*, **184**, 441.
30. Rammensee, H. G., Falk, K., and Rotzschke, O. (1993). *Annu. Rev. Immunol.*, **11**, 213.
31. Engelhard, V. H. (1994). *Annu. Rev. Immunol.*, **12**, 181.
32. van Bleek, G. M. and Nathenson, S. G. (1990). *Nature*, **348**, 213.
33. Nowak, M. A., May, R. M., Phillips, R. E., Rowland Jones, S., Lalloo, D. G., McAdam, S., *et al.* (1995). *Nature*, **375**, 606.
34. Goulder, P. J., Sewell, A. K., Lalloo, D. G., Price, D. A., Whelan, J. A., Evans, J., *et al.* (1997). *J. Exp. Med.*, **185**, 1423.
35. Chen, W., Khilko, S., Fecondo, J., Margulies, D. H., and McCluskey, J. (1994). *J. Exp. Med.*, **180**, 1471.
36. Townsend, A. R., Rothbard, J., Gotch, F. M., Bahadur, G., Wraith, D., and McMichael, A. J. (1986). *Cell*, **44**, 959.
37. Maryanski, J. L., Pala, P., Corradin, G., Jordan, B. R., and Cerottini, J. C. (1986). *Nature*, **324**, 578.
38. Pasternack, M. S., Verret, C. R., Liu, M. A., and Eisen, H. N. (1986). *Nature*, **322**, 740.
39. Czerkinsky, C., Andersson, G., Ekre, H. P., Nilsson, L. A., Klareskog, L., and Ouchterlony, O. (1988). *J. Immunol. Methods*, **110**, 29.
40. Lalvani, A., Brookes, R., Hambleton, S., Britton, W. J., Hill, A. V., and McMichael, A. J. (1997). *J. Exp. Med.*, **186**, 859.
41. Scheibenbogen, C., Lee, K. H., Stevanovic, S., Witzens, M., Willhauck, M., Waldmann, V., *et al.* (1997). *Int. J. Cancer*, **71**, 932.
42. Scheibenbogen, C., Lee, K. H., Mayer, S., Stevanovic, S., Moebius, U., Herr, W., *et al.* (1997). *Clin. Cancer Res.*, **3**, 221.
43. Miyahira, Y., Murata, K., Rodriguez, D., Rodriguez, J. R., Esteban, M., Rodrigues, M. M., *et al.* (1995). *J. Immunol. Methods*, **181**, 45.
44. Elliott, T., Cerundolo, V., Elvin, J., and Townsend, A. (1991). *Nature*, **351**, 402.
45. Townsend, A., Elliott, T., Cerundolo, V., Foster, L., Barber, B., and Tse, A. (1990). *Cell*, **62**, 285.
46. Elvin, J., Potter, C., Elliott, T., Cerundolo, V., and Townsend, A. (1993). *J. Immunol. Methods*, **158**, 161.
47. Tan, L., Andersen, M. H., Elliott, T., and Haurum, J. S. (1997). *J. Immunol. Methods*, **209**, 25.
48. Cerundolo, V., Elliott, T., Elvin, J., Bastin, J., Rammensee, H. G., and Townsend, A. (1991). *Eur. J. Immunol.*, **21**, 2069.
49. Ruppert, J., Sidney, J., Celis, E., Kubo, R. T., Grey, H. M., and Sette, A. (1993). *Cell*, **74**, 929.
50. Sette, A., Sidney, J., del Guercio, M. F., Southwood, S., Ruppert, J., Dahlberg, C., *et al.* (1994). *Mol. Immunol.*, **31**, 813.
51. Parker, K. C., Carreno, B. M., Sestak, L., Utz, U., Biddison, W. E., and Coligan, J. E. (1992). *J. Biol. Chem.*, **267**, 5451.
52. Olsen, A. C., Pedersen, L. O., Hansen, A. S., Nissen, M. H., Olsen, M., Hansen, P. R., *et al.* (1994). *Eur. J. Immunol.*, **24**, 385.
53. Fahnestock, M. L., Tamir, I., Narhi, L., and Bjorkman, P. J. (1992). *Science*, **258**, 1658.
54. Sykulev, Y., Brunmark, A., Jackson, M., Cohen, R. J., Peterson, P. A., and Eisen, H. N. (1994). *Immunity*, **1**, 15.
55. Kageyama, S., Tsomides, T. J., Sykulev, Y., and Eisen, H. N. (1995). *J. Immunol.*, **154**, 567.
56. Christinck, E. R., Luscher, M. A., Barber, B. H., and Williams, D. B. (1991). *Nature*, **352**, 67.
57. Luescher, I. F., Romero, P., Cerottini, J. C., and Maryanski, J. L. (1991). *Nature*, **351**, 72.

58. Schumacher, T. N., Heemels, M. T., Neefjes, J. J., Kast, W. M., Melief, C. J., and Ploegh, H. L. (1990). *Cell*, **62**, 563.
59. van der Burg, S. H., Ras, E., Drijfhout, J. W., Benckhuijsen, W. E., Bremers, A. J., Melief, C. J., *et al.* (1995). *Hum. Immunol.*, **44**, 189.
60. Deng, Y., Yewdell, J. W., Eisenlohr, L. C., and Bennink, J. R. (1997). *J. Immunol.*, **158**, 1507.
61. Sigal, L. J., Goebel, P., and Wylie, D. E. (1995). *Mol. Immunol.*, **32**, 623.
62. Zeh, H. J., Leder, G. H., Lotze, M. T., Salter, R. D., Tector, M., Stuber, G., *et al.* (1994). *Hum. Immunol.*, **39**, 79.
63. Stuber, G., Leder, G. H., Storkus, W. T., Lotze, M. T., Modrow, S., Szekely, L., *et al.* (1994). *Eur. J. Immunol.*, **24**, 765.
64. van der Burg, S. H., Visseren, M. J., Brandt, R. M., Kast, W. M., and Melief, C. J. (1996). *J. Immunol.*, **156**, 3308.
65. Pala, P., Bodmer, H. C., Pemberton, R. M., Cerottini, J. C., Maryanski, J. L., and Askonas, B. A. (1988). *J. Immunol.*, **141**, 2289.
66. Carreno, B. M., Anderson, R. W., Coligan, J. E., and Biddison, W. E. (1990). *Proc. Natl. Acad. Sci. USA*, **87**, 3420.
67. Kast, W. M., Brandt, R. M., Sidney, J., Drijfhout, J. W., Kubo, R. T., Grey, H. M., *et al.* (1994). *J. Immunol.*, **152**, 3904.
68. Springer, S., Doering, K., Cerundolo, V., Edwards, J., Skipper, J., and Townsend, A. (1998). *Biochemistry*, **37**, 3001.
69. Matsumura, M., Saito, Y., Jackson, M. R., Song, E. S., and Peterson, P. A. (1992). *J. Biol. Chem.*, **267**, 23589.
70. Gakamsky, D. M., Bjorkman, P. J., and Pecht, I. (1996). *Biochemistry*, **35**, 14841.
71. Levitsky, V., Zhang, Q. J., Levitskaya, J., and Masucci, M. G. (1996). *J. Exp. Med.*, **183**, 915.
72. Sigal, L. J., Berens, S., and Wylie, D. (1994). *J. Immunol. Methods*, **177**, 261.
73. Hill, A. V., Allsopp, C. E., Kwiatkowski, D., Anstey, N. M., Twumasi, P., Rowe, P. A., *et al.* (1991). *Nature*, **352**, 595.
74. Sette, A., Vitiello, A., Reherman, B., Fowler, P., Nayersina, R., Kast, W. M., *et al.* (1994). *J. Immunol.*, **153**, 5586.
75. http://bimas.dcrt.nih.gov:80/molbio/hla_bind
76. Hill, A. V., Elvin, J., Willis, A. C., Aidoo, M., Allsopp, C. E., Gotch, F. M., *et al.* (1992). *Nature*, **360**, 434.
77. Rovero, P., Riganelli, D., Fruci, D., Vigano, S., Pegoraro, S., Revoltella, R., *et al.* (1994). *Mol. Immunol.*, **31**, 549.
78. Hobohm, U. and Meyerhans, A. (1993). *Eur. J. Immunol.*, **23**, 1271.
79. Davenport, M. P., Ho Shon, I. A., and Hill, A. V. (1995). *Immunogenetics*, **42**, 392.
80. Parker, K. C., Bednarek, M. A., and Coligan, J. E. (1994). *J. Immunol.*, **152**, 163.
81. D'Amaro, J., Houbiers, J. G., Drijfhout, J. W., Brandt, R. M., Schipper, R., Bavinck, J. N., *et al.* (1995). *Hum. Immunol.*, **43**, 13.
82. Drijfhout, J. W., Brandt, R. M., D'Amaro, J., Kast, W. M., and Melief, C. J. (1995). *Hum. Immunol.*, **43**, 1.
83. Pamer, E. G., Harty, J. T., and Bevan, M. J. (1991). *Nature*, **353**, 852.
84. Rotzschke, O., Falk, K., Stevanovic, S., Jung, G., Walden, P., and Rammensee, H. G. (1991). *Eur. J. Immunol.*, **21**, 2891.
85. Oukka, M., Riche, N., and Kosmatopoulos, K. (1994). *J. Immunol.*, **152**, 4843.
86. Ressing, M. E., Sette, A., Brandt, R. M., Ruppert, J., Wentworth, P. A., Hartman, M., *et al.* (1995). *J. Immunol.*, **154**, 5934.
87. Connolly, J. M. (1994). *Proc. Natl. Acad. Sci. USA*, **91**, 11482.
88. Dong, T., Boyd, D., Rosenberg, W., Alp, N., Takiguchi, M., McMichael, A., *et al.* (1996). *Eur. J. Immunol.*, **26**, 335.

89. Vitiello, A., Sette, A., Yuan, L., Farness, P., Southwood, S., Sidney, J., *et al.* (1997). *Eur. J. Immunol.*, **27**, 671.
90. Townsend, A., Bastin, J., Gould, K., Brownlee, G., Andrew, M., Coupar, B., *et al.* (1988). *J. Exp. Med.*, **168**, 1211.
91. Tobery, T. W. and Siliciano, R. F. (1997). *J. Exp. Med.*, **185**, 909.
92. Levitskaya, J., Coram, M., Levitsky, V., Imreh, S., Steigerwald Mullen, P. M., Klein, G., *et al.* (1995). *Nature*, **375**, 685.
93. Niedermann, G., Butz, S., Ihlenfeldt, H. G., Grimm, R., Lucchiari, M., Hoschutzky, H., *et al.* (1995). *Immunity*, **2**, 289.
94. Elliott, T., Willis, A., Cerundolo, V., and Townsend, A. (1995). *J. Exp. Med.*, **181**, 1481.
95. Snyder, H. L., Yewdell, J. W., and Bennink, J. R. (1994). *J. Exp. Med.*, **180**, 2389.
96. Haurum, J. S., Arsequell, G., Lellouch, A. C., Wong, S. Y., Dwek, R. A., McMichael, A. J., *et al.* (1994). *J. Exp. Med.*, **180**, 739.
97. Michaelsson, E., Malmstrom, V., Reis, S., Engstrom, Å., Burkhardt, H., and Holmdahl, R. (1994). *J. Exp. Med.*, **180**, 745.
98. Abdel Motal, U. M., Berg, L., Rosen, A., Bengtsson, M., Thorpe, C. J., Kihlberg, J., *et al.* (1996). *Eur. J. Immunol.*, **26**, 544.
99. Jensen, T., Hansen, P., Galli Stampino, L., Mouritsen, S., Frische, K., Meinjohanns, E., *et al.* (1997). *J. Immunol.*, **158**, 3769.
100. Galli Stampino, L., Meinjohanns, E., Frische, K., Meldal, M., Jensen, T., Werdelin, O., *et al.* (1997). *Cancer Res.*, **57**, 3214.
101. Haurum, J. S., Tan, L., Arsequell, G., Frodsham, P., Lellouch, A. C., Moss, P. A., *et al.* (1995). *Eur. J. Immunol.*, **25**, 3270.
102. Ishioka, G. Y., Lamont, A. G., Thomson, D., Bulbow, N., Gaeta, F. C., Sette, A., *et al.* (1992). *J. Immunol.*, **148**, 2446.
103. Yellen Shaw, A. J., Laughlin, C. E., Metrione, R. M., and Eisenlohr, L. C. (1997). *J. Exp. Med.*, **186**, 1655.
104. Daly, K., Nguyen, P., Woodland, D. L., and Blackman, M. A. (1995). *J. Virol.*, **69**, 7416.
105. Vitiello, A., Potter, T. A., and Sherman, L. A. (1990). *Science*, **250**, 1423.
106. Valitutti, S., Muller, S., Cella, M., Padovan, E., and Lanzavecchia, A. (1995). *Nature*, **375**, 148.
107. Sykulev, Y., Joo, M., Vturina, I., Tsomides, T. J., and Eisen, H. N. (1996). *Immunity*, **4**, 565.
108. al Ramadi, B. K., Jelonek, M. T., Boyd, L. F., Margulies, D. H., and Bothwell, A. L. (1995). *J. Immunol.*, **155**, 662.
109. Oldstone, M. B., Lewicki, H., Borrow, P., Hudrisier, D., and Gairin, J. E. (1995). *J. Virol.*, **69**, 7423.
110. Mylin, L. M., Bonneau, R. H., Lippolis, J. D., and Tevethia, S. S. (1995). *J. Virol.*, **69**, 6665.
111. Gegin, C. and Lehmann Grube, F. (1992). *J. Immunol.*, **149**, 3331.
112. van der Most, R. G., Sette, A., Oseroff, C., Alexander, J., Murali Krishna, K., Lau, L. L., *et al.* (1996). *J. Immunol.*, **157**, 5543.
113. McHeyzer Williams, M. G. and Davis, M. M. (1995). *Science*, **268**, 106.
114. Eberl, G., Kessler, B., Eberl, L. P., Brunda, M. J., Valmori, D., and Corradin, G. (1996). *Eur. J. Immunol.*, **26**, 2709.
115. Cox, A. L., Skipper, J., Chen, Y., Henderson, R. A., Darrow, T. L., Shabanowitz, J., *et al.* (1994). *Science*, **264**, 716.
116. den Haan, J. M., Sherman, N. E., Blokland, E., Huczko, E., Koning, F., Drijfhout, J. W., *et al.* (1995). *Science*, **268**, 1476.
117. Henderson, R. A., Cox, A. L., Sakaguchi, K., Appella, E., Shabanowitz, J., Hunt, D. F., *et al.* (1993). *Proc. Natl. Acad. Sci. USA*, **90**, 10275.

118. Wang, W., Meadows, L. R., den Haan, J. M., Sherman, N. E., Chen, Y., Blokland, E., *et al.* (1995). *Science*, **269**, 1588.
119. Brichard, V., Van Pel, A., Wolfel, T., Wolfel, C., De Plaen, E., Lethe, B., *et al.* (1993). *J. Exp. Med.*, **178**, 489.
120. Coulie, P. G., Brichard, V., Van Pel, A., Wolfel, T., Schneider, J., Traversari, C., *et al.* (1994). *J. Exp. Med.*, **180**, 35.
121. Rammensee, H. G., Bachmann, J., and Stevanovic, S. (1997). *MHC ligands and peptide motifs*. Springer-Verlag, Heidelberg, Germany.
122. Kelly, A., Powis, S. H., Kerr, L. A., Mockridge, I., Elliott, T., Bastin, J., *et al.* (1992). *Nature*, **355**, 641.
123. Urban, R. G., Chicz, R. M., Lane, W. S., Strominger, J. L., Rehm, A., Kenter, M. J., *et al.* (1994). *Proc. Natl. Acad. Sci. USA*, **91**, 1534.
124. Salter, R. D. and Cresswell, P. (1986). *EMBO J.*, **5**, 943.
125. Anderson, K. S., Alexander, J., Wei, M., and Cresswell, P. (1993). *J. Immunol.*, **151**, 3407.
126. McIntyre, C. A., Rees, R. C., Platts, K. E., Cooke, C. J., Smith, M. O., Mulcahy, K. A., *et al.* (1996). *Cancer Immunol. Immunother.*, **42**, 246.
127. Gavioli, R., Zhang, Q. J., Marastoni, M., Guerrini, R., Reali, E., Tomatis, R., *et al.* (1995). *Biochem. Biophys. Res. Commun.*, **206**, 8.
128. Cerundolo, V., Alexander, J., Anderson, K., Lamb, C., Cresswell, P., McMichael, A., *et al.* (1990). *Nature*, **345**, 449.
129. Alexander, J., Payne, J. A., Murray, R., Frelinger, J. A., and Cresswell, P. (1989). *Immunogenetics*, **29**, 380.
130. Smith, K. D. and Lutz, C. T. (1996). *J. Immunol.*, **156**, 3755.
131. Smith, K. D., Mace, B. E., Valenzuela, A., Vigna, J. L., McCutcheon, J. A., Barbosa, J. A., *et al.* (1996). *J. Immunol.*, **157**, 2470.
132. Takiguchi, M., Kawaguchi, G., Sekimata, M., Hiraiwa, M., Kariyone, A., and Takamiya, Y. (1994). *Int. Immunol.*, **6**, 1345.
133. Cerundolo, V., Elliott, T., Elvin, J., Bastin, J., and Townsend, A. (1992). *Eur. J. Immunol.*, **22**, 2243.
134. Zhou, X., Momburg, F., Liu, T., Abdel Motal, U. M., Jondal, M., Hammerling, G. J., *et al.* (1994). *Eur. J. Immunol.*, **24**, 1863.
135. Crumpacker, D. B., Alexander, J., Cresswell, P., and Engelhard, V. H. (1992). *J. Immunol.*, **148**, 3004.
136. Reinholdsson Ljunggren, G., Franksson, L., Dalianis, T., and Ljunggren, H. G. (1993). *Int. J. Cancer*, **54**, 992.
137. Alexander Miller, M. A., Burke, K., Koszinowski, U. H., Hansen, T. H., and Connolly, J. M. (1993). *J. Immunol.*, **151**, 1.
138. Barnstable, C. J., Bodmer, W. F., Brown, G., Galfre, G., Milstein, C., Williams, A. F., *et al.* (1978). *Cell*, **14**, 9.
139. Brodsky, F. M., Parham, P., Barnstable, C. J., Crumpton, M. J., and Bodmer, W. F. (1979). *Immunol. Rev.*, **47**, 3.
140. Kornbluth, J., Spear, B., Raab, S. S., and Wilson, D. B. (1985). *J. Immunol.*, **134**, 728.
141. Ellis, S. A., Taylor, C., Hildreth, J. E., and McMichael, A. J. (1985). *Hum. Immunol.*, **13**, 13.
142. Yang, S. Y., Morishima, Y., Collins, N. H., Alton, T., Pollack, M. S., Yunis, E. J., *et al.* (1984). *Immunogenetics*, **19**, 217.
143. Rebai, N. and Malissen, B. (1983). *Tissue Antigens*, **22**, 107.
144. Parham, P. and Bodmer, W. F. (1978). *Nature*, **276**, 397.
145. Parham, P. and Brodsky, F. M. (1981). *Hum. Immunol.*, **3**, 277.
146. McMichael, A. J., Parham, P., Rust, N., and Brodsky, F. (1980). *Hum. Immunol.*, **1**, 121.

147. Berger, A. E., Davis, J. E., and Cresswell, P. (1982). *Hybridoma*, **1**, 87.
148. Foung, S. K., Taidi, B., Ness, D., and Grumet, F. C. (1986). *Hum. Immunol.*, **15**, 316.
149. Ellis, S. A., Taylor, C., and McMichael, A. (1982). *Hum. Immunol.*, **5**, 49.
150. Parham, P. and McLean, J. (1980). *Hum. Immunol.*, **1**, 131.
151. Parham, P. (1981). *Immunogenetics*, **13**, 509.
152. Haynes, B. F., Reisner, E. G., Hemler, M. E., Strominger, J. L., and Eisenbarth, G. S. (1982). *Hum. Immunol.*, **4**, 273.
153. Ozato, K., Mayer, N. M., and Sachs, D. H. (1982). *Transplantation*, **34**, 113.
154. Ozato, K., Hansen, T. H., and Sachs, D. H. (1980). *J. Immunol.*, **125**, 2473.
155. Ozato, K. and Sachs, D. H. (1980). *J. Immunol.*, **126**, 317.
156. Hammerling, G. J., Hammerling, U., and Lemke, H. (1979). *Immunogenetics*, **8**, 433.
157. Lemke, H., Hammerling, G. J., and Hammerling, U. (1979). *Immunol. Rev.*, **47**, 175.
158. Hammerling, G. J., Rusch, E., Tada, N., Kimura, S., and Hammerling, U. (1982). *Proc. Natl. Acad. Sci. USA*, **79**, 4737.
159. Hammerling, G. J., Lemke, H., Hammerling, U., Hohmann, C., Wallich, R.. and Rajewsky, K. (1978). *Curr. Top. Microbiol. Immunol.*, **81**, 100.
160. Brodsky, F. M., Bodmer, W. F., and Parham, P. (1979). *Eur. J. Immunol.*, **9**, 536.

Appendix 1

Class I MHC peptide motifs

Appendix 1 is intended only as a brief overview of the primary anchor motifs for a large range of human and murine class I MHC alleles. Such a list is due to change all the time, and therefore it is always wise to consult the literature for further details. Recently, an excellent review of this topic has been published by Rammensee and colleagues as a book format, and anyone with an interest in peptide binding should consult this book for further information on primary versus secondary anchors, examples of ligands and epitopes, as well as literature references (121).

	Position								
HLA-A1	1	2	3	4	5	6	7	8	9
			D						Y
			E						
HLA-A*0201	1	2	3	4	5	6	7	8	9
		L							L
		M							V
		I							I
HLA-A*0202 + 0204	1	2	3	4	5	6	7	8	9
		L							L
									V
HLA-A*0205	1	2	3	4	5	6	7	8	9
		L							L
		V							I
		I							V
HLA-A3	1	2	3	4	5	6	7	8	9
		L							F
		M							K
		V							Y

HLA-A11	1	2	3	4	5	6	7	8	9
		I	I						K
		F	L						
		V	M						
		Y	F						
			Y						
HLA-A24	1	2	3	4	5	6	7	8	9
		Y							F
		F							I
									L
HLA-A*2902	1	2	3	4	5	6	7	8	9
		E							Y
									L
HLA-A*3101	1	2	3	4	5	6	7	8	9
		L	F						R
		F	L						
		V	Y						
		Y	W						
		Q							
HLA-A*3302	1	2	3	4	5	6	7	8	9
		A							R
		I							
		L							
		V							
		F							
		Y							
HLA-A*6801	1	2	3	4	5	6	7	8	9
		V							R
		T							K
HLA-B*0701	1	2	3	4	5	6	7	8	9
		P							L
									F
HLA-B*0702	1	2	3	4	5	6	7	8	9
		P							L
HLA-B8	1	2	3	4	5	6	7	8	9
			K		K				L
					R				
HLA-B14	1	2	3	4	5	6	7	8	9
		R			R				L
		K			H				
HLA-B15	1	2	3	4	5	6	7	8	9
		Q							F
		L							Y
HLA-B*2702	1	2	3	4	5	6	7	8	9
		R							F
									Y
									I
									L
									W
HLA-B*2703 + 2704	1	2	3	4	5	6	7	8	9
		R							Y
									F
									L

	Position								
HLA-B*2705	1	2	3	4	5	6	7	8	9
		R							Y
									F
									L
									M
									I
									K
									R
HLA-B*3501	1	2	3	4	5	6	7	8	9
		P							Y
									F
									I
									L
									M
HLA-B*3503	1	2	3	4	5	6	7	8	9
		P							L
									M
HLA-B*3801	1	2	3	4	5	6	7	8	9
		H	D						F
			E						L
HLA-B*3901	1	2	3	4	5	6	7	8	9
		R							I
		H							L
									M
									V
HLA-B*3902	1	2	3	4	5	6	7	8	9
		K							L
		Q							
HLA-B40	1	2	3	4	5	6	7	8	9
		E	F						L
			I						W
			V						A
									M
									R
									T
HLA-B*4401	1	2	3	4	5	6	7	8	9
		E							Y
HLA-B*4402 + 4403	1	2	3	4	5	6	7	8	9
		E							Y
									F
HLA-B*4601	1	2	3	4	5	6	7	8	9
		M							Y
									F
HLA-B*5101	1	2	3	4	5	6	7	8	9
		A							F
		P							I
		G							
HLA-B*5102	1	2	3	4	5	6	7	8	9
		A							I
		P							V
		G							

HLA-B*5103	1	2	3	4	5	6	7	8	9
		A							I
		P							F
		G							V
HLA-B*5301	1	2	3	4	5	6	7	8	9
		P							I
									L
									F
									M
									W
HLA-B*5501 + 5502 + 5601	1	2	3	4	5	6	7	8	9
		P							A
HLA-B*5801	1	2	3	4	5	6	7	8	9
		A							F
		S							W
		T							
HLA-Cw*0102	1	2	3	4	5	6	7	8	9
		A							L
		L							
HLA-Cw*0301	1	2	3	4	5	6	7	8	9
									F
									I
									L
									M
HLA-Cw*0304	1	2	3	4	5	6	7	8	9
		A							L
									M
HLA-Cw*0401	1	2	3	4	5	6	7	8	9
		Y							L
		P							F
		F							M
HLA-Cw*0601 + 0602	1	2	3	4	5	6	7	8	9
									I
									L
									V
									Y
HLA-Cw*0702	1	2	3	4	5	6	7	8	9
									Y
									F
									L
H-2K^{b}	1	2	3	4	5	6	7	8	9
					Y			I	
					F			L	
								M	
								V	
H-2D^{b}	1	2	3	4	5	6	7	8	9
					N				I
									L
									M
H-2K^{d}	1	2	3	4	5	6	7	8	9
		Y							I
		F							L
									V

	Position								
H-2D^d	1	2	3	4	5	6	7	8	9
		G	P		R				I
					K				L
									F
H-2L^d	1	2	3	4	5	6	7	8	9
		P							F
		S							L
									M
H-2K^k	1	2	3	4	5	6	7	8	9
		E							I
									V

Appendix 2

Class I MHC alleles expressed by TAP-deficient cell lines

Allele	Cell line	Reference
HLA-A*0101	BM36.1	122, 123
HLA-A*0201	T2	124
HLA-A2	RMA-S-A2	125
HLA-A*0301	T2-A3	126
HLA-A3	RMA-S-A3	125
HLA-A*1101	T2-A11	127
HLA-Aw68	T2-Aw68	128
HLA-B*0701	T2-B7.01	129, 130
HLA-B*0702	T2-B7.02	131
HLA-B*2705	T2-B2705	125
HLA-B27	RMA-S-B27	125
HLA-B*3501	T2-B35	132
HLA-B*3501	BM36.1	122, 123
HLA-B*5101	T2	124
HLA-B*5301	T2-B53	76
H-2K^b	RMA-S	45
H-2K^b	T2-K^b	125, 129
H-2D^b	RMA-S	45
H-2D^b	T2-D^b	125, 133
H-2K^d	T2-K^d	60, 134
H-2K^d	RMA-S-K^d	T. J. Braciale, unpublished
H-2D^d	T2-D^d	135
H-2D^d	RMA-S-D^d	136
H-2L^d	T2-L^d	135, 137
H-2K^k	T2-K^k	P. Cresswell, unpublished

Appendix 3

Monoclonal antibodies specific for class I MHC molecules

Specificity	Name	Class	ATCC	Reference
Monomorphic HLA-A, -B, -C	W6/32	IgG_{2a}	HB 95	138
Monomorphic HLA-A, -B, -C	BB7.7	IgG_{2b}	HB 94	139
HLA-A locus	A131	IgG_1	–	140
HLA-B, -C, some A-alleles	MHM.5	IgG_1	–	141
HLA-B	4E	IgG_{2a}	–	142
HLA-B, -C	B1.23.2		–	143
HLA-A2, Aw69	PA2.1	IgG_1	HB 117	144
HLA-A2, Aw69	BB7.2	IgG_{2b}	HB 82	145
HLA-A2, -B17	MA2.1	IgG_1	HB 54	146
HLA-A3	GAP-A3	IgG_{2a}	HB 122	147
HLA-A11, -A24	A11.1 M	IgG_3	HB 164	148
HLA-A11, -A3	AUF 5.13		–	127
HLA-B7	BB7.1	IgG_1	HB 56	139
HLA-B27, -B7	ME 1	IgG_1	HB 119	149
HLA-B40, -A28, -B7	MB40.1	IgG_1	–	150
HLA-B7, -B40	MB40.2	IgG_1	HB 59	151
HLA-B5	4D12	IgG_1	HB 178	152
$H\text{-}2D^d$	34-4-20s	IgG_{2a}	HB 75	153
$H\text{-}2D^d$	34-2-12s	IgG_{2a}	HB 87	153
$H\text{-}2K^d$	31-3-4s		HB 77	153
$H\text{-}2K^d$, D^d	34-1-2s	IgG_{2a}	HB 79	153
$H\text{-}2L^d$	30-5-7s	IgG_{2a}	HB 31	154
$H\text{-}2L^d$, D^b	28-14-8s	IgG_{2a}	HB 27	154, 155
$H\text{-}2D^b$	B22.249	IgG_{2a}	–	156, 157
$H\text{-}2D^b$, D^d	27-11-13s	IgG_{2a}	–	155
$H\text{-}2K^b$	Y3	IgG_{2b}	HB 176	158
$H\text{-}2K^k$	H100-5	IgG_{2a}	–	157, 159
Human β_2-m	BBM.1	IgG_{2b}	HB 28	160

Appendix 4

Index peptides suitable for modification

Peptide	Protein	Allele
FLPSDYFPSV[a]	HBV core $_{18-27}$	A1
YLEPAIAKY[a]	Consensus peptide	
ILKEPVYGV[a]	HIV RT $_{476-484}$	A*0201
FLPSDYFPSV[a]	HBV core $_{18-27}$	
VYPLRPMTYK[a]	HIV nef $_{73-82}$	A3
KVFPYALINK[a]		
AVDLYHFLK[a]	HIV nef $_{84-94}$	A11
AYIDNYNKF[a]	Consensus peptide	A24
FAPGNYPAL[a]	Sendai virus NP $_{324-332}$	D^b, K^b
YSNENMDAM[a]	Influenza virus NP $_{366-374}$	D^b
YPHFMPTNL[a]	CMV $_{168-176}$	L^d, K^b
SIINFCKL[b]	OVA $_{257-264}$	K^b

[a] Peptides that are suitable for radioiodination.

[b] Suitable for chemical modification with thiol-reactive cross-linkers at position 6 (e.g. immobilization onto biosensor chip, fluorescination).

Chapter 6
The design, synthesis, and characterization of molecular mimetics

Ian T. W. Matthews
ChemOvation, The Yammond, North Heath Lane, Horsham RH12 5QE, UK.

1 Introduction

This chapter is concerned with practical approaches to the design, synthesis, and characterization of synthetic organic compounds that attempt to mimic bioactive natural molecules. The chapter will concentrate on the preparation of peptoids as peptidomimetics and discuss the design and synthesis of other amino acid/ peptide surrogates.

The ever increasing need for more and better drugs is now being driven by the development of new target identification sciences such as proteomics (1) and the now formidable high throughput screens which can, in some instances consume up to 100 000 compounds a day. This has lead to the need for new ways of synthesizing novel chemicals for screening and that these new ways should produce good lead compounds quickly to give the more innovative pharmaceutical companies an edge over their rivals.

Natural peptides and proteins have evolved to function either in an autocrine mode or a paracrine mode. Their modes of action can result in profound or subtle effects through high, medium, or low affinity interactions with extracellular or intracellular receptors. It has long been the goal of drug discovery scientists to mimic their biochemical behaviour in the form of low molecular weight, preferably orally available, easily synthesized organic compounds. There are many examples of an apparent deep understanding of the molecular interactions of proteins and peptides with their binding partners but relatively few examples of translating this information into a useful lead compound through conventional medicinal chemistry (2). This state of affairs is however changing, particularly with the development of a number of powerful yet relatively accessible molecular modelling and computer assisted design software packages (3, 4). These tools coupled to an increased ability to make rapidly compounds for fast screening means that drug discovery is moving into areas where more control can be exercised over the types of compounds that need to be made (5) and that these compounds should, it is hoped, make better leads.

If chemistry is to keep pace with developments in proteomics, screening, and computational innovations then it may need to break the link with traditional chemistries and progress the development of solid and liquid phase rapid synthesis methods.

2 Design, synthesis, and characterization of peptoid oligomers

2.1 Design of peptoid oligomers

As discussed briefly above, converting a natural peptide with good *in vitro* activity but poor biological stability and little oral availability can be difficult, particularly if this peptide is linear. Peptoids are an interesting class of peptide-like molecules that exhibit an increase in biological stability, oral absorption, and ability, in some instances to mimic the action of natural peptides (6). This class of compound comprises N-substituted glycine derivatives, an example of which is shown in *Figure 1*, and is compared to its natural peptide equivalent.

As can be seen in *Figure 1* the amino acid side chain has been moved from carbon to nitrogen in the peptoid example. There are a number of important factors associated with this move. It has the effect of making these compounds achiral, and thus the purification and synthetic development are easier. In addition, there are a large number of commercially available inexpensive building blocks that may be used directly as side chains in peptoid synthesis. These

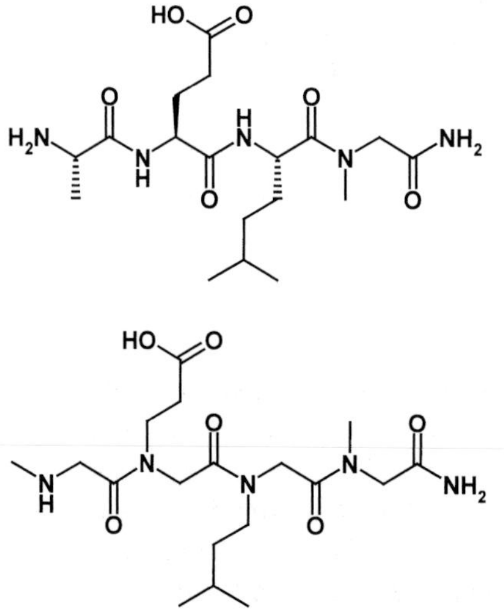

Figure 1 The top peptide represents the sequence Ala, Glu, Val, Sar amide (Sar is used as the linker to the resin for synthesis). The bottom sequence represents a peptoid equivalent. Note that the side chains of Ala, Glu, and Val have been moved along one from carbon to nitrogen thus producing an N-linked glycine oligomer.

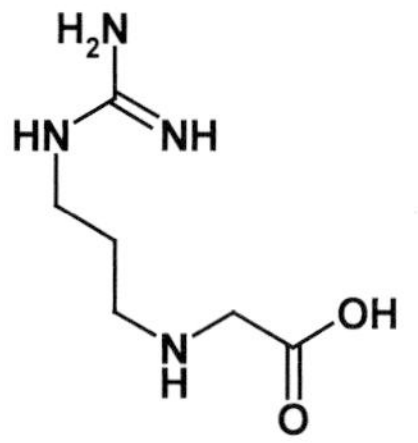

Figure 2 *N*-3-guanidinopropylglycine is an example of a surrogate for a natural amino acid, in this case arginine. This molecule will need its amino functions suitably protected prior to synthesis.

compounds can be made on solid phase and therefore the chemistry is easy to automate. An example of a natural amino acid surrogate building block that has been used in the construction of a peptoid library is the arginine replacement *N*-3-guanidinopropylglycine (7), *Figure 2*.

This arginine surrogate was used in 20 peptoid combinatorial libraries each containing 160 000 compounds. After multiple screenings and deconvolution to a single compound CGP64222 (8) was identified, the structure of which is shown in *Figure 3*. CGP64222 is an antiviral compound that inhibits the formation of the Tat/TAR-RNA complex with an IC^{50} of 12 nM and is one of the first examples of an antiviral compound specifically designed to inhibit a protein/RNA interaction.

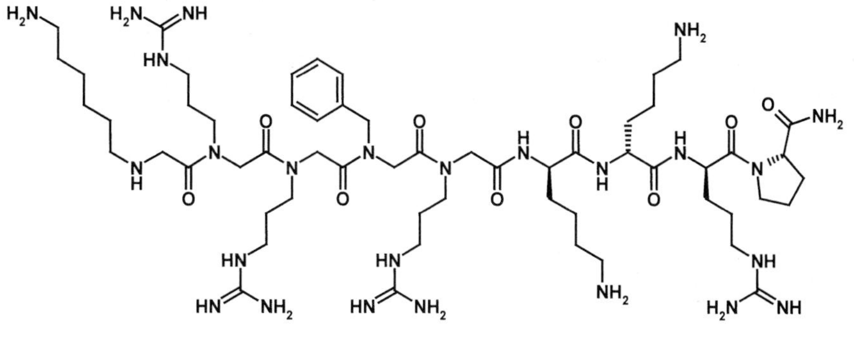

Figure 3 An example of a peptoid oligomer containing *N*-3-guanidinopropylglycine as a surrogate for arginine in the anti-HIV compound CGP64222.

The building blocks used in the libraries that lead to CGP64222 would probably preclude its oral absorption. Libraries can however, be constructed from building blocks that would not be inhibitory to absorption; keeping the molecular weights of the individual components below 500–600 Da would also help achieving this goal. Building blocks can be purchased and divided into a number of groups such as those that are hydrogen bond donors, hydrogen bond acceptors, and those that interact through hydrophobic or aromatic interactions. The use and selection of compounds within these different classes of reagents can have consequences for the diversity (9) of the libraries produced and the cost and ease of chemistry in making them.

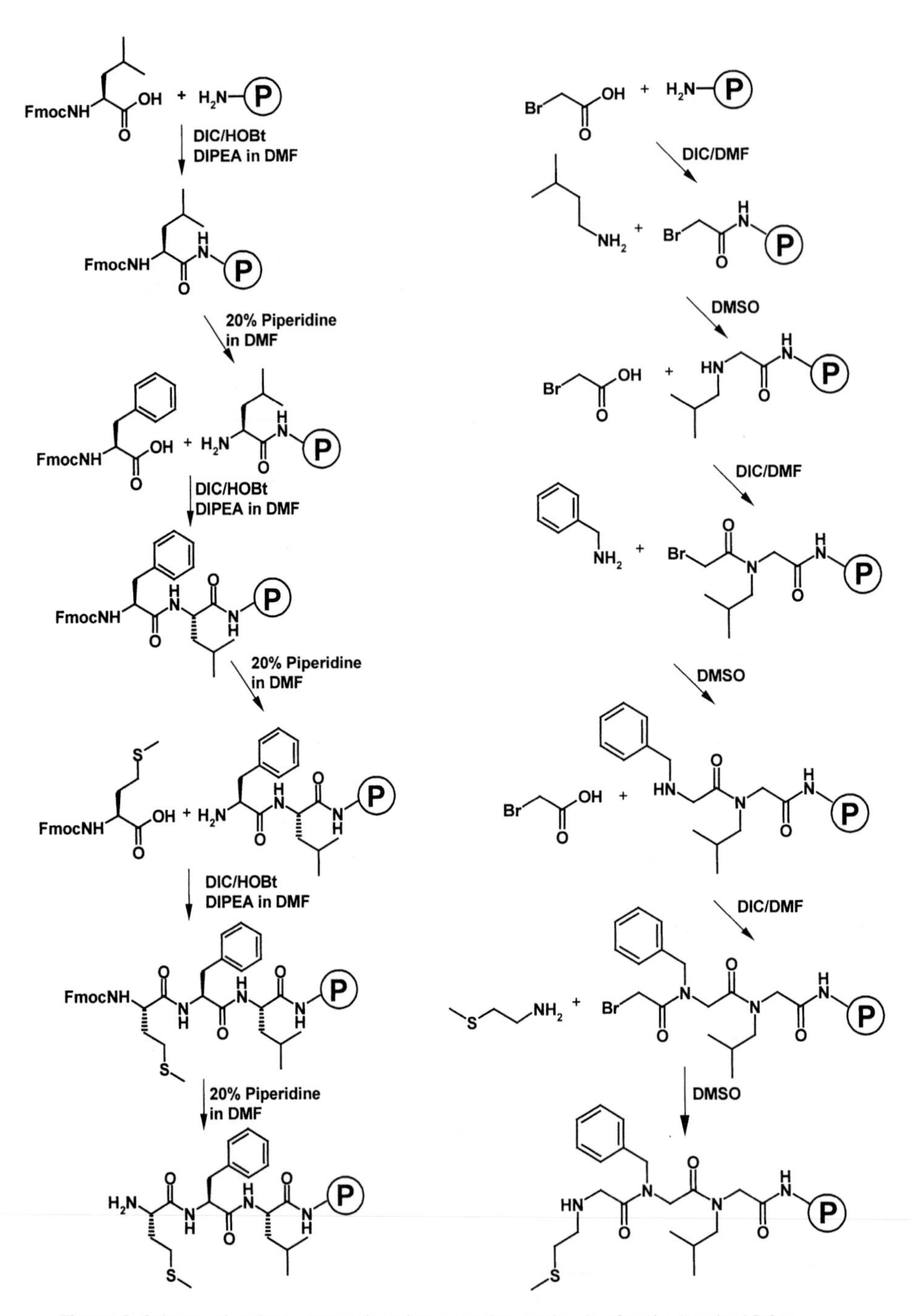

Figure 4 Scheme showing a comparison between the synthesis of resin- (marked P for polymer) bound peptide Leu, Phe, Met (*left*) and the synthesis of a resin-bound peptoid equivalent (*right*). In this instance the peptoid synthesis does not involve successive rounds of deprotection before elongation as in the case of the peptide synthesis. The final step (not shown here) will be cleavage from the resin (*Protocol 3*). Explanations for the abbreviations are given in *Protocol 1*. Diagonal arrows point to the product from the previous reaction.

2.2 The synthesis of peptoid oligomers

Depending on the building blocks chosen to form the peptoid oligomer there can be less of a need for amino protection than is the case with conventional peptide synthesis. An example of a synthetic route to a peptide and its hypothetical peptoid equivalent is illustrated in *Figure 4*.

The synthesis outlined can be very efficient with yields in excess of 90% per reaction. There is a great deal of knowledge associated with the solid phase synthesis of natural peptides due primarily to the limited number of building blocks (20 natural amino acids) and also to the large number of peptide sequences that have been made over the last three decades (10). The same degree of knowledge with regard to peptoid synthesis is less detailed primarily due to the large number of amino building blocks commercially available. Amino compounds that represent the natural amino acid side chains are not always commercially available. Synthesis of amine protected amine derivatized carboxylic acids are then required to be used as building blocks, such as in *Figure 5*. These peptoid monomers can then be used in either solution (11) or solid phase (12) synthesis of peptoid libraries.

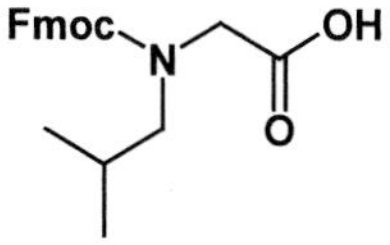

Fmoc N OH O

Figure 5 An example of an amino protected (Fmoc) amino derivatized (isobutyl) peptoid building block equivalent of Fmoc-leucine.

The solid phase can be resin or 96-well PIN format (13). The solid phase attachment point or linker can contain a number of different functionalities but the amine and carboxylic groups have been greatly used. The structure of one of the traditional linkers is shown in *Figure 6*.

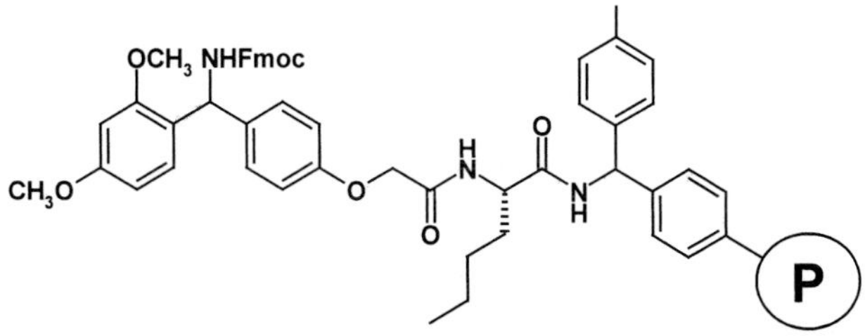

OCH_3 NHFmoc CH_3O O H N O O N H P

Figure 6 An Fmoc Rink amide MBHA linked resin (marked P for polymer). The Fmoc group is removed by treatment with 20% piperidine in dimethylformamide which releases the amino function for use as a handle in sequence synthesis.

Synthetic procedures with resin can be carried out in a number of different reaction vessels such as glass tubes, Eppendorf tubes, or in 96-deep well plates fitted with frits. The first three protocols outlined below contain procedures for resin conditioning, synthesis, and product cleavage from the resin in a 96-deep well format.

Protocol 1

Resin conditioning prior to synthesis

Equipment and reagents

- FlexChem 96-deep well synthesis and filtration system (Robbins Scientific Corporation, distributed by GRI, Molecular Biology)
- Vacuum pump (KNF, Laboport, BDH) and desiccator (BDH)
- Shaking platform (Heidolph titramax 100)
- Ultraviolet/visible 96-well spectrophotometer (i.e. Molecular Devices, Spectra Max 190)
- Deep well plates (Jones Chromatography)
- 20% piperidine (toxic) in dimethylformamide (Pip/DMF purchased separately form Aldrich, DMF, anhydrous grade)
- Isopropanol (Aldrich)
- Methanol (Aldrich)
- Fmoc Rink amide MBHA polystyrene resin (Novabiochem, typical loading of 0.3–0.8 mmole/g resin)

Method

1. Add Fmoc Rink amide MBHA resin (0.59 mmole/g, 50 mg, 29.5 μmole) to each of the FlexChem 96-deep wells and clamp into the bottom sealing cover with sealing cover gasket.
2. Add isopropanol (1 ml) to each well and leave the reaction block agitating for 30 min.
3. Place the reaction block in the filtration manifold and apply a vacuum to remove the isopropanol to waste.
4. Wash resin by addition of DMF (2 × 1 ml/well) and remove to waste.
5. Place the reaction block back into the bottom sealing cover and cleave Fmoc protecting group by addition of 1 ml of 20% piperidine in dimethylformamide to each well. Agitate the block for 30 min in a fume hood.
6. Place the reaction block in the filtration manifold containing a 96-deep well collection plate. Apply vacuum and collect cleavage solutions. Add a further aliquot of cleavage mixture (0.5 ml) to each well and repeat cleavage and isolation procedure.
7. If desired, dilute one or more of the solutions from the deep well plates with methanol and monitor at 300 nm. Extinction coefficient used for estimation of cleaved Fmoc adduct = 7800 $mol^{-1}dm^{3}cm^{-1}$.
8. Remove deep well collection plate from filtration manifold and replace with a waste vessel.
9. Wash resin with alternating dimethylformamide (3 × 1.5 ml) and isopropanol (3 × 1.5 ml) solutions. Remove final traces of dimethylformamide by washing with isopropanol (3 ×1.5 ml).
10. Dry resin in vacuum desiccator.

In the above deprotection step, the extent of Fmoc removal is generally 100%. Following this, *Protocol 2* outlines the synthesis of a simple dipeptoid. The most efficient chemistry is carried out with unhindered aliphatic amines. Hindered and aromatic compounds can require elevated temperatures or extensive purification.

Protocol 2

Synthesis of simple dipeptoid

Equipment and reagents

- FlexChem 96-deep well synthesis and filtration system (Robbins Scientific Corporation, distributed by GRI, Molecular Biology)
- Vacuum pump (KNF, Laboport, BDH) and desiccator (BDH)
- Shaking platform (Heidolph titramax 100)
- Deep well storage plates (Jones Chromatography)
- Dried, deprotected Rink resin (*Protocol 1*)
- A series of unhindered aliphatic primary amines (toxic, Aldrich, Maybridge, Lancaster, Fluka)
- Diisopropylcarbodiimide (DIC, toxic, Aldrich)
- Bromoacetic acid (Aldrich)
- Dimethyl sulfoxide (DMSO, Aldrich)
- Dimethylformamide (DMF, anhydrous grade, Aldrich)
- Isopropanol (Aldrich)

Method

1. Add DMF (1 ml) to each of the 96 resin filled wells and agitate for 30 min.
2. Place plate in filtration block and filter DMF under pressure to waste.
3. Add bromoacetic acid in DMF (41 mg, 295 μmole, 500 μl) to each of the DMF washed, resin filled wells followed by DIC (354 μmole, 44.7 mg, 55.4 μl) also in DMF (500 μl), and leave reaction agitating for 1 h at room temperature in a fume hood.
4. Wash resins with alternating DMF (5×1 ml) and DMSO (3×1 ml) solutions, making the final washes with DMSO (3×1 ml).
5. Displace bromine group by addition of unhindered primary aliphatic amines (0.885 mmole) in DMSO (500 μl) to each of the deep wells and leave agitating for 3 h at room temperature.
6. Wash with alternating rounds of isopropanol (3×1 ml) and DMF (3×1 ml), the final wash should be of DMF (2×1 ml).
7. Repeat bromoacetic acid coupling (step 3).
8. Wash as above (step 4).
9. Repeat with a second round of unhindered primary aliphatic amine reactions (step 5).
10. Wash the resin-bound dipeptoid as above (step 6), but with a final wash of isopropanol (3×1 ml).
11. Dry resin in vacuum desiccator.

The product from this synthesis will need to be presented for screening in the appropriate form. This may be on the resin for solid phase screening or cleaved off for solution screening. *Protocol 3* gives a general method for product cleavage from a Rink amide MBHA resin.

Protocol 3

Cleavage of dipeptoid sequence from resin

Equipment and reagents

- FlexChem 96-deep well synthesis and filtration system (Robbins Scientific Corporation, distributed by GRI, Molecular Biology)
- Vacuum pump (KNF, Laboport, BDH) and desiccator (BDH)
- Deep well storage plates (Jones Chromatography)
- Shaking platform (Heidolph titramax 100)
- Rotary deep well plate concentrator (Savant SC250DDA)
- 99% trifluoroacetic acid (Aldrich)
- Dichloromethane (Aldrich)
- Deionized, organics-free water (Millipore, Milli-RX 20)

Method

1. Prepare 95% trifluoroacetic acid (in a fume hood) by adding the required amount of TFA to pure water.
2. Add 95% TFA (1 ml) to each dry resin filled deep well and leave agitating for 1 h at room temperature in a fume hood (resin often becomes deeply coloured).
3. Filter the TFA solution into deep well plates and wash resin with dichloromethane (1×1 ml).
4. Place the deep well plate into Savant concentrator with plate balance and spin on the volatile solvent setting for 1.5 h and then for a further 2 h on the aqueous setting.
5. The residue in the plates can now be dissolved in DMSO or a suitable buffer for chemical characterization and screening.

2.3 Characterization of peptoids

There are a number of ways of characterizing organic molecules; peptoids hold no special difficulties and due to their possible natural peptide mimetic properties, increased information through specific biochemical characterization can be generated. The only really meaningful characterization is that the peptoid does what it was designed to do in some biological assay. Simply transferring a natural amino acid side chain from carbon to nitrogen does not automatically confer mimetic status.

Generally it is believed that synthesized compounds required for biochemical/biological screening need to be pure. This purity is usually proven by a combination of high performance liquid chromatography (HPLC), nuclear magnetic resonance (NMR), and mass spectroscopy (MS). Further characterization of peptoid building blocks can be by their ability to replace a natural amino acid or natural sequence. For instance, the successful replacement of a proline with the surrogate peptoid building block *N*-isobutylglycine maintains the triple helix stability in collagen-like structures (14). An example of the rational design of a low molecular weight peptoid based on a natural biologically active peptide is the synthesis of methyltryptophan derivatives as analogues of cholecystokinin. These peptoids have oral activity and potent anxiolytic properties (15).

3 Further oligomeric peptide mimetics

3.1 β-Peptides, β-peptoids, retro-peptoids, and amide surrogates

β-Peptides are a class of peptide analogues that are composed of β-amino acids (16). *Figure 7* shows the structure of a β-amino acid compared to an α-amino acid. Although considered non-natural, β-amino acids are found in natural compounds such as Taxol and β-lactam antibiotics. There are a number of interesting properties shown by β-peptides, one is that they adopt 3D structures (17) similar to natural peptide sequences and secondly they are resistant to enzymatic degradation (18). There is a great deal of effort being directed toward using β-peptides as scaffolds for the construction of large protein-like complexes.

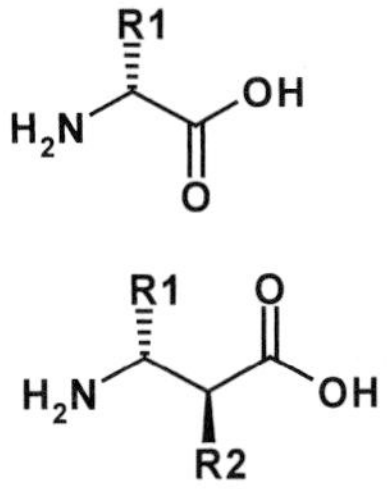

R1
H_2N
OH
O

R1 O
H_2N
OH
R2

Figure 7 Comparison of an α-amino acid with a β-amino acid. The β-amino acid has an extra carbon backbone and therefore they can have an additional side chain (R1 or R2).

There are a number of ways these molecules can be prepared or introduced into backbone structures. One method is to synthesize or purchase suitably protected β-amino acids (some are commercially available) and link these together through the use of conventional peptide chemistry. Another method is to make peptoid molecules with β-amino acids as the side chains, as shown in *Figure 8*.

A protocol for the synthesis of a peptoid structure with pendent β-peptides is given below and is carried out on Chiron Pins in 96-well format.

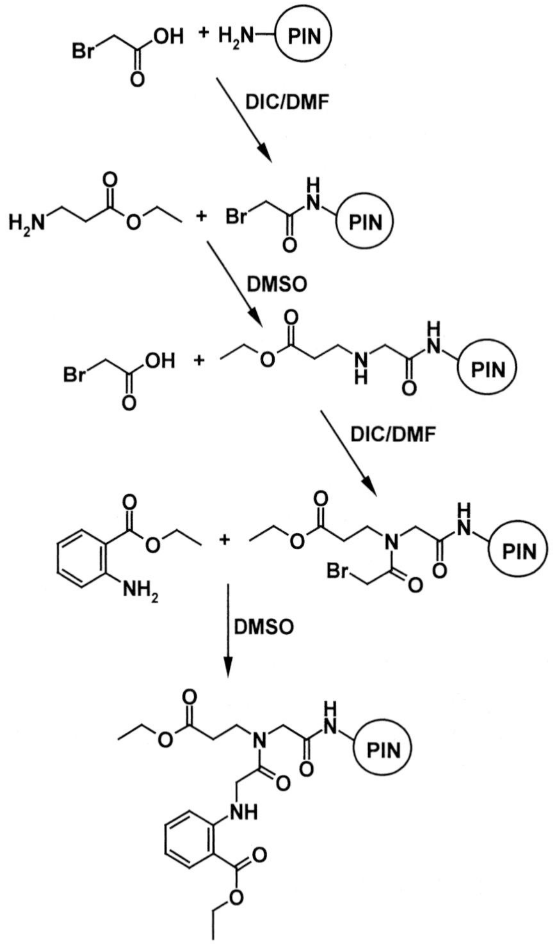

Figure 8 A PIN synthesis of a peptoid backbone with β-amino acid esters as pendent side chains (Protocol 4). The product can be released from the PINs as esters by trifluoroacetic acid treatment. Subsequent treatment with sodium hydroxide will liberate the free acids. Diagonal arrows point to the product formed from the previous reaction.

Protocol 4

Synthesis on Chiron Pins of a dipeptoid comprised of β-amino acid esters as the side chains (*Figure 7*)

Equipment and reagents

- Chiron 96 SynthPhase Multipin synthesis and washing system
- Polystyrene Pins with I- series crowns and Fmoc Rink amide linker (RAM)
- Shaking platform (Heidolph titramax 100)
- Vacuum pump (KNF, Laboport, BDH) and desiccator (BDH)
- Deep well storage plates (Jones Chromatography)
- β-Amino acid esters (Aldrich)

Protocol 4 continued

- Dimethylformamide (DMF, anhydrous grade, Aldrich)
- 20% piperidine/DMF (purchased separately from Aldrich)
- Methanol (Aldrich)
- Dichloromethane (Aldrich)
- Diisopropylcarbodiimide (DIC, Aldrich)
- Trifluoroacetic acid (Aldrich)

Method

1. Place crowns on pin stems and arrange in 96 format in pin holder.
2. Remove Fmoc by immersing mounted pins in 20% piperidine/DMF (50 ml) contained in a polypropylene wash-bath. Leave for 30 min at room temperature. The solution should completely cover the crowns.
3. Remove from bath and shake free of excess reagent. Wash in DMF (50 ml) for 5 min. Wash again by complete immersion of crowns and stems in methanol (200 ml) for 2 min and air dry.
4. Add DIC (2 M) in DMF (25 ml) to a solution of bromoacetic acid (2 M) also in DMF (25 ml) in a wash-bath and leave to react for 5 min.
5. Place pin holder with Fmoc deprotected amine pins in wash-bath and leave gently agitating for 4 h.
6. Wash pins in DMF (50 ml) for 5 min, wash pins in methanol (200 ml) for 2 min, finally wash pins in dichloromethane (200 ml) for 2 min, then air dry.
7. Add aliphatic amino esters dissolved in DMSO (0.2 M–2 M, 500 μl) to each deep well and leave agitating for 2.5 h.
8. Wash pins as in step 6.
9. Repeat steps 4 and 5.
10. Add aliphatic or aromatic (if final building block)[a] amino esters[a] in DMSO (0.2 M–2 M) to each well. Leave agitating for 2.5 h.
11. Wash pins as in step 6.
12. Place pin holder with pins in 96-deep well block containing 95% TFA (500 μl, see *Protocol 3*) and leave for 2 h at room temperature.
13. Place deep well plate containing cleaved compound and balance plate in Savant rotary evaporator and spin on volatile solvent setting for 1.5 h followed by 2 h on the aqueous setting.

[a] Secondary aromatic compounds (aniline derivatives) made as a result of step 10 are then quite unreactive to the next round of activated bromoacetic acid addition. If *t*-butyl ester derivatives of aliphatic or aromatic β-amino acids are used then these can be cleaved to the free acid by 95% TFA, this is generally not the case with linear aliphatic ester groups.

β-Peptoids are yet a further class of unique structures, recently exemplified in a simple yet highly effective synthesis (19). Similar to the original peptoids, the side chains have been moved to the nitrogen, as can be seen in *Figure 9*.

The chemistry of preparing these β-peptoids is amenable to solid phase synthesis and allows the preparation of many examples of this type of potential

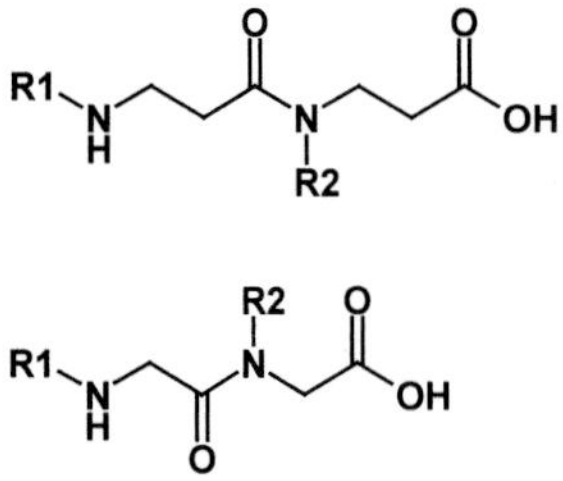

Figure 9 Structure of a β-peptoid (*top*) and peptoid (*bottom*) shows the extended nature of the backbone similar to the extension of peptides to β-peptides.

peptide mimetic, although, due to lack of reactivity, the aromatic amino compound aniline would not take part in the addition reaction needed to form such compounds.

Depending on the orientation of the backbone, peptides can be turned into peptoids as already seen, or, into retropeptoids which could be considered as the peptoid version of retropeptides. The orientation of the side chains in retropeptoids may appear to mimic natural peptides better than do peptoids (20) as can be seen in *Figure 10*.

The major aim of this area of study is to produce non-natural compounds that mimic natural ones but have fewer of the disadvantages. The search for amide

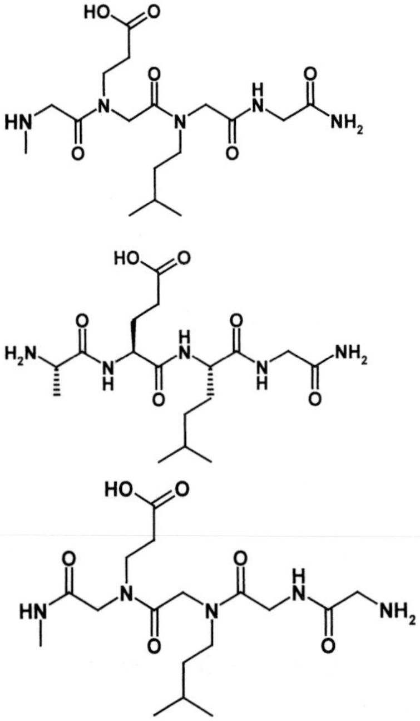

Figure 10 Representations of a peptoid (*top*), peptide (*middle*), and a retropeptoid (*bottom*). The side chains and carbonyl groups of the peptide and retropeptoid take up a similar orientation compared to the peptoid.

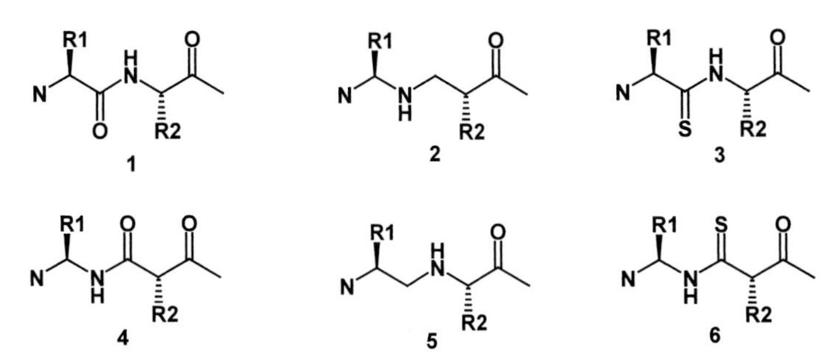

Figure 11 Examples of some amide (1) surrogates such as the retro-reduced bond (2), the thioamide bond (3), the retro-amide bond (4), the reduced amide bond (5), and the retro-thio amide bond (6).

bond replacements has also added to the possibilities for producing stable peptide-like compounds. Examples of these are given in *Figure 11* and comprise oxygen replacement, reverse bonds, and reduced bonds (21, 22). Many compounds with these replacements are more stable to enzymes but only a few retain significant biological activity.

Antigenic mimicry has been noted between an L-peptide sequence and its equivalent retro-inverso sequence (inverted chirality); both are recognized by the same antibody (23). Retro-inverso peptide analogues represent a major area of peptidomimetic synthesis and this backbone alteration has been introduced into a number of biologically active sequences (24).

4 Non-oligomeric peptide mimetics

4.1 Natural templates

One source of potential bioactive peptide mimetics is from already existing constrained natural molecules that can be used as templates on to which peptide-like non-natural structures can be placed. The amino acid sugar shown in *Figure 12* behaves as a dipeptide isostere and induces a β-turn in linear peptides (25).

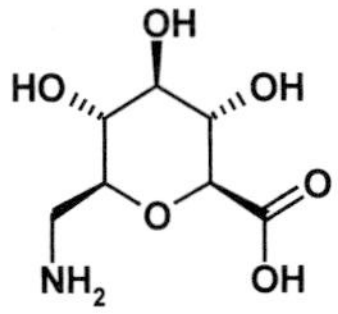

Figure 12 The sugar amino acid shown (synthesized from glucose) was used successfully as a GlyGly replacement in the middle of a peptide sequence.

A further sugar-based peptidomimetic, shown in *Figure 13*, was designed from a potent cyclic hexapeptide and has a number of unexpected actions other than the action it was designed for (26). Designed to act at the somatostatin receptor, some of the molecules with different side groups have antagonistic activity at other receptors.

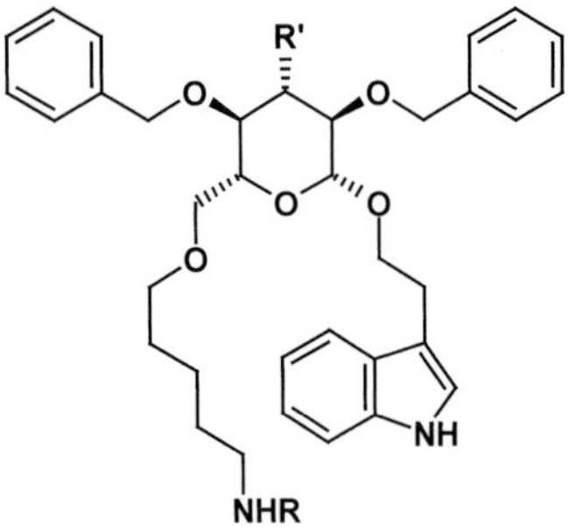

Figure 13 Designed to act at the somatostatin receptor, some of these molecules (R, R′ = different groups) had antagonistic activity at other receptors.

Constrained cyclic peptides are being used as templates for the design of protein mimetics (27). Large peptide sequences have been grafted on to templates as functional mimetics of proteins (28). An extreme result of work like this is the production of a non-natural compound that mimics the binding and functional properties of a monoclonal antibody (29).

4.2 Non-natural templates

A steroid nucleus has been used as a template to mimic a type 1 β-turn, but as can be seen from the activity of this structure (*Figure 14*) it is a weak mimetic of the RGD containing ligand.

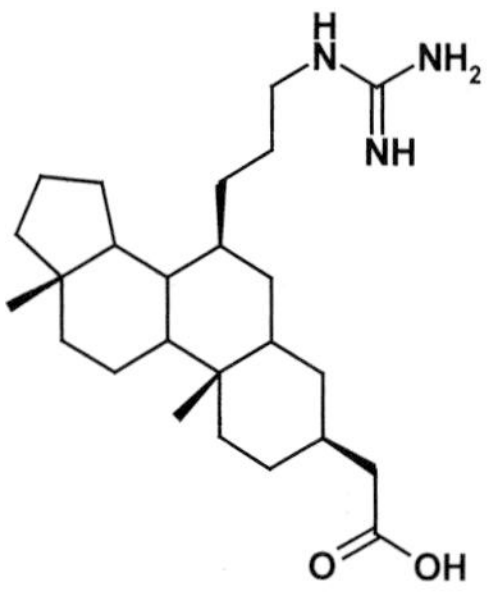

Figure 14 This molecule based on cyclopentanoperhydrophenanthrene exhibited an IC^{50} = 100 μM in a GP IIb/IIIa–fibrinogen receptor assay.

Again in the RGD field a benzodiazepine structure was produced from a constrained cyclic peptide and was one of the first examples of the synthesis of a highly potent peptidomimetic designed directly from the cyclic peptide conformation (30). The structures of these molecules, in *Figure 15*, show their similarities.

There are now companies that offer a growing number of non-natural building blocks. Some of these building blocks are considered to be 'turn' mimetics and can be introduced into growing chains. The more these compounds become available the more entry into combinatorial mimetic synthesis can begin, without the initial need for expensive time-consuming medicinal chemistry.

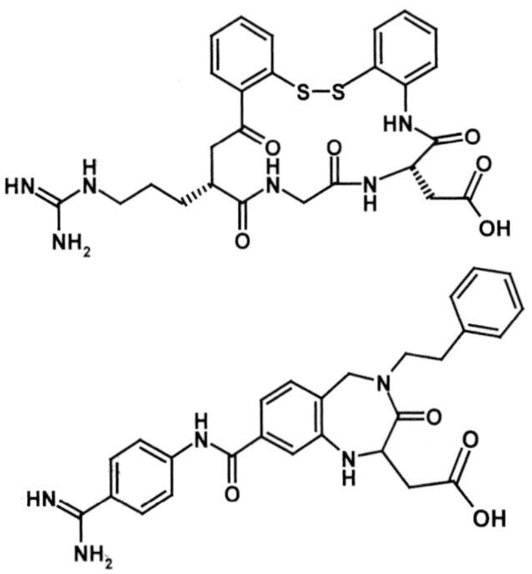

Figure 15 Non-natural benzodiazepine used as a template on to which are 'hung' the biologically important functionalities. In all assays tested both molecules showed low nM activity.

References

1. Mullner, S., Neumann, T., and Lottspeich, F. (1998). *Arzneimittelforschung*, **48** (1), 93.
2. Martin, S. F., Dorsey, G. O., Gane, T., Hillier, M. C., Kessler, H., Baur, M., *et al.* (1998). *J. Med. Chem.*, **41** (10), 1581.
3. Maddalena, D. J. (1998). *Exp. Opin. Ther. Patents*, **8** (3), 249.
4. Good, A. C. and Lewis, R. A. (1997). *J. Med. Chem.*, **40** (24), 3926.
5. Muller, G. and Giera, H. (1998). *J. Comput. Aided Mol. Des.*, **12** (1), 1.
6. Zuckermann, R. N., Martin, E. J., Spellmeyer, D. C., Stauber, G. B., Shoemaker, K. R., Kerr, J. M., *et al.* (1994). *J. Med. Chem.*, **37**, 2678.
7. Heizmann, G. and Felder, E. R. (1994). *Peptide Res.*, **7** (6), 328.
8. Hamy, F., Felder, E. R., Heizmann, G., Lazdins, J., Aboul-Ela, F., Varani, G., *et al.* (1997). *Proc. Natl. Acad. Sci. USA*, **94** (4), 3548.
9. Gillet, V. J., Willet, P., and Bradshaw, J. (1997). *J. Chem. Inf. Comput. Sci.*, **37**, 731.
10. Merrifield, R. B. (1966). *J. Am. Chem. Soc.*, **85**, 2149.
11. Kruijtzer, J. A. W. and Liskamp, R. M. J. (1995). *Tetrahedron Lett.*, **38** (38), 6969.
12. Zuckermann, R. N., Kerr, J. M., Kent, S. B. H., and Moos, W. H. (1992). *J. Am. Chem. Soc.*, **114**, 10646.
13. Geysen, H. M., Meloen, R. H., and Barteling, S. J. (1984). *Proc. Natl. Acad. Sci. USA*, **81**, 39998.
14. Feng, Y., Melacini, G., Taulane, J. P., and Goodman, M. (1996). *Biopolymers*, **39** (6), 859.
15. Horwell, D. C., Hughes, J., Hunter, J. C., Pritchard, M. C., Richardson, R. S., Roberts, E., *et al.* (1991). *J. Med. Chem.*, **34**, 404.
16. Iverson, B. L. (1997). *Nature*, **385**, 113.
17. Podlech, J. and Seebach, D. (1995). *Leibigs Ann.*, 1217.
18. Hintermann, T. and Seebach, D. (1997). *Synlett.*, 437.
19. Hamper, B. C., Kolodziej, S. A., Scates, A. M., Smith, R. G., and Cortez, E. (1998). *J. Org. Chem.*, **63**, 708.
20. Simon, R. J., Kania, R. S., Zuckermann, R. N., Huebner, V. D., Jewell D. A., Banville, S., *et al.* (1992). *Proc. Natl. Acad. Sci. USA*, **89**, 9367.
21. Campbell, M. M., Ross, B. C., and Semple, G. (1989). *Tetrahedron Lett.*, **30** (48), 6752.

22. Campbell, M. M., Ross, B. C., and Semple, G. (1989). *Tetrahedron Lett.*, **30** (15), 1997.
23. Guichard, G., Benkirane, N., Zeder-Lutz, G., Van Regenmortel, M. H. V., Briand, J. P., and Muller, S. (1994). *Proc. Natl. Acad. Sci. USA*, **91**, 9765.
24. Chorev, M. and Goodman, M. (1993). *Acc. Chem. Res.*, **26**, 266.
25. von Roedern, E. G. and Kessler, H. (1994). *Angew. Chem. Int. Ed. Engl.*, **33** (6), 687.
26. Hirschmann, R., Strader, C. D., Cascieri, M. A., Candelore, M. R., Nicolaou, K. C., Donaldson, C., *et al.* (1992). *J. Am. Chem. Soc.*, **114**, 9217.
27. Dumy, P., Eggleston, I. M., Nicula, E., and Mutter, M. (1998). http://www.ch.ic.ac.uk/ectoc.paper/56/
28. Mutter, M., Dumy, P., Garrouste, C. L., Mathieu, M., Peggion, S. P., Razaname, A., *et al.* (1996). *Angew. Chem. Int. Ed. Engl.*, **35** (13/14), 1482.
29. Saragovi, H. U., Fitzpatrick, D., Raktabutr, A., Nakanishi, H., Kahn, M., and Greene, M. L. (1991). *Science*, **253**, 792.
30. Ku, T. W., Ali, F. E., Barton, L. S., Bean, J. W., Bondinell, W. E., Burgess, J. L., *et al.* (1993). *J. Am. Chem. Soc.*, **115**, 8861.

Chapter 7

Generating monoclonal antibody probes and techniques for characterizing and localizing reactivity to antigenic determinants

Paul N. Nelson

Division of Biomedical Sciences, University of Wolverhampton, 62-68 Lichfield Street, Wolverhampton WV1 1DJ, UK.

1 Introduction

This chapter initially provides details of protocols useful for the generation of monoclonal antibody probes. Monoclonal antibodies (mAbs) in general, serve as powerful reagents in diagnostic applications and as tools for the investigation of macromolecules and cells (1, 2). However, it is the further scrutiny of their reactivity to desired target antigens, subsequently pursued in this chapter, which paves the way for the development of secure assay systems and reliable functional studies. The localization of a target epitope may be grossly identified though the combination of enzyme (or chemical) degradation studies, e.g. digestion of IgG to yield Fc and pFc′ fragments, and the analysis of mAb reactivity using techniques such as enzyme-linked immunosorbent assay, haemagglutination, and slot-blotting. Alternatively, *in situ* localization of cell surface or intracellular antigenic determinants may be achieved by immunocytochemical techniques.

The additional use of physicochemical techniques such as carbamylation and ultraviolet irradiation which perturb certain amino acids, can also provide evidence of key residues involved in mAb–epitope interactions (3). Of course underpinning many of these investigative techniques useful for localization studies is the prior separation of intact antigens and their fragments by SDS–polyacrylamide gel electrophoresis. The subsequent transfer of fragments to nitrocellulose membrane (Western blotting) again permits mAb reactivity studies. Evidently, other methodologies, e.g. for isolating cell membrane components and protein purification steps are required and it is suggested that these protocols are obtained from other sources (4–6). Ultimately, epitope localization can be further refined by epitope mapping studies which specifically highlights salient residues (7) (*Table 1*).

Overall monoclonal reagents readily permit standardization of reagent and technique. A monoclonal antibody being defined as an antibody of unique specificity, generated from the immortalization of a single plasma B cell *in vitro*. The method of generating a monoclonal antibody, offers the potential to provide an unlimited supply of reagent which can be of high titre and of reproducible quality. In contrast, polyclonal reagents, are limited in supply and generally of low titre, and may vary in quality from batch to batch.

Within this chapter, a number of references are given where techniques have been applied. In addition, useful addresses have been included and where appropriate, costings are shown to serve as a rough guide for a laboratory considering developing monoclonal reagents. For health and safety, it is assumed that the scientific investigator understands and is familiar with the codes of practice and safety advice within a laboratory for handling reagents, equipment, and spillages, i.e. COSHH assessment forms are duly completed before starting practical work.

2 Immunization strategies

Substances which induce an immune response are usually foreign to the individual and are termed antigens, e.g. tetanus toxoid in humans. In this case a large number of antibody molecules would be produced (i.e. a polyclonal antibody response); each antibody recognizing a different antigenic determinant or epitope on the protein. In effect the hybridoma technique permits the isolation and selection of a particular antibody for a clearly defined purpose. For the hybridoma technique, an antigen, be it protein, cell, or synthetic peptide (linear or multiple antigenic peptides: MAPs), is used to induce an immune response in a mouse; an important requirement being that the species employed for immunization is phylogenetically different from the species used to isolate the target antigen. Adjuvants, e.g. Freund's adjuvant can be employed to enhance the immune response by causing a 'depot' effect, i.e. allowing a slow release of antigen. Boosting is necessary to obtain a sufficient antibody response (evaluated by tail bleeds) and to switch isotype: gamma being preferred for monoclonal reagents. However,

Table 1 Comparison of monoclonal reactivity and epitope localization studies

mAb	Reactivity			
	ELISA	**HA/HAI**	**Western blot**	**Epitope mapping**
A57H[a]	pan-IgG Fc, pFc′	pan-IgG Fc, pFc′	Fc, pFc′ (IgG1)	378-AVEWESNGQPENNYK-392 *CH3 domain localization*
PNF69C[b]	pan-IgG	G3m(u)	pFc′ (IgG1) pFc′ (IgG3)	330-APIEKTISKAKGQPR-344 *CH2/3 domain localization*

[a] Refer to ref. 7.

[b] Refer to BSI abstracts: Nelson *et al.*, December 1995, Brighton, UK.

the *in vivo* strategy requires a substantial source of antigen (200–500 μg/mouse) whereas an *in vitro* immunization strategy (e.g. Cel-Prime, Immune Systems Ltd.) may utilize 2 × 30 μg doses of antigen. The latter could be particularly useful where the quantity of antigen is limited. In general, *in vivo* immunization remains the mainstay of most laboratories.

Protocol 1
Immunization

Equipment and reagents

- Glass syringe for immunization (easier for antigens emulsified in adjuvant)
- 1 ml plastic syringes
- Needles
- 0.5 ml Eppendorfs (for collecting bleeds)
- Microcentrifuge (e.g. IEC MicroMax, Life Sciences)
- Complete Freund's adjuvant (CFA) (Sigma P5881)
- Sterile PBS (Sigma D8537) pH 7.2
- Incomplete Freund's adjuvant (IFA) (Sigma F5506)
- Animals: Balb/c mice (allow three mice per group)
- Immunizations (13 week strategy)
- Maintenance and tail bleeds

Note: costs may vary between Home Office approved Biological Research Facilities and whether animals are bred in-house, i.e. carriage may be incurred.

Method

1 For a single immunization strategy, use three female Balb/c mice (six- to eight-week-old). Clearly label each mouse (1–3) by tail marking or using an ear puncher. House the mice in a single cage.

2 Immunize mice with appropriate antigen.
  (a) For each mouse emulsify 50–100 μg of protein in an equal volume of complete Freund's adjuvant using a glass syringe. Immunize via a subcutaneous (s.c.) route: one or two sites may be permitted with an injection volume of 100 μl per site.
  (b) For each mouse, grow 10^6 to 10^7 viable cells in culture medium and resuspended in 0.5 ml normal saline or PBS. Inject using a plastic syringe via an intraperitoneal (i.p.) route. No adjuvant is necessary.
  (c) Immunize each mouse as for protein although a linear peptide[a] should be linked to a carrier protein, e.g. keyhole limpet haemocyanin (KLH). Note that multiple antigenic peptides (MAPs)[b] do not require a carrier (8). Costing for three MAP peptides and one linear sequence (15- to 13-mers) including purification: approximately £1200.
  (d) Remove spleen cells from an unimmunized mouse and immunize *in vivo* according to manufacturer's instructions, e.g. Cel-Prime (Immune Systems Ltd.). Initial immunization of mice with crude antigen may well ensure an isotype switch to IgG (9).

Protocol 1 continued

3 Boost mice on week five, but with protein/peptide antigens emulsified in incomplete Freund's adjuvant. Premature boosting may decrease serum antibody titre (10).

4 Tail bleed each mouse during week six/seven. Collect venous blood (50–100 μl) from the tail vein of each mouse: in brief, restrain mouse and aseptically cut tip of tail and collect blood (applying Vaseline to the tail and gently squeezing towards the open end will aid in the delivery of blood). Allow sample to clot for 1 h at 37 °C and store overnight at 4 °C (this aids clot retraction). Spin sample at 6500 r.p.m. in a microcentrifuge for 5 min and collect serum by means of a fine Pasteur pipette. Dilute serum for appropriate assay system or store at −20 °C until required. A mouse pre-bleed may be used as a control, however normal mouse serum (e.g. Sigma M5905) is perfectly adequate.

5 Boost mice on week five and week eight and assess reactivity of tail bleed serum to target antigen (see ELISA; *Protocol 6*): bleeds may be taken on week six and weeks 9–11. Intact cells and proteins tend to give high titres after a second boost, whereas synthetic peptides may require additional boosts. However, note that continual boosting (including a pre-fusion boost) may cause sensitization and shock: the temperature of a mouse rapidly dropping from 37 °C to 31 °C. Always seek expert advice from trained BRF personnel as a hybridization may be the best option to save months of hard work. However mice usually recover after a few hours.

6 Three to five days prior to fusion, boost a selected mouse (based on tail bleed data) with antigen (50 μg i.p. and 50 μg i.v. in sterile saline/PBS) without adjuvant. It is suggested that boosting of peptide–KLH/MAP is performed only if the peptide has originally been purified and is soluble in solution. If uncertain, immunize i.p. as recommended for intact cells.

[a] Candidate peptides should be analysed (e.g. PC Gene) for hydrophilicity (12) and can be synthesized using the Merrifield solid phase method (11): hydrophilic 12- to 15-mers are potentially effective for immunization purposes. Generally a 30 μmole scale synthesis is sufficient for immunization and screening purposes. For a linear peptide, an N-terminal cysteine residue can be added to facilitate coupling (e.g. by using *N*-hydroxysuccinimide ester) to a carrier such as keyhole limpet haemocyanin. If using MAP peptides, an option is to synthesize 15 μmoles peptide for immunization and 15 μmoles biotinylated peptide for screening. In some cases, peptides may be difficult to dissolve and reconstitution requires trial and error using dilute acid, ammonium hydroxide, or dimethyl sulfoxide (note that material can always be recovered by freeze drying). Always ensure that lyophilized peptides warm up to room temperature and store peptide solutions at −20 °C in the short-term.

[b] For MAP peptides, the C-terminus of a peptide is synthesized on a branched poly-lysine core sequence, allowing eight peptide chains to be incorporated into each molecule. Because of the molecular weight (> 16 kDa), the compound is immunogenic. It is important that the crude MAP peptides are extensively dialysed (e.g. by using benzoylated dialysis tubing, Sigma D7884) prior to immunization so as to remove toxic substances remaining from their synthesis. It is stressed that MAP synthesis is not recommended for sequences immediate to the C-terminus of a molecule.

Isolated peptide may be linked to a carrier, e.g. keyhole limpet haemocyanin (KLH) using *N*-hydroxysuccinamide ester (13) in order to induce an immune response. In this method, a cysteine-containing peptide is rapidly linked to an activated carrier protein. Generally, a peptide, perhaps containing an N-terminal cysteine, is linked to one carrier, e.g. KLH for immunization and an alternative carrier for hybridoma screening, e.g. bovine serum albumin (BSA).

Protocol 2

KLH conjugation

Equipment and reagents

- Magnetic stirrer and flea
- Dialysis tubing (e.g. Visking Size 9)
- Sephadex G25
- UVicord 280 nm
- Synthetic peptide (e.g. synthesized by Alta Bioscience, University of Birmingham)
- 10 mM sodium phosphate buffer pH 7.2
- 50 mM sodium phosphate buffer pH 6.0
- PBS pH 7.5
- Keyhole limpet haemocyanin (KLH) (Sigma H2133)
- *m*-maleimodo-benzoyl-*N*-hydroxysuccinamide ester (MBS) (Sigma M2786)
- Dimethylformamide (Sigma D4254)

Method

1. Dissolve 4 mg KLH in 250 μl of 10 mM sodium phosphate buffer pH 7.2.
2. Dissolve 0.7 mg MBS in a minimal quantity of dimethylformamide. Then add drop-wise to the KLH solution with constant stirring (i.e. use a magnetic stirrer and flea) and allow to react for 30 min at room temperature.
3. Equilibrate a Sephadex G25 column with 50 mM sodium phosphate buffer pH 6.0 and pass reaction mix through the column. The KLH-MBS elutes as the breakthrough fraction and is monitored at 280 nm.
4. Dissolve 5 mg synthetic peptide in 1 ml PBS pH 7.5.
5. Mix peptide solution with KLH-MBS (ensure reaction remains between pH 7.0–7.5) and stir for 3 h at room temperature.
6. Dialyse against PBS to remove unlinked peptide (preparation may be lyophilized if required).

3 Hybridization and culture of hybridomas

The principle of the hybridoma technique lies in the fact that cells possess two pathways of nucleotide biosynthesis: the *de novo* (normal) pathway and the salvage pathway which uses an enzyme called hypoxanthine-guanine phosphoribosyl transferase (HGPRT). Mouse myeloma cells lines are selected (generally in 8-azaguanine) which have a defective gene for HGPRT: these cells are unable to utilize the salvage pathway for purine synthesis. Mouse B cells from an immunized animal are fused with myeloma HGPRT negative cells to produce a hybrid cell or hybridoma: the B cell providing genes for antibody production and HGPRT, and the myeloma cell providing genes for 'immortality'. By using a selective medium such as 'HAT' containing hypoxanthine, aminopterin (which blocks the *de novo* pathway), and thymidine, the net result is that unfused myeloma cells die. This is because both pathways are no longer functioning: should unfused myeloma survive, the latter would rapidly outgrow the newly formed hybridomas. Over a few days, unfused B cells die but the hybridomas proliferate since their salvage pathway enzyme is provided by the original B cell. Typically, an agent such as polyethylene glycol (< 50%, w/v), is used to fuse the plasma membranes of adjacent myeloma and antibody secreting cells to form a hybridoma. Evidently because of the abnormal number of chromosomes, some may be lost, e.g. those responsible for immunoglobulin heavy or light chains, or drug selection, leading to unstable hybridomas.

Generally, hybridomas are grown in RPMI 1640 medium with L-glutamine, containing 10% fetal bovine serum (normally containing very low concentrations of IgG that could interfere in assays or purification). Other supplements include, penicillin (inhibits cell wall synthesis: effective against Gm positive bacteria for two days) and streptomycin (effective against Gm negative bacteria for four days). Fused cells may be grown in 96-well culture plates and hybridoma supernatants (containing antibody in the order of 1–60 μg/ml) are screened and selected for antibody reactivity in a desired assay system. It is important to carefully select at the earliest possible stage so as to avoid unnecessary expansion of useless hybridomas, i.e. a robust initial screening method is desired and ruthless nature required! Often an initial screening method can be designed and optimized using mouse tail bleeds. It is vital that culture plates are visually inspected for hybridoma growth (a cluster of cells is normally observed) and any contaminated wells treated immediately with sodium hydroxide. Overall, hybridomas appear 10–14 days post-fusion, but beware of slow growing hybridomas (25–30 days post-fusion); the latter are often very stable and yield useful reagents (14). It is often worthwhile waiting for a given hybridoma to be at least 50% to 75% confluent before screening: usually 100–150 μl can be aseptically removed for testing, e.g. in enzyme-linked immunosorbent assay or immunocytochemistry. Selected hybridomas can then be expanded to 24-well plates: at least 1 ml of culture supernatant is available at this stage for characterization, e.g. reactivity in other systems and isotype determination. Once selected, expand to a limited extent (e.g. using 25 cm^2 and 75 cm^2 flasks) to provide at least three to five vials for cryopreservation. However, a hybridoma must be recloned to ensure monoclonality and at all stages mAb reactivity should be assessed to ensure that a hybridoma line is stable. Subsequently, bulk production of a mAb can be achieved using expanded surface culture flasks, a capillary cell harvester, or as ascites (yield 1–25 mg/ml). Whilst mAbs should be tested in the applicable system chosen for its use, it is sensible to screen supernatants in a number of assay systems. On occasions, mAbs exhibit assay restriction (15), performing well in some systems but not others. This approach maximizes the chances of obtaining a mAb which may prove useful for purification purposes, probing antigenic determinants, or be of commercial value. As a rough guide, allow at least £4000 for a single fusion: in general a couple of fusions are necessary. This figure includes the immunization strategy (mice, peptides, plus carrier if required) with media and supplements ~ £500, fetal bovine serum ~ £300, disposable plastics ~ £1500, mAb purification and conjugation ~ £250. For screening by ELISA, blotting, and immunocytochemistry, allow £1000 for each technique whilst screening by haemagglutination may offer a cheaper alternative.

Protocol 3

Hybridization and subsequent propagation of hybridomas

Equipment and reagents

- Two sets of sterile scissors and forceps
- Sterile Petri dish (90 mm, e.g. BDH triplevent 402/0066/02)
- Syringes: 20 ml (BDH 406/0375/15), 10 ml (BDH 406/0375/14), 1 ml (BDH 406/0375/11)
- Needles: 21G × 1″ (BDH 406/0379/01), 25G × 1″ (BDH 406/0379/03)
- Sterile transfer pipettes (e.g. Sarstedt 86.1172.001)
- Sterile bottles (Nalgene 100 ml, 500 ml)
- CO_2 incubator (Heraeus BB6060), automatic CO_2 cylinder change over unit
- Inverted microscope (e.g. Olympus CK2 with adjustable stage)
- Biological Safety Cabinet Class II (Biohit)
- MSE bench centrifuge (Centaur II)
- Cryotube 1.8 ml (Nalge Nunc Int. 375418), 50 ml sterile polypropylene tubes (Falcon 2098)
- Multichannel pipette: 25–200 μl (dedicated for cell culture use, e.g. Anachem Autoclavable 200-8A)
- Sterile 96-well flat-bottom tissue culture plates (Sarstedt 83.1835), 24-well plates (Sarstedt 83.1836)
- Tissue culture flasks: 75 cm^2 (Sarstedt 83.1811.001 with phenolic style cap), 175 cm^2 (Sarstedt 83.1812.001)
- Cell scraper 25cm (Sarstedt 83.1830, Falcon 3086)
- Pipet-aid (Anachem)
- Sterile graduated pipettes 1 ml (Sarstedt 86.1251.001), 10 ml (Sarstedt 86.1254.001), 25 ml (Sarstedt 86.1685.001)
- Fetal bovine/calf serum (Sigma F7524): batches of FCS may be compared by supplementing basic medium (e.g. at 10%) with test serum and culturing myeloma cells (seeded at $< 1 \times 10^5$ cells/ml) for 5–7 days. Growth curves are then obtained by plotting the viable/non-viable cell number against consecutive days in culture. Always use heat inactivated serum which is achieved by immersing a bottle of FCS in a water-bath at 56 °C for 35 min. When cool, aliquot into sterile containers and store at −20 °C until required.
- Myeloma cells: a number of compatible mouse myeloma cell lines are available, e.g. NSO (laboratory of Molecular Biology, Cambridge), a non-secreting variant of MOPC 21 cell line, and SP-2, AG8, NS1 (purchased from Immune Systems Ltd.)
- Basic medium: to 485 ml RPMI 1640 (Sigma R0883), add 5 ml L-glutamine (200 mM, Sigma G-7513), 5 ml penicillin–streptomycin solution (10 000 U/ml penicillin, 10 mg/ml streptomycin in 0.9% NaCl, Sigma 0718), and 50 μl gentamycin solution (Sigma G1272, 10 mg/ml gentamycin in deionized water). Gentamycin is added for hybridization purposes, but is optional for the growth of established hybridomas.

Protocol 3 continued

- General maintenance medium (10% FCS): as for basic medium but add 50 ml of FCS to a total volume of 435 ml RPMI plus supplements
- 8-Azaguanine myeloma medium (containing 10% FCS): to 430 ml RPMI 1640, add 5 ml L-glutamine, 5 ml penicillin–streptomycin, and 50 ml FCS. Reconstitute one vial of 8-azaguanine (50 $\times$, 6.6 $\times$ 10^{-3} M, Sigma A5284) with 10 ml of spare RPMI 1640 and add this volume to the medium.
- Maintenance HAT medium: to 425 ml RPMI 1640, add 5 ml L-glutamine, 5 ml penicillin–streptomycin, 50 μl gentamycin, and 50 ml FCS. Reconstitute one vial of HAT media supplement (50 $\times$, 5 $\times$ 10^{-3} M hypoxanthine, 2 $\times$ 10^{-5} M aminopterin, 8 $\times$ 10^{-4} M thymidine, Sigma H0262) with 10 ml of spare RPMI 1640 and add this volume to the medium.
- Maintenance HT medium: as for HAT medium except add 10 ml of reconstituted HT media supplement (50 $\times$, 5 $\times$ 10^{-3} M hypoxanthine, 8 $\times$ 10^{-4} M thymidine, Sigma H0137)
- Polyethylene glycol, PEG: melt 8 mg PEG (PEG 1500 BDH, Sigma P7777) in a water-bath and add 10 ml of pre-warmed RPMI 1640 plus L-glutamine and filter sterilize. Alternatively PEG may be autoclaved and sterile medium added immediately to obtain a 40% (w/v) solution (following fusion conditions of Galfre *et al.*, 1977) (16). Other protocols may utilize 50% (w/v) PEG solution (6, 7) and solutions can be readily purchased, e.g. Polyethylene Glycol Hybri-Max solution (Sigma P7181), PEG 1500

Method

1. Recover myeloma cells from liquid nitrogen storage in general maintenance medium. One to two weeks prior to fusion, grow in 8-azaguanine medium. Perform fusion when cells are in a logarithmic phase of growth. Myeloma cells can be removed by gently tapping flask or preferably using a cell scraper. Prior to and on the day of fusion, count cells: require a ratio of 10^7 viable myeloma cells: 10^8 spleen cells. Collect myeloma cells in a sterile 50 ml polypropylene tube and maintain at 37 °C immediately prior to fusing.
2. **Fusion** (method based on Galfre *et al.*, 1977) (16). Perform in a Class II sterile cabinet:
   (a) Place a few drops of basic medium in a sterile Petri dish.
   (b) Fill $\times$ 3 10 ml syringes with basic medium and attach a 25G needle. Maintain at RT.
   (c) Fill a 20 ml syringe with 20 ml basic medium and attach a 21G needle. Maintain at RT.
   (d) Fill a 20 ml syringe with 21 ml basic medium and attach a 21G needle. Maintain at 37 °C.
   (e) Place 24 ml general maintenance medium in a sterile 50 ml tube and maintain at 37 °C.
   (f) Fill a 1 ml syringe with 0.8 ml PEG and attach a 25G needle. Maintain at 37 °C.

Protocol 3 continued

(g) Aseptically remove spleen: place cervically dislocated mouse in a supine position. Dab fur on right-hand side with methanol and cut skin with sterile scissors and forceps: tear open to allow removal of spleen. Using a second set of sterile instruments, cut open the peritoneal membrane, remove spleen, and place in a sterile Petri dish. Remove any adhering fat.

(h) Perforate spleen with multiple holes using the 25G needle attached to the 10 ml syringe(s) and perfuse gently (30 ml is usually sufficient). The cells will collect in the Petri dish and the spleen will become translucent as perfusion continues. Alternative methods include: wrapping the spleen in sterile gauze (or using a sterile cell dissociation sieve, e.g. Sigma CD-1) and gently rubbing the spleen with the plunger of a syringe to release cells. At the end of the day, viable and not crushed spleen cells are required! Collect cells in a sterile Universal and perform a viability count using 0.4% trypan blue (e.g. Sigma T8154) dye exclusion.

(i) Spin NSO and spleen cells separately (2000 r.p.m., 5 min) and discard supernatant.

(j) Resuspend each pellet with 10 ml of basic medium: use the 20 ml syringe (step 2c) for simplicity. Add both cell suspensions together in a sterile 50 ml polypropylene tube and mix gently by hand. Spin at 1850 r.p.m. for 7 min.

(k) Remove supernatant and gently flick pellet. Then allow pellet to warm up in hand for 1 min.

(l) Add PEG solution (step 2f) over 1 min (i.e. 0.3 ml/15 sec) and shake tube continuously.

(m) Allow suspension to warm up in hand for 1 min.

(n) Add 21 ml basic medium as follows; 1 ml over 1 min (0.25 ml/15 sec), then 20 ml over 5 min (2 ml/30 sec). Again the syringe with needle (step 2d) provides an easily controllable system.

(o) Spin suspension at 1500 r.p.m. for 15 min and discard the supernatant.

(p) Resuspend cells in 24 ml general maintenance medium (step 2e) and then add one drop (approximately 30 μl) to each well of six 96-well sterile plates (suitably labelled). Maintain plates at 37 °C in a 5% CO_2 gassed incubator (100% humidity).

(q) After 24 h (day 2) add 150–175 μl HAT medium to each well (medium may be poured into a sterile aluminium tray and dispensed using a multichannel pipette with sterile tips.

(r) On day 6/7, change the medium, i.e. remove 150 μl using a sterile Pasteur attached to a suction pump with suitable trap. Dispense 150 μl of maintenance HAT medium (add 100 μl of 1 M NaOH to any contaminated wells). Check cells daily and replace medium every 2/3 days.

3 **General maintenance of hybridomas**. After 10–14 days, hybridomas may be visible under the inverted microscope and often the culture medium is yellow. Generally 150 μl of culture supernatant is removed for testing, e.g. in ELISA. Once appropriate hybridomas have been selected, transfer from a 96-well plate to one well of a 24-well plate. Ensure that some cells remain in the 96-well plate as a

Protocol 3 continued

back-up (some hybridomas initially grow poorly in 24-well plates but this may be overcome by initially coating wells with feeder cells: derived from a normal mouse spleen).[a] Once 75% confluent, transfer to a 25 cm^2 flask and then to 75 cm^2 flask.

4 **Cloning of hybridomas**, e.g. by limiting dilution eliminates non-producers and contaminating hybridomas. Count viable hybridomas and prepare at least 3.2 ml of the following: 10 cells/ml, 3 cells/ml, and circa 1.6 cells/ml in HT medium. Pipette 100 μl of each dilution into 32 wells of a sterile 96-well microtitre. The latter accommodates all three seeding dilutions which roughly equates to one cell per well, one cell per three wells, and one cell every six wells respectively. Note that because of the low cell density, not all hybridomas will grow and it is suggested that plates are seeded with feeder cells[a] two days prior to cloning. Check growth of hybridomas under an inverted microscope, and select wells containing a single colony. The process should be repeated to ensure monoclonality. Verify class/subclass of mAb using an isotype kit (e.g. Sigma ISO-2).

5 **Freezing down hybridomas**. Prepare freezing down medium: composed of 90% FCS[b], and 10% DMSO (e.g. Sigma C6295). Freeze aliquots (e.g. 5 ml) at −20 °C until required. Alternatively use a commercial freezing down reagent (Sigma C6295 or Gibco 11101). Collect cell suspension (approx. 10^{6-7} cells may be collected from one 75 cm^2 flask) into a 50 ml polypropylene tube. Centrifuge at 1000 r.p.m. for 10 min and decant supernatant containing relevant mAb. With a small drop of supernatant remaining, resuspend cells by gently flicking the side of the container. Add 1 ml of freezing down medium (maintained on ice) and ensure that cells are fully resuspended. Place cell suspension into a sterile cryotube (colour coded for easy reference) which is resting on an ice-bath (a polystyrene box with ice will suffice). Insert into a glycerol bath (−32 °C) for 1 h, before being place in liquid nitrogen. Alternatively, place cryotube in a suitable polystyrene box with tray inset and leave at −20 °C for 1 h, −70 °C overnight, and then transfer the following day to a Dewar flask containing liquid nitrogen. Note that smaller numbers of hybridomas, e.g. from 24-well microtitre plates may be successfully preserved: simply remove most of the supernatant and gently scrape cells of the surface with a pastette. Add freezing down medium and place in a cryotube. This may provide peace of mind for important hybridomas, or when the workload becomes excessive.

6 **Recovery of hybridomas**. Remove hybridoma from liquid nitrogen and rapidly thaw cells using warm water (utilize a 37 °C water-bath). Transfer cells using a pastette (or 1 ml pipette) into 10 ml RPMI 1640 plus L-glutamine in a sterile Universal and spin at 1500 r.p.m. for 5 min. Discard supernatant into a beaker of Virkon. Gently resuspend pellet in 1 ml of general maintenance medium and aseptically remove 30–50 μl to perform a viability count using trypan blue. Add required number/volume of cells to propagate hybridoma line in culture flasks containing HT medium.

Protocol 3 continued

7 **Propagation of hybridomas**: expand cloned hybridomas in HT medium from a 96-well plate, to a 24-well plate, to a 25 cm^2 flask, and finally a 75 cm^2 flask. Further expansion can be achieved using expanded tissue culture flasks, hollow-fibre technology (e.g. IBS Integra Biosciences; Harvest Mouse, Serotec; Technomouse) or as ascites. For the latter, inject mice (i.p.) with 0.5 ml Pristane (2-,6-,10-,14-tetramethylpentadecane, Aldrich Chemical Co. Ltd.) 3–60 days prior to use. Inject 2.5–5 $\times 10^6$ hybridoma cells in 0.5 ml PBS per mouse. After 10–14 days, collect ascitic fluid, e.g. 1–10 ml and place into a sterile Universal containing heparin anticoagulant. Freeze peritoneal ascitic cells as for hybridomas (or 'passage' into other mice) and verify[c] and purify mAb from ascitic fluid.

[a] Feeder cells: remove spleen from an unimmunized mouse as described in step 2(h) using a cell dissociation sieve. Resuspend in approximately 10 ml general maintenance medium (sufficient for × 2 96-well plates). Add 50 μl/well and place in 37 °C CO_2 incubator for 48 h prior to use.

[b] Note that in order for hybridoma cell cultures to be utilized in the USA, North American fetal bovine serum is required. Evidently this will increase the cost of production as compared to using European serum.

[c] Ascites may be verified for immunoglobulin using zone electrophoresis: apply 1 μl of ascitic fluid to a trough of a pre-poured agarose plate (Corning). Using a Corning Electrophoresis apparatus and a high resolution buffer (Gelman Sciences Inc.), run for 35 min and stain gel for 15 min in a solvent of ethanol, distilled water, glacial acetic acid (45:45:10) containing 0.6% (w/v) naphthalene black (BDH). Subsequently destain in the same solvent but containing the proportions 31.25:56.25:12.5. Read strips using a laser densitometer (LKB 2202 Ultrascan).

4 Monoclonal purification

Methods include ammonium sulfate precipitation, gel filtration, and ion exchange chromatography (5–7), although a rapid and more convenient method is affinity chromatography. The latter may utilize Protein A, or Protein G cross-linked to Sepharose. The latter binds to the Fc region of mouse γ1, γ2a, γ2b, γ3, and α isotypes and is stable over a wide pH range: pH 7.2 for binding and pH 2.3 for eluting purposes. The yields of mAb obtained from 20 ml culture supernatant (using a 1 ml HiTrap Protein G Column; Pharmacia) ranges from 183–250 μg of immunoglobulin from hybridomas PNQ312D, PNG312G (9, 15). Evidently larger column volumes will be utilized for bulk preparation of purified mouse immunoglobulin. For an IgM isotype, alternative methods are required, e.g. Prosep-Thiosorb-M (Bioprocessing Ltd.), rProtein L (Actigen Ltd.) according to manufacturer's instructions. In general γ immunoglobulins are preferable as monoclonal reagents, since IgM antibodies are prone to degradation, particularly following freezing and thawing.

Protocol 4

Monoclonal antibody purification

Equipment and reagents

- 1.5 ml Eppendorf tubes, rack
- Sterilin tubes
- Varaible pipette (1–200 μl, 400–1000 μl)
- 10 ml syringes
- UV spectrophotometer, e.g. Amersham Pharmacia Biotech Ultrospec III
- For bulk production: peristaltic pump, glass columns, UV flow cell and chart recorder, fraction collector
- HiTrap Protein G column (1 ml, Amersham Pharmacia Biotech 17-040-03)
- Culture supernatant
- Binding buffer: 20 mM sodium phosphate buffer pH 7.0
- Eluting buffer: 0.1 M glycine–HCl pH 2.7
- Neutralizing buffer: 1 M Tris
- 20% ethanol
- Distilled water

Method

1. For small scale preparation, e.g. from 20 ml culture supernatant: label Eppendorf tubes 1–40 ensuring that the 1 ml mark is clearly identified. Add 75 μl of Tris buffer (which neutralizes eluted fractions) from tube 27 onwards. Fill syringes with appropriate buffers and expel any air bubbles.
2. Flush a 1 ml Protein G column with 5 ml distilled water to expel 20% ethanol (storage buffer).
3. Equilibrate column with 3 ml of 20 mM sodium phosphate buffer pH 7.0 and collect fractions (i.e. 1–3 in this case and continue to collect thereafter and monitor fractions using a UV spectrophotometer set at 280 nm).
4. Add 20 ml culture supernatant: it is advisable to centrifuge samples prior to application (e.g. 2000 r.p.m., 5 min) to remove any debris that might block a column.
5. Equilibrate column with 8 ml of 20 mM sodium phosphate buffer pH 7.0.
6. Add 8 ml of 0.1 M glycine–HCl pH 2.7 to elute antibody. At this stage ensure that eluted fractions are well mixed with neutralizing buffer. In general the antibody will elute over two or three fractions. Assess concentration and pool accordingly, aliquot (50–100 μl volumes), and store at −20 °C.
7. Equilibrate column with 5 ml of 20 mM sodium phosphate buffer pH 7.0 and finally 3 ml of 20% ethanol if column is to be stored.

5 Monoclonal conjugation

Conjugation of an enzyme marker to a mAb may be desired to direct probing of antigenic determinants, e.g. in ELISA system or perhaps where a capturing reagent is also of mouse origin. This approach may well save the expense of commercial conjugates and reduce the steps (and time) required for a particular assay. Clearly other labels (17) can be employed which may provide enhanced sensitivity, e.g.

biotinylation of immunoglobulin together with streptavidin-HRP/alkaline phosphatase conjugate systems.

Protocol 5

Horseradish peroxidase (HRP) conjugation of immunoglobulins

Equipment and reagents

- Dialysis membrane
- Borosilicate tubes
- HRP (Sigma P8375)
- 0.1 M sodium periodate (Sigma solid S1878)
- Sodium borohydride (Sigma S9125)
- 1 mM sodium acetate buffer pH 4.4
- 0.2 M sodium carbonate buffer pH 9.5
- 0.01 M sodium carbonate buffer pH 9.5
- Distilled water

Method

1. Dissolve 4 mg HRP in 1 ml distilled water and then add 200 μl of fresh 0.1 M sodium periodate.
2. Stir mixture for 20 min at room temperature and dialyse against 1 mM sodium acetate buffer pH 4.4 overnight at 4 °C.
3. Raise pH of mixture to 9.0–9.5 by adding 20 μl of 0.2 M sodium carbonate buffer pH 9.5 prior to conjugation.
4. Dissolve or dialyse 8 mg of purified immunoglobulin in 1 ml of 0.01 M sodium carbonate buffer pH 9.5.
5. Mix immunoglobulin solution to HRP solution and stir for 2 h at room temperature.
6. Add 100 μl of fresh sodium borohydride (4 mg/ml in distilled water) and allow to stand for 2 h at room temperature.
7. Dialyse conjugate 0.1 M sodium borate buffer pH 7.4. Aliquot mAb-HRP preparations and store at 4 °C.

6 Techniques for screening and characterizing mAb reactivity to antigenic determinants

Stringent evaluation of a mAb and its target epitope is necessary because some antibodies exhibit assay restriction: a given mAb performing well in some assays whilst being poor or ineffective in others (2, 14). The variation in reactivity between assay systems may be attributed to changes in epitope display since denaturation may be induced by physical, or chemical procedures used to immobilize antigen. In addition, a given mAb may exhibit multispecificity as the antibody combining site may recognize more than one antigenic determinant. Therefore testing a given mAb in a range of assay systems, alludes to its potential applications and limitations: generally and within fixed limits it is possible to achieve the desired specificity of a suitable mAb (14).

6.1 Enzyme-linked immunosorbent assay (ELISA)

ELISA provides a convenient method for ascertaining reactivity of a given mouse monoclonal antibody to intact antigen, fragments, and peptides. The design of an indirect ELISA requires the immobilization of antigen onto a solid phase support, e.g. a 96-well formatted polystyrene plate. Following the blocking of potentially free sites on the plastic (with milk protein or albumin), a mAb or 'primary antibody' is allowed a requisite time to bind to the target antigen. Residual antibody is then removed by a washing step and the bound mAb detected with a 'secondary' antibody conjugate: usually a polyclonal reagent, e.g. goat anti-mouse immunoglobulin labelled with an enzyme such as horseradish peroxidase (HRP). Again following incubation of this reagent, residual conjugate is removed by washing the plate. Finally, the bound conjugate is revealed by adding a chromogenic substrate which gives rise to coloured product: the latter being measured in a plate reader and signalled as an optical density (OD) value. Evidently there are a number of varieties on this basic design: direct-, inhibition-, and capture-ELISA, each possessing individual nuances (2, 14). These are further complicated by the plethora of revealing/signalling systems which are available such as enzyme (HRP or alkaline phosphatase), biotinylated, or radiolabelled secondary antibodies and visible or luminescent products. Ultimately, the ELISA system chosen must be practical and offer a high signal-to-noise ratio. In addition the system must provide the required sensitivity, specificity, and a low coefficient of variation of replicate results between ELISA plates.

Overarching all these variables are key factors which must be resolved in order to achieve satisfactory and meaningful results.

(a) The type of plate requires consideration: a number of companies (e.g. Dynex Technologies) provide immunoassay plates with differing binding capacities for proteins, synthetic peptides, and glycopeptides (18). Most antigens may be coated at 1–10 μg/ml in alkaline buffer, however, coating antigens employing a chemical fixating agent, e.g. glutaraldehyde, may denature the epitope.

(b) Putative target antigens should be coated at equal molar concentrations, e.g. if investigating mAb reactivity to molecules of differing molecular weight such as IgG Fc and pFc′ fragments. This simple rule avoids the potential bias in mAb reactivity towards lower molecular weight substances which would occur if antigens were coated at equal mass concentrations, i.e. 1 μg/ml.

(c) The blocking of non-specific binding sites on the ELISA well surface should be tested empirically: common reagents include milk powder, bovine serum albumin, and gelatin. However, plates used for testing of hybridoma supernatants which contain 10% FCS may not require blocking. Non-specific binding of antibody reagents is also aided by incorporating a detergent, e.g. 0.05% Tween 20 in diluents and washing buffer. Note that whilst most cellular ELISAs do not recommend this detergent because of cell lysis, it is worth investigating individual cell lines: washing human umbilical vein endothelial cells (Huvec) with 0.05% Tween 20 has no deleterious effect on antibody reactivity and helps reduce background levels (P. N. Nelson and C. A. Brown, personal communication).

(d) The conjugated reagent (19), e.g. an anti-GAM for mouse monoclonal screening, should be absorbed by affinity chromatography to remove cross-reacting antibodies.

(e) The chromogenic substrate together with stopping reagent, may enhance the detection limit of an immunoassay: for HRP the enzyme substrates can be ranked; tetramethylbenzidine (TMB) > *o*-phenylenediamine dihydrochloride (OPD) > 2,2′-azino-bis (3-ethylbenzthiazoline-6-sulfonic acid) (ABTS) which give blue (yellow after stopping), yellow (orange after stopping), and green soluble reaction products respectively measured at 450 nm, 492 nm, and 405 nm.

Protocol 6

Enzyme-linked immunosorbent assay

Equipment and reagents

- Flat-bottomed 96-well polystyrene microtitre plates, e.g. Sarstedt Microtest Plate: Flat Well Cat. 82.1581; Dynex Technologies: Immulon 1 (M129AI), Immulon 2 (M129AII), Immulon 4 (M129AIV); Life Sciences (Greiner MP01-1)
- ELISA plate strips: may be more preferable if the full 96-well plate is not routinely required
- ELISA plate reader (e.g. Labsystems Multiskan MS Type 352) linked to computer with appropriate software, e.g. Labsystems Genesis
- Plate washer, e.g. Labsystems Multiwash (Model 8070-06)
- Multichannel pipette
- 0.15 M phosphate-buffered saline (PBS) pH 7.2
- Blocking buffer: 2% (w/v) milk powder (Marvel) in PBS; alternatively use bovine serum albumin, caesin, gelatin
- 0.05 M carbonate buffer pH 9.6
- PBS/Tween (0.05%): use Tween 20 (polyoxyethylenesorbitan monolaurate, Sigma P7949)
- Substrate buffer: 0.1 M citrate buffer pH 4.0 or 0.05 M citrate–phosphate buffer pH 5.0; use substrate buffer as recommended by manufacturer or optimize empirically to obtain a desired signal from the positive control over a set period of time, e.g. 30 min
- Peroxidase-conjugated goat anti-mouse immunoglobulin (GAM), affinity isolated and absorbed against human immunoglobulins and fetal calf serum (Dako P0447); guideline for dilution, 1:2000 for ELISA
- Hydrogen peroxide 30% (w/w) solution (Sigma H1009)
- Substrates tablets,10 mg: ABTS (Sigma A9941), OPD (Sigma P8287)

A. Evaluation of mouse tail bleeds

1 Apply 100 μl of antigen[a] (1–10 μg/ml) in carbonate buffer to individual wells of a flat-bottomed microtitre plate and incubate overnight at 4 °C (100 ng/well is generally sufficient for proteins, 500–1000 ng/well for non-biotinylated peptides). Placing a

Protocol 6 continued

microtitre plate against a dark background will aid in sample application. Cover plate with lid and use a plastic sandwich box containing moist tissue to act as a humidity chamber for all steps. For biotinylated peptides, initially coat microtitre plate (100 μl/well) with streptavidin (e.g. Sigma S4762,) at 5 μg/ml in carbonate buffer overnight at 4 °C. Block plate as in step 2 and then add 100 μl/well of biotinylated peptide at 1 μg/ml in PBS/Tween. Place plate on a shake table and incubate for 1 h at room temperature and proceed to step 3.

2 Block plate by applying 150 μl/well of 2% Marvel solution in PBS (alternatively 2% BSA in PBS). Incubate at room temperature (or 37 °C) for 1 h.

3 Wash plate (× 3) in PBS/Tween and shake the plate dry. Invert the plate on paper towelling to remove excess fluid (alternatively use a plate washer ensuring that the dispensing/collecting assembly does not touch the bottom of the well).

4 Add 100 μl of mouse tail bleed diluted 1/100, 1/1000, 1/10 000, and 1/100 000 in PBS/Tween. Add samples in duplicate and include normal mouse serum as a control. Add PBS/Tween to four wells to assess conjugate reactivity. Incubate at 37 °C for 2 h. Wash plate (× 3) in PBS/Tween and shake the plate dry.

5 Apply 100 μl of polyclonal anti-mouse (GAM) HRP-conjugated antibody at the recommended dilution in PBS/Tween and incubate for 1 h at 37 °C.

6 Wash plate (× 5) in PBS/Tween and shake the plate dry.

7 Apply 100 μl of the substrate ABTS (10 mg in 20 ml citrate buffer plus 5 μl hydrogen peroxide; 30%, v/v) to each well. Prepare substrate immediately prior to use.

8 Allow the colour reaction to develop for an appropriate time, e.g. 30 min (determine empirically for a particular antigen/antibody system) at room temperature and stop reaction by adding 50 μl/well of 1% (w/v) SDS (Sigma L-4509) in distilled water. It is recommended to leave the plate for 20 min, since the OD may increase over the first 5–10 min after stopping a reaction, after which the OD value plateaus. Read the optical density (OD) of individual wells at 414 nm using an automatic plate reader.

B. Screening of hybridomas

1 Apply 100 μl of antigen (1–10 μg/ml) in carbonate buffer to individual wells of a flat-bottomed microtitre plate and incubate overnight at 4 °C as stated in part A, step 1. Alternatively use method for biotinylated peptide.

2 Add 100 μl of hybridoma supernatant (neat) and incubate at 37 °C for 2 h. Wash plate (× 3) in PBS/Tween and shake the plate dry. For a positive control, include a mouse tail bleed, e.g. diluted 1/1000 (ascertains that assay system is working). Negative controls may include HAT medium and supernatant derived from myeloma cells: at this stage, the isotype of the mAb is unknown and it is therefore unrealistic to compare supernatants of known hybridomas of similar isotype.

3 Apply 100 μl of polyclonal anti-mouse (GAM) HRP-conjugated antibody at the recommended dilution in PBS/Tween and incubate for 1 h at 37 °C.

Protocol 6 continued

4 Wash plate (× 5) in PBS/Tween and shake the plate dry.

5 Proceed as stated in part A, steps 7 and 8. Alternatively apply 100 μl of the substrate *o*-phenylenediamine (10 mg in 25 ml citrate-phosphate buffer plus 5 μl hydrogen peroxide; 30%, v/v) to each well. Prepare substrate immediately prior to use.

6 Allow the colour reaction to develop in the dark for 30 min, and stop by adding 25 μl/well of 20% sulfuric acid.

7 Read the optical density (OD) at 492 nm using an automatic plate reader.

C. Characterization of purified monoclonal antibodies

1 Apply 100 μl of antigen (protein 100 ng, peptide 100–500 ng per well, or equivalent molar concentrations) in carbonate buffer to individual wells of a flat-bottomed microtitre plate and incubate overnight at 4 °C. Use a plastic sandwich box containing a moist tissue to act as a humidity chamber for all steps.

2 Apply 150 μl of 2% BSA in PBS (alternatively use 2% Marvel solution) to individual wells for the blocking stage. Incubate at 37 °C for 1 h.

3 Wash plate (× 3) in PBS/Tween and shake the plate dry.

4 Add 100 μl of test antibody, diluted in PBS/Tween and incubate at 37 °C for 1 h. Use purified mAbs at concentrations ranging from 1–25 μg/ml.

5 Wash plate (× 3) in PBS/Tween and shake the plate dry.

6 Apply 100 μl of polyclonal anti-mouse (GAM) HRP-conjugated antibody at the recommended dilution in PBS/Tween (alternatively use isotype-specific conjugated antibody at the supplier's recommended dilution) and incubate for 1 h at 37 °C.

7 Wash plate (× 5) in PBS/Tween and shake the plate dry.

8 For substrate use Option 1/Option 2 as in part B.

D. Capture (sandwich) ELISA

1 Apply 100 μl of purified monoclonal antibody[b] (10 μg/ml well) in carbonate buffer to individual wells of a flat-bottomed microtitre plate and incubate overnight at 4 °C. Use a plastic sandwich box containing a moist tissue to act as a humidity chamber for all steps.

2 Apply 150 μl of 2% BSA in PBS (alternatively use 2% Marvel solution) to individual wells for the blocking stage. Incubate at 37 °C for 1 h.

3 Wash plate (× 3) in PBS/Tween and shake the plate dry.

4 Add 100 μl of test protein at appropriate concentrations (0.001–10 μg/ml) diluted in PBS/Tween and incubate at 37 °C for 1 h.

5 Wash plate (× 3) in PBS/Tween and shake the plate dry.

6 Apply 100 μl of monoclonal HRP-conjugate (optimal dilution must be estimated empirically) and incubate for 1 h at 37 °C.

Protocol 6 continued

7 Wash plate (× 5) in PBS/Tween and shake the plate dry.

8 For substrate use as in part A or part B.

[a] A cell suspension of adherent cells (e.g. Huvec) at 5×10^4 cells/ml may applied (100 μl) to individual wells of a tissue culture treated plate and allowed to adhere overnight. Alternatively polyvinyl plates (Falcon 3912, Becton Dickinson) may be used. Other cells may require fixation (6, 7), e.g. using an isotonic solution of 2% paraformaldehyde.

[b] The capture ELISA format is useful where the epitope is masked or altered by direct physical absorption.

6.2 Haemagglutination

Haemagglutination provides an alternative method for investigating mAb reactivity to target epitopes which may be masked or denatured by the passive absorption onto plastic ELISA plates. The technique relies on the specific interaction and agglutination (i.e. cross-linking) of antibody with antigen which may be coated onto the surface of red blood cells. Invariably, the test antibody is serially diluted (generally doubling dilutions) across a microtitre plate containing 'U'- (or 'V'-) shaped wells which also contain both positive and negative controls. A suspension of red blood cells with antigen non-covalently using chromium chloride or tannic acid (6, 7), or covalently attached, e.g. by glutaraldehyde, is then added to each well. The dilution of an antibody ultimately gives rise to an end-point, after which agglutination is no longer visible: agglutination being visible as a mat of red cells at the bottom of the well. Conversely, a red pellet or dot is evident of non-agglutination; in this case the red cells simply settle at the lowest point of the well (tilting of a 'V'-shaped plate gives a characteristic 'comma' in non-agglutinated wells).

In haemagglutination (HA), antibodies can be compared in parallel with the same antigen, thus providing a titre. The latter represents a measure of the antigen-binding capacity of an antibody and is generally taken as the end-point value. Conversely, a haemagglutination inhibition (HAI) system, can be used to determine the concentration of an antigen in solution which specifically inhibits the homologous agglutinating system. Hence this technique readily allows the comparison of mAb reactivity to isolated fragments to aid in epitope localization studies (2). Usually to obtain maximum sensitivity in HAI, an antibody is used at a dilution comparable to twice the minimum haemagglutinating dose.

Protocol 7

Haemagglutination

Equipment and reagents

- Protein sensitization of SRBC: extensively dialyse protein solution against 0.15 M NaCl to remove phosphate ions: add purified protein solution (10 ml; 1–10 mg/ml) to a visking tubing (initially placed in boiling water for 5 min), seal (by knotting or clips), and insert tubing into a 2 litre beaker of saline containing a magnetic stirrer. Stir for 4 h at 4 °C, replace saline, and stir again for 4 h. Finally replace saline and stir overnight. For sensitization, add the equivalent of 0.3 mg protein to 10% packed SRBC and mix carefully. Add 'matured' $CrCl_3$ at 0.1 mg/ml dropwise (one drop equivalent to 30 μl) with continual mixing with a vortex. The number of drops must be worked out empirically: suggest use six drops (180 μl) initially. Wash down walls of tube with 1 ml sterile saline and incubate mixture overnight at 4 °C. Wash sensitized SRBC in 3 ml sterile PBS and centrifuge at 1000 r.p.m. (Sartorius MSE) for 5 min. Repeat wash step and then resuspend in 4 ml Hepes-buffered RPMI 1640 pH 7.4 containing 0.1% bovine serum albumin. Prior to use, prepare a 0.33% cell suspension of sensitized SRBC (termed SRBC-Ag), i.e. take 2 ml of SRBC-Ag and add 13 ml Hepes-buffered RPMI 1640.
- Single and multichannel pipettes
- U-bottomed microtitre plates (Sterilin, Birmingham)
- Chromium chloride solution: prepare a 0.1% (w/v) stock solution of chromium chloride (Sigma C1896) in isotonic saline and allow to stand at room temperature for three to four weeks. Neutralize the solution to pH 5.0 by dropwise addition of 1 M NaOH (do not allow pH to rise to pH 6.0) and allow the solution to mature for five months at room temperature.
- Sheep red blood cells: dilute SRBC (in heparin) 1:1 in 0.15 M NaCl and centrifuge at 2000 r.p.m. (Sartorius MSE) for 5 min. Remove supernatant and repeat this step five times. Pack SRBCs by centrifuging at 2600 r.p.m. for 10 min. Dilute in 0.15 M NaCl to give a 10% SRBC suspension. Immediately prior to sensitization, pipette 1 ml of a 10% SRBC suspension into a sterile glass tube (1 cm × 7 cm) and centrifuge at 1700 r.p.m. for 5 min. Remove supernatant.

A. Direct haemagglutination (HA)

1. Double dilute antibody (25 μl) in diluent (Hepes-buffered RPMI 1640 containing 2% (v/v) fetal calf serum) in a U-shaped microtitre plate.
2. For a typical 96-well plate, i.e. 8 rows (A–H) by 12 columns (1–12), pipette 25 μl diluent into wells 2–12 of a single row, e.g. A.
3. Pipette 50 μl antibody solution, e.g. hybridoma culture supernatant into well 1 of row A (for a monoclonal ascites, commence from an initial dilution of 1 in 10). Remove 25 μl and add to well 2 of row A, mix (by repetitive withdrawal and expulsion of fluid), and continue to doubly dilute across the plate until column 11 of row A. Discard the excess 25 μl in column 11.

Protocol 7 continued

4 Add one drop (30 μl) of SRBC-Ag to wells 1–12 of row A. Column 12 in this case serves as a negative control since it only possesses diluent and SRBC-Ag. Note that a doubling dilution series may well need to be continued along row B for a high titred antibody: in this case a negative control could be ascribed to row B, column 12.

5 Tap plate gently and leave for 2 h at room temperature.

6 Read the agglutination pattern (agglutination visible as a mat of coated SRBCs; absence of agglutination evidenced by a dot of coated SRBCs).

B. Haemagglutination inhibition (HAI)

1 Determine the minimum haemagglutinating dose (MHD) from an HA titration: essentially this is the last well to give a clear mat of agglutinated SRBCs. For an example, let us say this occurs in well 10.

2 Determine the 2 MHD. In our example this would be well 9 and the antibody would be diluted accordingly, e.g. if initial antibody was used neat (well 1), the dilution would be $\times 2^8$ (i.e. for wells 2–9) or 1/256, in a diluent of Hepes-buffered RPMI 1640 containing 2% (v/v) fetal calf serum.

3 For a typical 96-well plate, i.e. 8 rows (A–H) by 12 columns (1–12): pipette 25 μl diluent into wells 2–12 of a single row, e.g. A.

4 Pipette 50 μl of protein solution (1 mg/ml) into well 1 of row A. Remove 25 μl and add to well 2 of row A, mix (by repetitive withdrawal and expulsion of fluid), and continue to doubly dilute across the plate until column 10 of row A. Discard the excess 25 μl in column 10.

5 Pipette 25 μl of diluted antibody solution (2 MHD) into wells 1–11.

6 Incubate for 15 min at room temperature.

7 Add one drop (30 μl) of SRBC-Ag to wells 1–12 of row A. Tap plate gently and leave for 2 h at room temperature. Column 12 in this case serves as a negative control possessing only diluent and SRBC-Ag (should give a dot) and column 11 serves as a positive control possessing diluent, antibody, and SRBC-Ag (should give a clear mat of agglutinated SRBCs). Note that for some antibodies an inhibition pattern may well continue along row B: in this case a the negative and positive controls could be ascribed to row B, column 11 and 12 respectively.

8 Read the haemagglutination inhibition pattern (the inhibition titre defined as the last well providing a dot of coated SRBCs).

6.3 Slot-blotting

Slot-blotting permits the evaluation of mAb reactivity to antigens/fragments immobilized on a membrane, e.g. nitrocellulose. In brief, a slot-blot filtration manifold (e.g. Hoefer PR600) is used to apply samples (arranged in a 2 × 12 array) under vacuum. Volumes ranging from 50 μl to 1 ml may be applied which for initial purposes should contain at least 100 ng of protein. Once the antigen is blotted onto a nitrocellulose membrane (NCM), the latter is blocked and then incubated with mAb, followed by conjugate, as described for the ELISA system. However in contrast to the latter, the substrate used provides an insoluble coloured product, e.g. blue/black coloured deposit for the HRP substrate 4-chloro-1-naphthol. Alternatively, enhanced sensitivity may be achieved by using an HRP chemiluminescent system, e.g. ECL (Amersham) which necessitates biotinylated secondary antibody and HRP-conjugated streptavidin. In general, slot-blots are preferable for accurate densitometric measurements as compared to dot-blots. However, it is worth bearing in mind that crude dot-blots can be effective and achieved by directly pipetting antigen onto a nitrocellulose sheet in a minimal volume, e.g. 10 μl. The absorption of antigen onto nitrocellulose membrane can always be verified by staining a sample blot with a protein stain such as Ponceau S solution.

Protocol 8

Slot-blotting

Equipment and reagents

- Slot-blot filtration manifold (e.g. Hoefer PR600)
- Liquid trap: Buchner flask with horizontal vent to vacuum pump (and vertical vent, i.e. through a rubber bung to slot-blot manifold); a vacuum gauge or bleed valve is useful in the vacuum line
- Nitrocellulose: 0.45 μm (GRI EP4HYB0010), 0.2 μm if using synthetic peptides (GRI WP2HY320F)
- Variable pipettes for delivering 10 μl, 50–100 μl, 500–1000 μl
- 90 mm Petri dish
- Light-proof cassette (18 × 24 cm) and Hyperfilm-ECL (Amersham RPN 2103, 18 × 24 cm)—if using an ECL system
- PBS/Tween (0.05%)
- PBS/Marvel (2%, w/v)
- Peroxidase-conjugated goat anti-mouse immunoglobulin (GAM), affinity isolated and absorbed against human immunoglobulins and fetal calf serum (Dako P0447); guideline for dilution, 1:1000–1:2000 for immunoblotting
- Substrate buffer: 50 mM Tris–HCl pH 7.4, 150 mM NaCl
- Substrate: 4-chloro-1-naphthol (4C1N) (Sigma C 6788, 30 mg tablets)
- Dissolve one tablet in 1 ml methanol and make up to 20 ml in substrate buffer, 0.01% (v/v) H_2O_2 (add H_2O_2 last and prior to use, since substrate will start to change colour if left for a prolonged period of time)
- Ponceau S solution (Sigma P7170): 0.1% (w/v) in 5% acetic acid

Method

1 Pre-cut membrane slightly smaller than the membrane support block, but covering all the slots. Always handle nitrocellulose membranes using gloves otherwise extraneous protein (yours!) will be left on the membrane. Label with pencil and/or cut one corner for easy identification and orientation.

2 Soak in appropriate buffer for 2 min and carefully place the membrane, i.e. ensure

Protocol 8 continued

that it is flat, on the membrane support block. Align the top section of the block and tighten screws to seal the unit. Attach to vacuum pump.

3. Apply vacuum according to the manufacturer's instructions (e.g. 13–25 cm Hg) and then switch off pump. Load samples (50 μl–1 ml) into appropriate wells but avoid forming bubbles. Turn on vacuum again and pull the solution through the membrane.
4. Apply 1 ml of buffer to sample wells and pull solution through (38–50 cm Hg): this step can be repeated two or three times.
5. Whilst still applying a vacuum, remove the top block and lift off the membrane with forceps and place on clean, dry filter paper. Turn off vacuum pump.
6. Cut and trim membrane accordingly, e.g. for verification of bound protein or to enable membrane to fit into a suitable incubation chamber, e.g. a 90 mm plastic Petri dish.
7. Block membrane with 2% Marvel solution for 30 min.
8. Rinse membrane twice in PBS/Tween and apply mAb: dilute purified antibody in PBS/Tween and use at 1–25 μg/ml. If using a 90 mm Petri dish, 20 ml is sufficient to cover a 2 × 6 array. Place lid on Petri dish during all incubation steps.
9. Incubate for 1 h at 37 °C.
10. Wash membrane for 2 min in PBS/Tween and repeat step two or three times.
11. Add antibody conjugate[a] diluted in PBS/Tween and incubate for 1 h at 37 °C.
12. Wash membrane as in step 10.
13. Add substrate solution and leave for 20–30 min. Blue/black slots will be visible where mAb has bound to antigen. Briefly wash membrane in distilled water. Dry membrane between filter paper and scan/photograph pattern.

[a] For chemiluminescent detection: incubate with biotinylated anti-mouse immunoglobulin (Dako E0413), for 1 h, wash, and then incubate with streptavidin–HRP (Amersham RPN 1231). Finally use the Amersham ECL system (RPN 2106) as directed by manufacturer and develop autograph. The pattern can then be visualized using imaging technology (e.g. BioGene Ltd., Gel documentation system).

6.4 Immunocytochemistry

Immunocytochemistry provides a simple method for detecting the cellular localization and distribution of antigen. Following fixation and permeabilization of a tissue/cell preparation, a mAb is used to locate the target epitope which in turn is revealed by a secondary antibody conjugated to enzyme, e.g. horseradish peroxidase (HRP), alkaline phosphatase (AP), or fluorochrome, e.g. fluoroisothiocyanate (FITC), tetramethylrhodamine (TRITC). Alternatively, secondary antibodies may be biotinylated and then revealed using an HRP/AP-streptavidin system. Immunocytochemistry is also versatile in that initial treatment of serial sections with enzymes, e.g. *N*-glycosidase F, protease, prior to adding primary antibodies allows an investigator to establish whether the target is protein or glycoprotein in composition (15).

Within a particular assay, the concentration of the primary antibody (ranging from 1–30 μg/ml) and time of incubation must be assessed empirically and will depend on the mAb affinity and the abundance of target epitopes. Furthermore, meticulous care must be taken in blocking and all washing steps to avoid excessive background staining: particularly unwelcome for immunofluorescent staining of sections. Evidently, background reactivity of secondary antibodies can be minimized by using affinity purified and cross-absorbed reagents (i.e. against antibodies of the species under investigation). In all tests, appropriate controls must be included: the use of a positive control, e.g. an anti-HLA (ABC) antibody, not only ensures that the assay system is functional but permits comparison of mAb reactivity, i.e. +++ (strong), ++ (moderate), + (weak), – (absent) (15). Furthermore, negative controls should not only include tissue/cell section with antibody conjugate but also an irrelevant mAb of similar isotype. Note that a proportion of mAbs will function on frozen sections but not on formalin-fixed tissue (15).

It should be appreciated that certain tissues possess endogenous peroxidase/alkaline phosphatase activity and therefore the label chosen should accommodate this knowledge (20, 21). For an HRP system, diaminobenzidine (DAB) provides a brown insoluble end-product which is easily observed under a light microscope (the application of a counterstain will also enhance tissue morphology). DAB is also insoluble in alcohols (other substrates require an aqueous mountant, e.g. aminoethylcarbazole) and permits the mounting of sections in an organic solvent-based medium. This methodology gives a permanent record of sections with little deterioration of the coloured product over time. In contrast the signal derived from immunofluorescence is easily quenched and short-lived. Consequently slides should be photographed immediately. However, a particular advantage of immunofluorescence is the possibility of co-localizing antigens using primary antibodies directly conjugated to alternative fluorochromes, e.g. FITC-mAb1 and TRITC-mAb2.

Protocol 9

HRP immunocytochemistry

Equipment and reagents

- Cryostat, e.g. Reichert-Jung Cryocut E (Leica)
- Paintbrush (for cleaning microtome stage)
- Diamond pen (for marking slides)
- General microscope, Zeiss MC63
- Coplin jars for washing slides and substrate/product development
- Microscope glass slides (Merck Ltd.): 72 × 26 mm Gold Star KTH380 (with single frosted end); microscope glass coverslips: No.1 22 × 40, 22 × 64
- Multispot microscope slides (C. A. Hendley) H-005 white 4-spot slides

Protocol 9 continued

- Microscope glass slides (BDH): 72 × 26 × 1–1.2 mm thick (406/0188/22); glass coverslips 22 × 32 mm, thickness No.1 (406/0188/22)
- OCT compound BDH 36160: embedding medium for frozen tissue specimens
- TBS buffer: 0.05 M Tris-buffered saline pH 7.6 – for 1 litre dissolve 6.1 g Trizma base, 8.0 g NaCl in distilled water and adjust to pH 7.6 with HCl (alternatively PBS may be used)
- Working buffer: 3% BSA (albumin Fraction V, A-9647, Sigma) in TBS or PBS
- Conjugate: goat anti-mouse immunoglobulin HRP conjugate (GAM, antibody absorbed) (Dako P0447)
- DePeX mounting medium (BDH product 36125 4D)
- Hydromount (National Diagnositics HS-106) or Aquamount Mountant (BDH 36086)
- Harris haematoxylin solution modified (Sigma HHS-16)
- Diaminobenzidine tetrahydrochloride (DAB) medium: dissolve 3 g in 50 ml TBS, stir for 20 min, filter, and dispense into 2 ml aliquots. Store at −20 °C. Prior to use, dissolve 2 ml aliquot in 300 ml TBS. Block endogenous HRP by adding 180 mg sodium azide and activate with 300 μl hydrogen peroxide (30%). Alternatively use DAB tablets (Sigma D5905) or Sigma Fast Enhancer tablets (D0426) according to manufacturer's instructions.
- Pronase (NBS Biologicals: use at 50 μg/ml), protease (Sigma P-0390: use at 50 μg/ml), neuraminidase (sialidase) (Roche Molecular Biochemicals: use at 50 μg/ml)
- *N*-glycosidase F; cleaves asparagine bound *N*-glycans (Roche Molecular Biochemicals: use at 1 U/ml), *O*-glycosidase; cleaves serine or threonine bound *O*-glycans (Roche Molecular Biochemicals: use at 10 mU/ml)

Method

1. Cut fresh tissue from surgical specimens/post-mortems into appropriately sized blocks, for snap-freezing and storage in cryo-vials (under liquid nitrogen), and for immediate sectioning. For the latter, place one or two drops of OCT embedding medium onto a pre-labelled circular 10 mm cork tile and adhere specimen: this may be accomplished by placing the cork mat plus specimen in a plastic beaker of isopentane, cooled by immersion in liquid nitrogen (if liquid nitrogen not available snap-freeze specimen onto cork mat in dry ice).
2. Allow cryostat to cool (may require at least 2–3 h) to optimal cutting temperature, e.g. guinea-pig heart −20 °C, human osteoclastoma −30 °C. Adhere the specimen onto the microtome block with OCT and cut 5–8 μm serial sections[a] of tissue.
3. Fix cryostat sections (single-spot, or multi-spot glass microscope slides), in acetone at −20 °C for 10 min, allow to air dry for 15–30 min, and then store individually wrapped in silver foil, at −20 °C for short-term (one month) or at −70 °C for longer periods of time (deterioration of epitopes may occur with time).
4. Prior to using a section,[b] thaw for 20 min at room temperature and unwrap slide.
5. Block tissue section in buffer containing 3% BSA for 30 min. The following optional step should be used for investigating protein or carbohydrate determinants. For proteases; add 50 μl of required enzyme solution for 2, 6, 12 min at 37 °C. For glyco-

Protocol 9 continued

sidases; add 50 μl of required enzyme solution for 30–60 min at 37 °C remembering to include a positive and negative control for each antibody and section.

6. Tip off excess medium, and with paper towelling, carefully wipe dry the zone around the tissue section. When using multi-spot slides *make sure to carefully dry between tissue sections*: this prevents mixing of solutions by capillary action, i.e. when different primary antibodies are used (scoring around a plain slide with a diamond pen may also help retain medium).
7. Add 35–50 μl of primary antibody (approx. 1–30 μg/ml): for hybridoma supernatant use neat or dilute 1/2 to 1/10, and for crude mAb ascites dilute 1/100–1/5000 in 3% BSA in TBS/PBS, and incubate in a moist chamber at room temperature for 30–60 min.
8. Tip off excess medium and wash slides by immersing sections in TBS/PBS: the number of washes may be determined empirically to reduce background staining although two washes (for 3 min each) in fresh medium is generally adequate. Carefully wipe dry the zone around the tissue section.
9. Add 35–50 μl of optimally diluted secondary antibody,[c] e.g. polyvalent anti-mouse HRP-conjugated reagent (e.g. for Dako P0447 suggested working dilution 1/100) and incubate in a moist chamber at room temperature for 30 min.
10. Wash slides as in step 8.
11. Incubate sections for up to 8 min in TBS/PBS containing diaminobenzidine (DAB). Check progress of DAB incubation so as to prevent excessive background staining.[d]
12. Rinse in tap-water to remove excess chromogen. Counterstain section with haematoxylin for 20–60 sec and rinse in tap-water. An alternative counterstain may be employed when using Sigma Fast Enhancer since the latter generates an intense brown/black product.
13. Dehydrate sections in graded alcohols, e.g. 3 min respectively in 70% IMS (industrial methylated spirit),[e] 90% IMS, absolute alcohol, then xylene (× 2 changes), and mount with coverslip using DePeX (these steps should be performed in a fume hood). Alternatively, mount slides directly using an aqueous medium, e.g. Hydromount (dries in 15 min).
14. Allow to air dry for 15–30 min before visualizing section under a light microscope.

[a] Alternatively sections may be fixed in formalin–saline solution (10% formalin), paraffin-embedded, and dewaxed for immunostaining using standard procedures (21, 22). Note that formalin fixation may destroy some antigenic determinants. In some cases, demasking of antigens prior to addition of primary antibody may be necessary, e.g. using a target retrieval solution (Dako S1700).

[b] Note that pre-treatment of slides with Vectabond (see manufacturer's recommendations), or poly-L-lysine, in combination with baking for 30 min at 55–60 °C may ensure adhesion of tissue. Sections may also be prepared as tissue imprints, e.g. by lightly dabbing material onto 4 × 1 cm multi-spot glass microscope slides (14). Cells may also be prepared as cytospins (in particular non-adherent cells) or pre-grown on glass coverslips, or using a single or multiwell chamber

Protocol 9 continued

slide (Falcon 8 chamber culture slide 4108, Becton Dickinson). If sections are unfixed, perform step 2 and air dry for 10 min before immunostaining.

[c] A conjugate may be diluted in buffer employing 1–2% normal serum from the species in which the tissue section is derived in order to minimize reactivity to endogenous immunoglobulins. For alkaline phosphatase staining use anti-mouse Ig AP conjugate and substrate, e.g. Vector Red SK-5100 (Vector Laboratories).

[d] Reduction of endogenous peroxidase may be inhibited by the addition of sodium azide to the chromogenic substrate solution. Reduction of alkaline phosphatase activity may be inhibited by 1 mM levamisole solution.

[e] Industrial methylated spirit provides a cheaper alternative as compared to diluting absolute alcohol.

Protocol 10

Immunofluorescence immunocytochemistry

Equipment and reagents

- Fluorescence microscope: filters for excitation of fluorescein (absorption maximum 494 nm, emission maximum 518 nm), tetramethylrhodamine (absorption maximum 555 nm, emission maximum 580 nm)
- Biotinylated goat anti-mouse IgG (H+L), affinity purified (BA-9200 Vector Laboratories); recommended concentration range: 2–10 μg/ml
- Fluorescein–streptavidin (SA-5001 Vector Laboratories); recommended concentration range: 5–30 μg/ml
- Avidin/biotin blocking kit (SP-2001 Vector Laboratories)
- Vectashield mounting medium for fluorescence microscopy H-1000
- Nail polish

Method

1. Perform procedures as described in *Protocol 9*, steps 1–4.
2. Place one drop of avidin D onto section and incubate for 15 min at room temperature.
3. Tip off excess medium and wash slides by immersing sections in 0.5% BSA in PBS (× 2 washes for 3 min each).
4. Repeat as in step 2 with one drop of biotin, incubate for 15 min at room temperature.
5. Tip off excess medium and wash slides as in step 3.
6. Add 35–50 μl of primary antibody (as in *Protocol 9*, diluted in 0.5% BSA in PBS) and incubate in a moist chamber at room temperature for 30–60 min.
7. Tip off excess medium and wash as in step 3.
8. Add 35–50 μl of optimally diluted biotinylated anti-mouse immunoglobulin and incubate for 30 min.

Protocol 10 continued

9 Wash as in step 3.

10 Add 35–50 μl of optimally diluted streptavidin–FITC and incubate for 30 min at room temperature.

11 Wash as in step 3.

12 Mount section with a minimal amount of fluoromount and with coverslip. Seal edges of coverslip with a transparent nail polish (e.g. Boots No.17) and allow to dry for at least 10 min before visualizing using a fluorescence microscope set at an appropriate wavelength.[a]

[a] For storage, wrap slides with tin foil and store at 4 °C. For an alternative method to the above, follow *Protocol 9* but use a diluted fluorochrome-conjugated secondary antibody, e.g. rabbit polyvalent FITC-conjugated anti-mouse immunoglobulin (Dako F261) and fluoromount medium (BDH 36098).

7 Physiochemical and chemical modification, pepsin digestion

Physiochemical modification and chemical modification of proteins may provide details of particular residues which maintain epitope integrity (3), e.g. carbamylation of a protein with potassium cyanate specifically abrogates lysine residues. In contrast, UV irradiation particularly effects cysteine and aromatic residues: a low intensity ultraviolet source (366 nm + 254 nm) may be used to induce oxidative stress which is essentially mediated by oxygen-free radicals (ROS) which attack proline, cysteine, and aromatic residues. For the latter, oxidation of tyrosine and tryptophan generates phenolic and fluorescent kyneurine derivates respectively whilst oxidation of cysteine generates thiyl radicals. Evidently, the speed in which protein/epitope integrity is lost depends on the relative composition of labile to non-labile residues. Some epitopes may be conformational (i.e. dependent on the interaction of single or multiple chains) and may be disrupted following mild reduction and alkylation (dithiothreitol can be used as a reducing agent, whilst iodoacetamide is used to block free sulfydryl groups). Ultimately, fragments of proteins may be generated using enzymes which may aid in gross epitope localization studies: pepsin hydrolyses N-region peptide bonds of tyrosine and phenylalanine residues. Clearly an appreciation of the amino acid composition of a molecule under investigation is useful in predicting the number and size of fragments. Other proteolytic enzymes may also be used, e.g. trypsin: hydrolyses C-region peptide bonds of arginine and lysine residues and the chemical reagent cyanogen bromide cleaves peptide bonds on the carboxyl side of methionine residues. Evidently successful assessment of the localization of antigenic determinants will be dependent on applicable separatory techniques (5, 6) and immunoassays (3).

Protocol 11
Carbamylation

Reagents
- Potassium thiocyanate (Sigma P3011)
- 1 M acetic acid
- 0.1 M sodium phosphate buffer
- PBS pH 7.4

Method
1. Add solid potassium thiocyanate to a protein solution (5 mg/ml) in 0.1 M sodium phosphate buffer to give a final concentration of 0.5 M.
2. Maintain the pH between 8.0–8.5 during the addition of potassium cyanate by adding dropwise 1 M acetic acid.
3. Allow the reaction to proceed for 72 h at room temperature.
4. Dialyse modified protein against PBS.

Protocol 12
UV irradiation

Equipment and reagents
- Matched quartz cuvettes 1 cm^2 in cross-section
- PBS pH 7.4
- 366 nm + 254 nm UV irradiation source (Anderman & Co.)

Method
1. Dilute protein to 1 mg/ml in PBS and decant into matched quartz cuvettes.
2. Irradiate protein solution at a distance of 6 cm from a 366 nm + 254 nm UV irradiation source (this light source provides an average intensity of 10.5 watts/cm^2 and 17 watts/cm^2 respectively at 1 metre).
3. Remove from UV source at set time intervals: 30 min, 1 h, 2 h, 6 h, 12 h, and 24 h.

Protocol 13
Mild reduction and alkylation

Equipment and reagents
- Water-bath, Saran Wrap, borosilicate tubes
- Dithiothreitol (Sigma D5545)
- Iodoacetamide (Sigma I6125)
- 0.05 M Tris–HCl buffer pH 8.5
- 0.14 M saline

Protocol 13 continued

Method

1. Dilute protein (e.g. to 10 mg/ml) in Tris–HCl buffer.
2. Add dithiothreitol to a final concentration of 0.05 M and incubate at room temperature for 30 min.
3. Add iodoacetamide to a final concentration of 0.1 M and incubate at room temperature for 30 min.
4. Dialyse solution exhaustively against saline.

Protocol 14

Pepsin digestion

Equipment and reagents

- Borosilicate tubes, dialysis tubing
- Pepsin (Sigma P6887)

Method

1. Dialyse protein solution, e.g. 4 ml (20–25 mg/ml) against 0.1 M sodium acetate buffer pH 4.5. Use three changes of buffer (2 litre volumes) to ensure the optimum pH.
2. Add 1 mg pepsin for every 100 mg protein placed in a borosilicate tube. Seal the tube with Saran Wrap and incubate at 37 °C.
3. Monitor digestion over 2–24 h by removing aliquots (50–100 μl) which may be analysed using HPLC and/or SDS–PAGE. As an example, pepsin digestion of IgG1 will yield a pFc′ fragment after 2 h and the reaction may go to completion after 24 h.
4. Terminate reaction by adding 5 M Tris to raise the pH > 6.0.

8 Sodium dodecyl sulfate polyacrylamide gel electrophoresis (SDS–PAGE) and Western blotting

SDS–PAGE provides a robust and simple analytical tool for the separation of proteins with high resolution (22). Furthermore, when combined with Western blotting, it enables target epitopes to be localized to specific fragments, e.g. mAb A57H (2) exhibits reactivity to human IgG in ELISA and HA, and shows reactivity to an IgG pFc' fragment in Western blotting. Analytical polyacrylamide gels are prepared by polymerizing acrylamide with a bifunctional cross-linking agent (e.g. diallyltartardiamide or methylene bisacrylamide): the gel possessing a pore size inversely proportional to the concentration (w/v) of acrylamide (usually fixed) and to the ratio of acrylamide to cross-linking agent. Consequently, analytical gels can be modified in regard to their pore size and therefore induce a sieving effect on proteins and their corresponding fragments. Thus within an electric field, proteins may be separated by charge and size. However by incorporating SDS in a gel (and sample buffer), denaturation of proteins occur: SDS binds and unfolds proteins (ratio of SDS to protein 1.4:1) producing rod-shaped molecules of net negative charge which migrate to the anode on electrophoresis. Consequently any charge on a protein molecule is masked by the SDS and thus a protein is fractionated according to its molecular size/weight. In addition to an analytical gel (e.g. 5–15% acrylamide), a stacking gel of larger pore size (i.e. 3% acrylamide) is prepared above and in juxtaposition to the analytical gel. This facility permits the concentration of samples in a narrow band and subsequent resolution of proteins components in the analytical gel.

Polymerization of acrylamide gels requires free radicals: achieved by the addition of an initiator (ammonium persulfate) and a catalyst (tetramethylenediamine, TEMED). However it is important to note that gel mixtures should be degassed prior to adding ammonium persulfate: whilst a small amount of oxygen is required initially, the continued presence of oxygen prevents polymerization.

An estimate of the molecular weight of an unknown protein may be determined by electrophoresing protein markers of known molecular weight at the same time. This is simply achieved by applying samples to adjacent wells (generally 1–10 μg of protein per well) in the stacking gel (prepared by inserting a comb of appropriate tooth size during gel polymerization). Following electrophoresis, sample constituents give rise to discrete bands which can be visualized by staining, e.g. by Coomassie Brilliant Blue (sensitivity 0.5–1 μg/band) or silver stain (sensitivity is in the order of 1–5 ng/band). By plotting the logarithm ($\log_{10}$) of molecular weight against the relative mobility (R_f) of protein standards, an almost linear relationship is obtained from which the molecular weight of an unknown component can be interpreted.

SDS–PAGE readily permits the investigation of macromolecules which may also be reduced into individual protein/peptide components though disruption of disulfide bonds, e.g. by using 2-mercaptoethanol. Furthermore, membrane components can be solubilized in detergents (6) and analysed in SDS–PAGE, whilst native structures may be investigated in non-denaturing PAGE (i.e. no SDS treatment of proteins prior or during electrophoresis). Evidently, vertical slab gels may require 2–4 h to run whilst mini-gels (e.g. Mighty Small, Pharmacia Biotech) offer rapid separation (30–60 min). The subsequent transfer of resolved proteins to an immobilizing media such as nitrocellulose paves the way for probing antigenic determinants with mAbs. In essence, transfer is effected by a buffer tank transfer system or a semi-dry transfer system. The latter may be advantageous for the rapid and simultaneous multiple transfer of proteins. However, it is important to note that not all mAbs will function in Western blotting. Evidently the initial denaturation of proteins by SDS will destroy conformational epitopes and thus mAb reactivity will be biased to those antibodies recognizing linear determinants.

For membranes which will ultimately be visualized by radioactive or luminescent probing, the transfer of proteins may be ascertained initially by staining with India ink (Pelican). Ponceau S red staining of a single track may be a useful option if an enzyme substrate system is to be used for other lanes: this simple verification of effective transfer is worthwhile to reduce needless waste of immunoprobes, amplification reagents, and time.

Protocol 15

Sodium dodecyl sulfate polyacrylamide gel electrophoresis

Equipment and reagents

- Glass plates 20 cm × 15 cm
- Scalpel, gloves, ruler, forceps
- Acrylamide (Sigma A8887), diallytardiamide DATD (Aldrich 15686, 99+% Gold Label) or *N′N′*-methylene bisacrylamide (Sigma M7279, purity > 98%), tetramethylenediamine TEMED (Sigma T9281), SDS (BDH 10807 3J), Tris (BDH 10315 4M), ammonium persulfate (Sigma A3678), glycine (Sigma G8898), sucrose (BDH 10274), glycerol (Sigma G8773), water saturated butan-2-ol (Sigma B1888), bromophenol blue (Sigma B8026), 2-mercaptoethanol (Sigma M7154), Coomassie Brilliant Blue R-250 (BDH 44328), acetic acid (BDH 27013 BV), methanol (BDH 10158 5A), isopropanol (Sigma I-9516)
- Prepare the following solutions and store at 4 °C in dark bottles (shelf-life approx. four weeks):
- Solution A (3 M Tris–HCl, TEMED pH 8.6): dissolve 36.6 g Tris, 48 ml of 1 M HCl, 0.23 ml TEMED, and make up to 100 ml distilled water
- Solution B (acrylamide, DATD): dissolve 28.8 g acrylamide, 0.735 g DATD, make up to 100 ml distilled water, and filter using Whatman No. 1 or a 0.45 μm Millipore filter
- Solution C (1 M Tris–HCl pH 7.0, SDS): dissolve 19.2 ml of 1 M Tris–HCl pH 7.0 (121.14 g Tris in 900 ml distilled water, adjusted to pH 7.0 with concentrated HCl, and make up to 1 litre), 0.8 ml of 20% SDS (i.e. 20 g in 1 M Tris–HCl pH 7.0), 0.05 ml TEMED
- Sealing agarose (1%): dissolve 1 g agarose in 100 ml of 0.375 M Tris–HCl pH 8.9 (45.43 g Tris in 900 ml distilled water, adjusted to pH 8.9 with concentrated HCl, and make up to 1 litre), containing 0.1% SDS in 0.375 M Tris–HCl pH 8.9
- Electrode buffer (0.025 M Tris–glycine buffer): dissolve 6 g Tris, 28.8 g glycine, 10 ml of 20% SDS in distilled water, and make up to 2 litres distilled water
- Ammonium persulfate (0.14% solution): dissolve 0.14 g in 100 ml distilled water (prepare a fresh batch prior to use)
- Stacking gel (3% acrylamide): mix 2.55 ml solution B, 3 ml solution C, 6.45 ml distilled water, 12 ml ammonium persulfate
- Sample buffer (2 ×): for a non-reducing buffer, mix 2 ml of 20% SDS, 1 ml of 1 M Tris pH 7.0, 1 ml of 60% sucrose (i.e. 6 g in 10 ml distilled water), 15 ml distilled water, and 100 μl bromophenol blue solution. This is prepared by adding 10 ml glycerol to 10 ml electrode buffer in a Sterilin Universal, and adding 1 g of bromophenol blue powder. A reducing buffer consists of the above components plus 1 ml of 2-mercaptoethanol.
- Staining solution (Coomassie blue): dissolve 0.3 g Coomassie blue in 1 litre solvent, 100 ml glacial acetic acid, 250 ml isopropanol, 650 ml distilled water
- Destain solution: mix 50 ml glacial acetic acid, 100 ml isopropanol, 850 ml distilled water; add 30 ml methanol to 100 ml destain solution in order to obtain the original size of gel

Protocol 15 continued

Method

1. Clean glass plates using detergent, water, methanol, and then acetone.
2. Grease side spacers (1/2 the width) and clamp plates together with bulldog clips.
3. Heat sealing agarose until clear and seal inner sides of side spacers and base spacer (if required). Allow 40 min.
4. Prepare analytical gel[a] of appropriate percentage: as a rough guide use a 15% gel for proteins of 150 000, 7–10% for proteins of 14 000–100 000, 5% for proteins of 10–60 000 Daltons.

	5%	7%	10%	12.5%	15%
Solution A (ml)	5	5	5	5	5
Solution B (ml)	7.05	9.9	14.3	17.7	21.2
20% SDS (ml)	0.2	0.2	0.2	0.2	0.2
Distilled water (ml)	17.25	14.4	10	6.6	3.1
Amm. persulfate (ml)*	10	10	10	10	10

*Ammonium persulfate is only added after above solutions have been mixed and degassed.

5. Pour analytical gel and layer 1 cm of distilled water gel (alternatively use isobutanol) on the top surface by dripping it into the mould using a 5 ml syringe: remove excess water prior to adding the stacking gel. Allow to set.
6. Prepare stacking gel, pour, insert appropriate comb, and allow to set.
7. Prior to adding electrode buffer, remove the base spacer and ensure that any air bubbles are removed (achieved by using a 10–20 ml syringe and needle bent at into a 'V' shape and depressing plunger to expel electrode buffer).
8. Add 50 μl of protein solution to 50 μl of sample buffer[b] in a 0.5 ml Eppendorf and boil for 2 min. Apply an appropriate volume, e.g. 20 μl into each well: about 1–2 μg of protein or 10 μg if overloading required to assess sample purity.
9. Run at constant voltage 150 volts for 3–4 h until the bromophenol blue is approx. 1 cm from the anodic (lower buffer chamber) end of the gel. Perform electrophoresis at room temperature since SDS precipitates in the cold.
10. Remove glass plates from acrylamide gel. This may be accomplished by teasing the edges of one plate away from the other with a spatula. Gently slide gel into staining tank (note that low percentage acrylamide gels are easily damaged).
11. Place in Coomassie blue[c] for 2 h[d] and then destain to remove excessive background staining. Determine the relative mobilities[e] of proteins and preserve gel by employing a gel dryer, e.g. Bio-Rad Model 583 and/or photographically record using a gel documentation system, e.g. Biogene.

[a] Acrylamide/diallyltartardiamide DATD (or *N'*,*N'*-methylene bisacrylamide) are neurotoxins although the polymerized product is non-toxic. Stock solutions should be stored at 4 °C and if the acrylamide should crystallize, warm the solution gently (< 40 °C); stock solutions may polymerize explosively if rapidly heated.

Protocol 15 continued

[b] The sample buffer described, includes sucrose which increases the density of sample preparation and prevents mixing of adjacent wells and the upper buffer solution. It is important that two adjacent wells are left blank if reduced and non-reduced samples are electrophoresed on the same gel. Note that if a protein sample contains a high salt concentration > 0.2 M (especially potassium), dialyse against 0.1% NaCl.

[c] Silver staining provides an increase in sensitivity for detecting proteins in polyacrylamide gels and can be visualized within 1 h (e.g. silver stain plus kit, Bio-Rad). Coomassie blue stained gels can be destained in 40% methanol/10% acetic acid and subsequently silver stained.

[d] For protein electroelution, stain gel with Coomassie blue for a minimum period, e.g. < 10 min and destain to obtain a visible protein band. Carefully cut out the band of interest and insert into dialysis tubing containing a Tris–glycine buffer plus 0.1–1% (w/v) SDS (0.1 M sodium phosphate buffer pH 7.8 plus 0.1% SDS). Displace most of the fluid and air bubbles, and seal the dialysis tubing. Finally submerge the dialysis sac in a submarine electrophoresis tank, e.g. Pharmacia GNA-100 containing the same buffer and run at 25–100 V (20–150 mA) for 3–20 h and then momentarily (e.g. 30 sec) reverse the polarity to remove any protein on the dialysis membrane. Remove the gel and any fragments (decant and centrifuge if necessary) and dialyse against a desired buffer, or distilled water containing SDS (if required for solubility) for freeze-drying.

[e] Molecular weight markers either pre-stained (Sigma SDS-7B: molecular weights 25–180 000) or standard (Sigma SDS-6H: molecular weights 30–200 000 or SDS-7: molecular weights 14–70 000) may be used for comparison. Running at least two markers (and/or a personally designed set) containing some common components, e.g. BSA, OVA, will ensure correct identification of standard markers and consequently the molecular weight of the protein band of interest. Both Sigma SDS-6H, SDS-7 contain bovine albumin (66 000), egg albumin (45 000), and erythrocyte carbonic anhydrase (29 000). Note that heavily glycosylated proteins bind less SDS and therefore estimations will be inaccurate. A given glycosylated protein will migrate more slowly that its non-glycosylated counterpart. The R_f value of a band is obtained by dividing the migration distance from top of the analytical gel to the centre of the protein band by the migration distance of the tracking dye, e.g. bromophenol blue (again from the top of the analytical gel).

Protocol 16

Electroblotting (tank transfer system)

Equipment and reagents

- Electroblot EC215 transfer system (E-C Apparatus Corporation)
- Whatman 3MM filter paper
- Transfer buffer: 25 mM Tris, 192 mM glycine, 20% (v/v) methanol pH 8.3
- Nitrocellulose membrane (NCM) or polyvinylidene fluoride membrane (PVDF), Immobilon-P Cat. IPVH 15150, Millipore (distributed though Sigma P2563)

Method

1 Follow Protocol 15 up to step 10 and tease apart glass plates, but ensure that the acrylamide gel remains on one plate.

Protocol 16 continued

2 Remove stacking gel (and any residual agarose) from the analytical gel and place a sheet of membrane (NCM/PVDF) adjacent to the analytical gel (ensure air bubbles are removed from between gel and membrane). Notch top left-hand corner of gel and membrane to facilitate orientation of membrane after transfer.

3 Carefully remove gel and NCM sandwich and insert into the clip of the transfer system; with the nitrocellulose membrane adjacent to the anode. Use Whatman No. 1 filter paper (pre-soaked in transfer buffer) for packing between the Scotch Brite pads to ensure that the gel and NCM remain in contact.

4 Close the grid, place phenolic grid over electrodes, put assembly into transfer tank, and turn on circulation unit.

5 Blot for 1–2 h at 0.6 mA (approx. 25 volts); for an overnight blot, set at 0.1–0.2 mA (approx. 10–20 volts). The transfer should be conducted at < 5 watts to minimize heating of the buffer and to avoid excessive bubble formation.

Protocol 17

Electroblotting (semi-dry transfer system)

Equipment and reagents

- Sartorius Semi-dry Electroblotter SM 17556 (alternatively Hoefer Semi-dry system)
- Anode buffer 1: 0.3 M Tris, 20% (v/v) methanol pH 10.4
- Anode buffer 2: 25 mM Tris, 20% (v/v) methanol pH 10.4
- Cathode buffer: 25 mM Tris, 40 mM 6-amino-*n*-hexanoic (caproic) acid, 20% methanol pH 9.4 (note: glycine can be substituted for 6-amino-*n*-hexanoic)

Method

1 Follow Protocol 15 up to step 10 and tease apart glass plates, but ensure that the acrylamide gel remains on one plate.

2 Place two layers of filter paper (16 × 16 cm) soaked in anode buffer 1 (30 sec) on the anodic graphite plate followed by one layer of filter paper soaked in anode buffer 2.

3 Next place an equal size sheet of membrane wetted with distilled water onto the layers of filter paper and then the analytical acrylamide gel. Notch gel and membrane to facilitate orientation of membrane after transfer and remove all air bubbles by carefully using a roller/pipette.

4 Place one layer of filter paper soaked in the cathode buffer on top of the acrylamide gel and finally the dialysis membrane soaked in water. This complete assembly constitutes a single transfer stack. Further transfer stacks (up to six) can be accommodated if required.

Protocol 17 continued

5 Finally place two layers of filter paper soaked in cathode buffer on top of the membrane.

6 Position cathode plate lid and connect electroblotter to power supply (6–8 V/cm, 0.8–1.2 mA/cm^2) for 1–2 h at room temperature.

Protocol 18

Probing of membranes

Reagents

- PBS/Tween
- Ponceau S red (Sigma P7170)

Method

1 Wash membrane twice in PBS/Tween 20 (0.05%, v/v) for 5 min. Alternatively remove one track and wash in PBS/Tween.

2 Immerse membrane in 0.1% (v/v) black India ink (Pelican) in PBS/Tween for < 30 min at room temperature (the use of a rocking platform ensures even staining). Alternatively immerse in Ponceau stain (1% (v/v) in PBS) for 10–15 min.

3 Wash membrane two to three times in PBS/Tween for 5 min.

Protocol 19

Detection: enzyme–substrate deposition

Equipment and reagents

- See *Protocol 8*

Method

1 Incubate membrane in 3% (w/v) BSA (alternatively bovine haemoglobin, BDH) in PBS for 1 h to block remaining protein binding sites. Note that dried Immobilon-P, following electrophoretic transfer does not require blocking (see Millipore User Guide P15372).

2 Wash blot twice in 1% BSA/PBS for 5 min.

3 Incubate membrane with monoclonal antibody (e.g. 10^{-3} dilution of ascites) in 1% BSA/PBS for 2 h at room temperature (alternatively overnight at 4 °C).

4 Wash blot five times in PBS/Tween.

5 Incubate for 1–2 h at room temperature with HRP-conjugated anti-mouse immunoglobulin and continue as for slot-blotting (*Protocol 8*) for enzyme substrate deposition or ECL chemiluminescence.

Protocol 20
Detection: radioactive

Reagents

- Anti-mouse Ig (GAM), F(ab')2 fragment, ^{125}I-labelled species-specific whole antibody (from sheep), code IM 1310, Amersham

Method

1. Incubate membrane in 3% (w/v) BSA (alternatively bovine haemoglobin, BDH) in PBS for 1 h to block remaining protein binding sites.
2. Wash blot twice in 1% BSA/PBS for 5 min.
3. Incubate membrane with monoclonal antibody (10^{-3} dilution of ascites) in 1% BSA/PBS for 2 h at room temperature (alternatively overnight at 4 °C).
4. Wash blot five times in PBS/Tween.
5. Incubate for 1–2 h at room temperature with ^{125}Iodine anti-mouse immunoglobulin (an activity of 0.1 μCi/ml of incubation medium is adequate).
6. Wash extensively in PBS/Tween, dry NCM, and insert into a light-proof cassette against an X-ray film (Kodak X-Omat RP). Securely close cassette. Exposure times may require empirical investigation, e.g. overnight at room temperature may prove sufficient. Develop X-ray film in Kodak D-19 developer for 4 min under red light in dark-room. Rinse in tap-water and fix ('Unifix' fixer) for 4 min. Wash in tap-water and dry film.

9 Protein determination

9.1 UV absorption

The concentration of a given immunoglobulin (or protein) in solution, may be ascertained via its absorption of ultraviolet (UV) light. In using a spectrophotometer, e.g. Pharmacia Ultrospec III/Pye Unicam SP500 set at 280 nm, a protein solution (in a quartz cuvette) will maximally absorb UV light: more specifically the aromatic residues such as tryptophan and tyrosine, and to some extent phenylalanine. In general, the spectrophotometer is referenced or 'blanked' using the buffer employed to dissolve/dilute the protein. A convenient approach is to use matched quartz cuvettes, e.g. 1 cm^2 in cross-section, containing reference and protein solutions respectively. In using this particular method, the absorbance value obtained can be divided by the extinction coefficient $E^{1\%}{}_{1\ cm}$ recorded for a given protein (defined as the absorbance of a 1% solution of a protein in a 1 cm light path) to give a concentration (5, 6). The main advantages of this approach include its simplicity along with the fact that it is non-destructive in the short-term and that protein samples are recoverable. However the protein solution must be pure. Absorbance at 215 nm or 230 nm can be used for monitoring synthetic peptides which may not contain aromatic residues.

Remember that an $E^{1\%}_{1\ cm}$ represents a value obtained for 1 g/100 ml (or 10 mg/ml) solution. It is therefore more convenient to use the extinction coefficient which has simply been divided a factor of 10, thus giving a protein solution in mg/ml.

Example: $E^{1\%}_{1\ cm}$ for human IgG is 14.3

Concentration of IgG sample = 0.50/1.43 (absorbance obtained at 280 nm)
= 0.35 mg/ml

Protocol 21

UV absorption

Equipment and reagents

- 100 ml volumetric flask
- 1.5 ml Eppendorfs
- Variable pipettes, e.g. 1 ml, 50–200 μl
- Matched quartz cuvettes (1 cm^2 in cross-section)
- UV spectrophotometer, e.g. Amersham Pharmacia Biotech Ultrospec III /Pye Unicam SP500 Series 2
- PBS (or buffer in which protein is dissolved/diluted)
- Protein (e.g. albumin, bovine A-2153, Sigma): as an example prepare a 10 mg/ml solution and prepare concentrations of 0.1, 0.25, 0.5, 1.0, 2.5 mg/ml

Method

1. Centrifuge protein solution at 1000 r.p.m. (Sigma 302 bench centrifuge) for 5 min to remove any suspended particles.
2. Turn on spectrophotometer at least 10 min prior to use.[a] Set at 280 nm.
3. Blank by pipetting 1.0 ml buffer into a cell (hold by frosted glass sides) and remove any fluid from the external sides of a cell (cuvette) with tissue. Orientate cuvette correctly and insert into spectrophotometer (if only a single cell is available: use this for blanking and for determining absorbance).
4. Read absorbance[b] which should be within 0.1–1.5 units. If an absorbance is > 2.0, dilute an aliquot of stock solution 1/2, 1/10.[c]
5. Calculate protein concentration as above[d,e,f] (method is not effective if the protein is in suspension, e.g. cell membranes: see fluorescamine method). For BSA (weighed) solutions at 0.1, 0.25, 0.5, 1.0, 2.5 mg/ml the following concentrations were derived using the $E^{1\%}_{1\ cm}$: 0.16, 0.27, 0.47, 0.95, and 2.07 mg/ml. This method overestimated and underestimated at 0.1 and 2.5 mg/ml respectively.

[a] Follow standard operating procedure for selected spectrophotometer, i.e. turn on appropriate lamp if necessary: deuterium 186–335 nm, tungsten 335–1000 nm.

[b] Steps 3 and 4 in addition can be repeated at 260 nm: the absorbance ratio 260/280 gives an indication of contamination (e.g. from nucleic acids) and should be below 0.6.

[c] If the protein solution requires diluting to obtain an absorbance value < 2.0, remember to correct the value obtained by multiplying by the dilution factor. Alternative cells, i.e. 2 mm path length may be used, only in this case the absorbance must be multiplied by a factor of 5.

Protocol 21 continued

[d] The following formula may be used if extinction coefficients are unknown: 1.55 × absorbance (280 nm) − 0.77 × absorbance (260 nm). Some useful extinction coefficients include: $E^{1\%}{}_{1\,cm}$ 6.7, bovine serum albumin; 1.22, Ig Fc; 1.38, IgG pFc'.

[e] Certain substances (e.g. detergents Nonidet P-40 and Triton X-100, nucleic acids, and certain lipids) may interfere with protein estimations since they absorb in the UV range. In addition, the preservative sodium azide absorbs strongly at 215–230 nm. If collecting fractions, e.g. eluate from affinity chromatography, a 'bubble test' (i.e. shaking a tube vigorously) may be used to determine whether protein is present (a froth generally appears) and protein concentrations can then be determined as above.

[f] Sensitivity is in the order of 0.1 mg/ml: clearly interpretation of a protein concentration is dubious at or below 0.1 mg/ml. For greater sensitivity, a colorimetric method should be used, e.g. Folin phenol method (modified from Lowry: Johnstone and Thorpe 1987, sensitivity approx. 5 μg/ml) or the Pierce system (bicinchonic acid: does not form precipitates with detergents and sensitivity is in the order of 1 μg/ml).

9.2 Fluorescence

The use of a fluorescamine assay (23) to determine protein concentration (effectively a measure of nitrogen content) is particularly useful for the assessing the protein concentration of a cell membrane preparation.

Protocol 22

Fluorescence

Equipment and reagents

- Fluorimeter Model 8-9: turn on 30 min prior to use
- Albumin (bovine serum, globulin free, Fraction V; A-7638 Sigma)
- Fluorescamine (F-9878 Sigma)
- Sodium borate (BDH): 0.1 M sodium borate buffer pH 9.2

Method

1. Prepare BSA standards 0, 1, 5, 10, 50, 100 μg/ml in distilled water.
2. Dilute sample 1/10, 1/50.
3. Add 0.5 ml of amine (i.e. protein solution to be tested and standard preparations) to 0.75 ml of 0.1 M sodium borate buffer pH 9.2.
4. Add 0.25 ml fluorescamine (0.2 μg/μl in acetone) and vortex.
5. Read fluorescence (fluorimeter) within 5 min. As an example, for a titration series of BSA 100, 50, 10, 5, 1, and 0 μg/ml, the following intensity units (× 10 setting) were obtained: +503, +316, +051, +009, −004, −012.

Acknowledgements

The protocols presented in this chapter have been gained from practical experience with colleagues gained in a number of laboratories. My grateful thanks to the following: Frank Hay, Alan Johnstone, Terry Poulton, Andy Soltys, Shantha Perera, Tim Chambers, Matt Walker, Margaret Goodall, Roy Jefferis, and Dorothy McDonald.

References

1. Blottiere, H. M., Daculsi, G., Anegon, I., Pouezat, J. A., Nelson, P. N., and Passuti, N. (1995). *Biomaterials*, **16**, 497.
2. Nelson, P. N., Fletcher, S. M., De Lange, G. G., Van Leeuwen, A. M., Goodall, M., and Jefferis, R. (1990). *Vox. Sang.*, **59**, 190.
3. Nelson, P. N., Goodall, M., and Jefferis, R. (1994). *Immunol. Invest.*, **23** (1), 39.
4. Rose, N. R., de Macario, E. C., Folds, J. D., Lane, H. C., and Nakamura, R. M. (1997). *Manual of clinical laboratory immunology* (5th edn). American Society for Microbiology, Washington DC.
5. Johnstone, A. and Thorpe, R. (1987). *Immunochemistry in practice* (2nd edn). Blackwell Scientific Publications, Oxford.
6. Hudson, L. and Hay, F. C. (1980). *Practical immunology* (2nd edn). Blackwell Scientific Publications, London.
7. Nelson, P. N., Westwood, O. M., Jefferis, R., Goodall, M., and Hay, F. C. (1997). *Biochem. Soc. Trans.*, **25**, 373S.
8. Tam, J. P. and Zavala, F. (1989). *J. Immunol. Methods*, **124** (1), 53.
9. Nelson, P. N. and Chambers, T. J. (1995). *Hybridoma*, **14**, 91.
10. Hanly, W. C., Artwohl, J. E., and Bennett, B. T. (1995). *ILAR J.*, **37** (3), 98.
11. Atherton, E. and Sheppard, R. C. (1989). *Solid phase peptide synthesis: a practical approach.* IRL Press.
12. Kyte, J. and Doolittle, R. F. (1982). *J. Mol. Biol.*, **157**, 105.
13. Liu, F. T., Zinnecker, M., Hamoaka, T., and Katz, D. H. (1979). *Biochemistry*, **18** (4), 690.
14. Nelson, P. N., Pringle, J. A. S., and Chambers, T. J. (1991). *Calcif. Tissue Int.*, **49** (5), 317.
15. Nelson, P. N., Fletcher, S. M., MacDonald, D., Goodall, D. M., and Jefferis, R. (1991). *J. Immunol. Methods*, **138**, 57.
16. Galfre, G., Howe, S. C., Milstein, C., Butcher, G. W., and Howard, J. C. (1977). *Nature*, **266**, 550.
17. Sambrook, J., Fritsch, E. F., and Maniatis, T. (ed.) (1989). *Molecular cloning, a laboratory manual* (2nd edn). Cold Spring Harbor Laboratory Press.
18. Stevens, P. W., Hansberry, M. R., and Kelso, D. M. (1995). *Anal. Biochem.*, **225**, 197.
19. Portsmann, B., Porstmann, T., Nugel, E., and Evers, U. (1985). *J. Immunol. Methods*, **79**, 27.
20. Sternberger, L. A. (1986). *Immunocytochemistry* (3rd edn). John Wiley and Sons, New York.
21. Chilosi, M., Gilioli, E., Lestani, M., Menestrinz, F., and Fiore-Donati, L. (1988). *J. Pathol.*, **156**, 251.
22. Weber, K. and Osborn, M. (1969). *J. Biol. Chem.*, **244**, 4406.
23. Udenfriend, S., Stein, S., Bohlen, P., Dairman, W., Leingruhei, W., and Weigel, M. (1972). *Science*, **178**, 680.

Chapter 8

Epitope mapping of carbohydrate binding proteins using synthetic carbohydrates

U. J. Nilsson and G. Magnusson
Organic Chemistry 2, The Lund Institute of Technology, University of Lund, PO Box 124, S-221 00 Lund, Sweden.

1 Introduction

Specific recognition of carbohydrates by proteins is involved in a number of important biological phenomena. For example, carbohydrates may act as cell surface antigens, e.g. in the ABO, Lewis, P, and I blood group systems or as tumour-specific antigens; they control growth and tissue development in an embryo, and are involved in fertilization; they regulate nerve growth and neural cell adhesion; and they direct the recruitment of leukocytes to damaged tissue (1, 2). Furthermore, carbohydrates often act as recognition sites for pathogenic bacteria and viruses (3). As a consequence, detailed knowledge on a molecular level concerning the recognition mechanisms and driving forces would open up possibilities to specifically interfere with, or promote, biological phenomena stemming from carbohydrate–protein interactions (i.e. development of therapeutics that target carbohydrate-binding proteins) (4).

Characterization of the carbohydrate ligand of carbohydrate-binding proteins can be divided into three levels of refinement. First the oligosaccharide sequences involved in binding are determined, then the critical monosaccharide residues required for binding are investigated, and finally the submolecular details of the interaction between the carbohydrate and the protein (i.e. steric and electronic complementarity through hydrogen bonds, electrostatic interactions, hydrophobic interactions, etc.) are mapped out. The third level of refinement, characterization of submolecular details of interaction (5), is usually referred to as the *epitope mapping*. The focus of this chapter will be on how systematic synthesis and biological evaluation of libraries/collections of chemically modified carbohydrates can be used to map the epitopes of carbohydrate-binding proteins. Section 2 discusses why certain carbohydrate analogues are synthesized for use as epitope mapping probes, Section 3 describes methods to perform binding studies, while Section 4 exemplifies how data obtained from binding studies can be interpreted in terms of submolecular interactions.

2 Synthetic carbohydrate analogues

A rough indication of which monosaccharides that constitute an absolute prerequisite for binding (i.e. the 'key sugars' for lectins and enzymes or 'immunodominant sugars' for antibodies) (6) by the protein can be obtained by screening against a large number of natural glycoconjugates (glycolipids, glycoproteins, and soluble oligosaccharides). Ideally, when the 'key sugars' have been identified, X-ray crystallography of the protein–carbohydrate complex would reveal the submolecular details of interaction, such as how the 'key sugars' are interacting and if any monosaccharide residues flanking the 'key sugars' are contributing to the interaction with the protein. Unfortunately, very few carbohydrate–protein complexes have been crystallized and, in the absence of an X-ray structure, alternative approaches have to be considered to further refine the picture of a molecular interaction. Evaluation of synthetic carbohydrate analogues as epitope probes is an efficient and often employed approach. Carbohydrate analogues include fragments of natural oligosaccharides, deoxy, deoxyfluoro, methoxy, deoxyamino, carboxy, thio, and conformationally restricted analogues. The chemical synthesis of saccharide analogues and fragments has been reviewed as parts of a chapter on the synthesis of neoglycoconjugates (7). The systematic synthesis of modified saccharides for use as probes in binding studies is a relatively time-consuming task. However, saccharide analogues are generally stable, except for deoxyfluoro analogues, if stored under dry and cold conditions. The labour put into chemical synthesis will thus be richly rewarded, since a library of synthetic saccharides can be used for many years in epitope mapping of various proteins.

2.1 Oligosaccharide fragments

The importance of the individual monosaccharide residues for the recognition of an oligosaccharide by a protein can be probed with synthetic fragments of the oligosaccharide. Such fragments are generally synthesized by conventional organic synthetic methodology, which has been the subject of several exhaustive reviews (8, 9). An example of a complete library of synthetic di → pentasaccharide fragments (10, 11) of the natural glycolipids belonging to the P blood group antigens, is shown in *Figure 1*. These compounds were used for mapping of *inter alia* the binding epitope of the adhesin protein expressed on the surface of pathogenic *Escherichia coli* bacteria, colonizing the human urinary tract (12, 13).

2.2 Deoxy analogues

The use of monodeoxy saccharide analogues as inhibitors of carbohydrate-recognizing proteins will reveal which of the hydroxyl groups in the natural substrate that are necessary for complex formation. Such a hydroxyl group has been termed a 'key polar group' by R. U. Lemieux (5), the pioneer of epitope mapping of carbohydrate-recognizing proteins (antibodies and plant lectins). We later synthesized all the monodeoxy galabiosides (14–16) (*Figure 2*) and used them

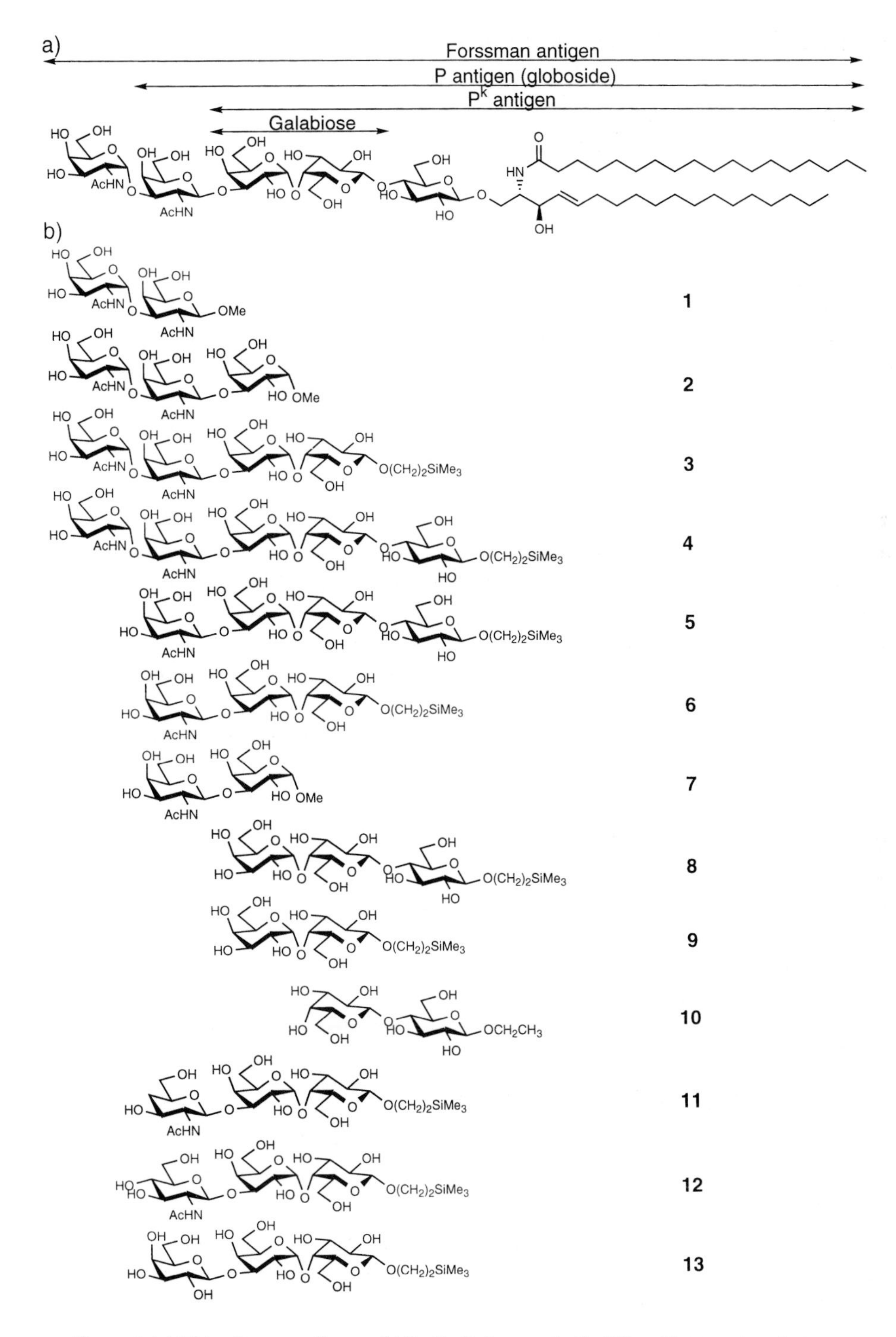

Figure 1 (a) P blood group antigens. (b) Synthetic fragments (**1–10**) and fragment analogues (**11–13**) of the Forssman pentasaccharide useful for epitope mapping of proteins recognizing P blood group antigens.

for mapping the hydrogen bonding network between galabioside residues and various galabiose-recognizing proteins of bacterial origin (see Section 4).

2.3 Deoxyfluoro analogues

Monodeoxyfluoro analogues reveal the directionality of hydrogen bonds necessary for complex formation, since the fluoro group can act as a proton acceptor but not as a donor in a hydrogen bond. A case where a monodeoxy analogue is inactive, but the corresponding deoxyfluoro analogue is active, implies that the hydroxyl group in the parent natural saccharide is *accepting* a proton in a hydrogen bond to the protein. Monodeoxyfluoro galabiosides (15, 16) useful for mapping out the hydrogen bonding networks between galabiosides and proteins are shown in *Figure 2*.

2.4 *O*- and *C*-methyl analogues

O-Methylated saccharides have also been used to determine the directionality of hydrogen bonds, since the methoxy group can act as a proton acceptor but not as a donor in a hydrogen bond (analogously to the fluoro group). In addition, *O*-methyl analogues have also been used to detect those saccharide hydroxyl groups that are situated in the periphery of the combining site and thus do not interact strongly with the protein (17). Removal of such a hydroxyl group is not expected to affect complexation, while *O*-methylation will abolish complexation due to increased steric crowding.

C-Methyl analogues of saccharides (i.e. saccharides where a hydroxyl group has been replaced by a methyl group; cf. *Figure 2*, compound **25**) are also useful in epitope mapping. The hydroxyl and methyl groups are similar in size, but

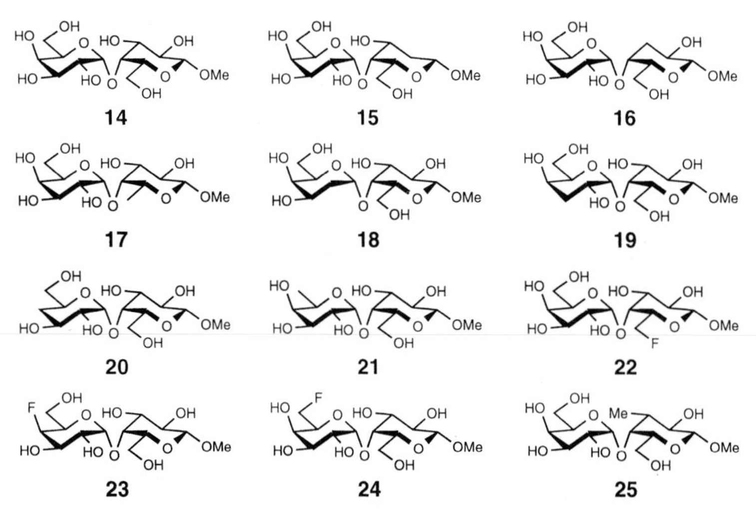

Figure 2 Monodeoxy, monodeoxyfluoro, and *C*-methyl analogues (**15–25**) of methyl β-D-galabioside (**14**) useful for mapping of hydrogen bonding networks with galabiose-recognizing proteins.

have very different polarity. A *C*-methyl saccharide analogue can therefore be used to investigate if a certain hydroxyl group in the parent saccharide is recognized by a polar or non-polar pocket in the protein.

2.5 Amino and carboxy analogues

Deoxyamino and carboxy analogues are used to detect the presence of charged amino acid residues in the combining site of the protein. A deoxyamino analogue may show enhanced affinity if the amino group is able to form a charge–charge interaction with a glutamate or aspartate of the protein. Analogously, a carboxy analogue forming a charge–charge interaction with lysine or arginine residues displays enhanced affinity for the protein. *Figure 3* shows amino and carboxy galabioside analogues (18) useful for mapping of charge–charge interactions and hydrogen bonding networks between galabiosides and proteins.

2.6 Conformationally restricted or altered analogues

Conformationally restricted saccharide analogues may be obtained by introducing methylene acetals bridging two hydroxyl groups of a saccharide (19, 20) or by preparing lactams as analogues of ganglioside lactones (21). Small changes in the conformation of a saccharide can be induced by replacing interglycosidic oxygen atoms with other atoms, such as sulfur (22). Observing activity or inactivity of conformationally restricted/altered analogues may shine light on requirements on the spatial arrangements of hydroxyl groups interacting with the protein. *Figure 3* shows a thiodisaccharide galabioside (22) analogue and a methylene acetal-bridged galabioside analogue (20), which are useful for studying conformational aspects on interactions between galabiosides and proteins.

2.7 Combinatorial synthesis of carbohydrate libraries

The approaches described in Sections 2.1–2.6 are essentially examples of a traditional medicinal chemistry methodology, where rational design and multistep synthesis provides a library of well-defined analogues (such as the galabiosides **14–30**), which answer specific questions when evaluated in a biological assay. Recently emerged, and conceptually different, random approaches based on

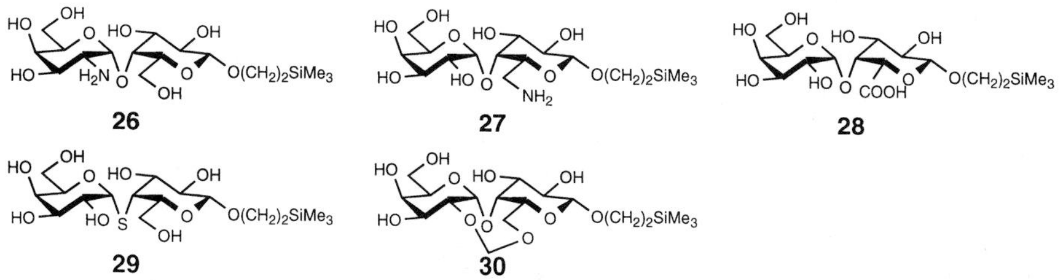

HO OH O HO OH HO H_2N O O $O(CH_2)_2SiMe_3$ OH **26**

HO OH O HO OH HO HO O O $O(CH_2)_2SiMe_3$ NH_2 **27**

HO OH O HO OH HO HO O COOH O $O(CH_2)_2SiMe_3$ **28**

HO OH O HO OH HO HO S O $O(CH_2)_2SiMe_3$ OH **29**

HO OH O HO OH HO O O O O $O(CH_2)_2SiMe_3$ **30**

Figure 3 Amino and carboxy analogues (**26–28**) of 2-(trimethylsilyl)ethyl β-D-galabioside (**9**) useful for mapping of electrostatic interactions with galabiose-recognizing proteins. A thiodisaccharide (**29**) analogue and a conformationally restricted methylene acetal-bridged analogue (**30**) of 2-(trimethylsilyl)ethyl β-D-galabioside (**9**) useful for studying conformational aspects of galabiose bound by a protein.

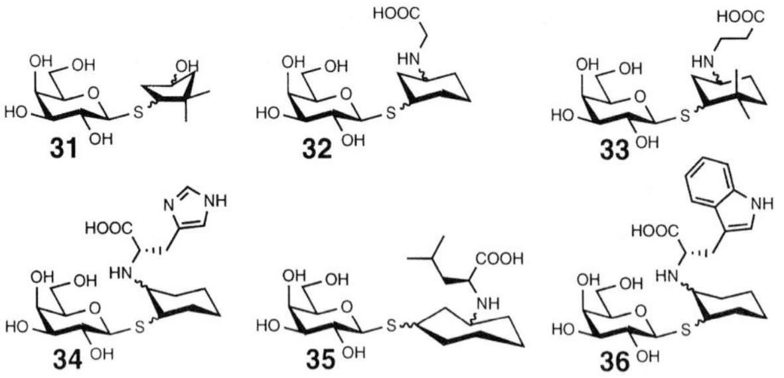

Figure 4 Members of a 1-thio-β-D-galactoside library useful for exploring space and chemical funtionality in and surrounding a combining site of a galactose-recognizing protein.

screening chemical libraries, prepared via combinatorial synthesis (23), can be regarded as tools for mapping the epitope of carbohydrate binding proteins. In combinatorial chemistry a large number of random and structurally diverse compounds are synthesized simultaneously. The collection or library of compounds obtained are subsequently screened for activity against a target protein. The identification and characterization of an active compound will provide information on what structures fit into the combining site of the protein. One approach is screening of oligosaccharide libraries against a target protein, which may reveal novel saccharides that are recognized (24). Another approach is to attach diverse non-carbohydrate structures at varying positions on a retained 'key sugar' and thus exploring the space and chemical functionalities in and surrounding the combining site (24, 25). *Figure 4* shows members of a library of 1-thio-β-D-galactosides carrying diverse non-carbohydrate structures useful for exploring a combining site of a galactose-recognizing protein.

3 Measuring carbohydrate–protein interactions

A complete review of the literature concerning epitope mapping of carbohydrate-binding proteins using synthetic carbohydrates is beyond the scope of this chapter. Instead, in order to illustrate how mapping of submolecular details is performed using chemically modified carbohydrates, our investigations on the binding specificities of human uropathogenic *E. coli* bacteria will be used as representative examples of how binding assays and data interpretation can be performed. Most uropathogenic *E. coli* strains associated with pyelonephritis (severe infection in the kidney) carry hair-like appendages called P pili (26). An αGal*p*(1–4)βGal*p* (galabiose) recognizing adhesin protein, PapG, is located at the tip of the P-pilus (12) and mediates adhesion to uroepithelial cells presenting αGal*p*(1–4)βGal*p*-containing glycolipids. The PapG adhesins of different pyelonephritic *E. coli* strains display small variations in their binding to various cells (27). The carbohydrate binding epitopes of two PapG adhesins, $PapG_{J96}$ and $PapG_{AD110}$, expressed by two different clinical isolates, J96 and AD110, were mapped using a library of saccharide analogues (**1–30**) in haemagglutination inhibition (28, 29) ($PapG_{J96}$ and $PapG_{AD110}$) and competitive ELISA (30) ($PapG_{J96}$) experiments. The haemagglutination inhibition and ELISA, described in Sections 3.1–3.2, are *competitive* assays, i.e. carbohydrate analogues are used in inhibition of a protein binding to an immobilized receptor. Methods for the *direct* measurement of carbohydrate recognition exist and will be briefly discussed in Section 3.3.

3.1 Haemagglutination inhibition

3.1.1 Principles

Aggregation of erythrocytes by species presenting carbohydrate-binding proteins (bacteria, viruses, toxins, antibodies, etc.) is termed haemagglutination (HA). Bacterial adhesins, such as the *E. coli* PapG adhesins mentioned above, bind to saccharides presented on the surface of erythrocytes thus causing haemagglutination, which results in the formation of a diffuse sheet of clumped cells. Different erythrocytes present different saccharides on their surface and HA experiments should therefore be performed with erythrocytes presenting saccharides that are recognized by the protein of interest. The erythrocytes will sediment (to form a sharp red pellet) in the presence of an analogue which inhibits the adhesin (and consequently haemagglutination), thus enabling identification of potent inhibitors (31). This phenomenon is termed haemagglutination inhibition (HAI). The inhibitor concentration giving 50% inhibition of haemagglutination is referred to as the IC_{50} value, which is obtained by performing the HAI with a dilution series of the inhibitor. Plastic particles coated with an active saccharide have been used as an alternative to erythrocytes (32). *Protocols 1* and *2* outline detailed standard procedures for a haemagglutination reaction and a haemagglutination inhibition using an *E. coli* strain carrying $PapG_{J96}$ adhesin and human blood group A^+ erythrocytes (which present the P antigen, *Figure 1a*).

Protocol 1

Haemagglutination reaction (35)

Equipment and reagents

- ACD Solution B Vacutainer tubes (Becton Dickinson)
- Bacterial strain HB101/pPAP5: the plasmid pPAP5 contains the entire *pap* gene cluster cloned from the human urinary tract *E. coli* isolate J96 (33); transformation of pPAP5 into the non-piliated laboratory *E. coli* host strain HB101 confers on the resulting strain HB101/pPAP5 the ability to produce P pili (34)
- Microtitre plate with pointed well bottoms (Costar)
- Blood group A^+ human erythrocytes
- Phosphate-buffered saline (PBS): 120 mM NaCl, 2.7 mM KCl, 10 mM phosphate-buffer salts pH 7.4
- CFA: 20 g/litre casamino acids, 3 g/litre yeast extract, 1 mM $MgSO_4$, 0.1 mM $MnCl_2$, and 20 g/litre agar

A. Bacterial suspension

1 Grow the HB101/pPAP5 bacteria overnight at 37 °C on the solid medium CFA.

Protocol 1 continued

2 Passage the cells several times to ensure activation of the transcription of the *pap* operon.
3 Suspend bacteria[a] in PBS to an A_{540} of 1.0.
4 Spin down 1 ml of the suspension in a microcentrifuge for 1 min.
5 Suspend the pellet in 100 μl PBS.

B. Erythrocyte suspension

1 Introduce erythrocytes into ACD Solution B Vacutainer tubes in order to prevent clotting.
2 Wash erythrocytes in PBS by repeated mixing and centrifugation until the supernatant is transparent.[b]
3 Suspend erythrocytes in PBS to an A_{640} of 1.9.

C. Haemagglutination assay and determination of titre

1 Dispense 25 μl PBS/well in a microtitre plate.
2 Add 25 μl bacterial suspension to the first well. Mix. Make serial dilutions by transferring 25 μl from the first well to the second well, mixing, transferring 25 μl from the second well to the third well, and so on.
3 Add 25 μl erythrocyte suspension to all wells.
4 Mix suspension gently by tapping the side of the plate.
5 Cover with Parafilm and place at 4 °C.
6 Read the haemagglutination (HA) titre as the maximum dilution of bacteria which yields 50% haemagglutination.[c]

[a] Fresh, overnight culture.
[b] No lysed cells.
[c] In the absence of haemagglutination the erythrocytes will sediment in the pointed well bottoms and form a small and distinct red pellet. 50% haemagglutination is seen as incomplete pellet formation.

Protocol 2

Haemagglutination inhibition (HAI)

Equipment and reagents

- Microtitre plate with pointed well bottoms (see *Protocol 1*)
- Bacterial suspension (see *Protocol 1*)
- Erythrocyte suspension (see *Protocol 1*)
- Carbohydrate analogues as inhibitors
- PBS (see *Protocol 1*)

Method

1 Dilute the bacterial suspension to a titre of 64, i.e. to a bacterial density which gives 50% haemagglutination after 64-fold dilution.

Protocol 2 continued

2. Dissolve each inhibitor in PBS to 100 mM solutions.
3. Dispense 25 μl PBS/well in a microtitre plate.
4. Add 25 μl inhibitor solution to the first well. Mix. Make serial dilutions by transferring 25 μl from the first well to the second well, mixing, transferring 25 μl from the second well to the third, and so on. Remove and discard 25 μl from second last well. The very last well will contain no inhibitor.
5. Add 25 μl bacterial suspension to all wells.
6. Incubate 15 min at room temperature.
7. Add 25 μl erythrocyte suspension to all wells.
8. Mix suspension gently by tapping the side of the plate.
9. Cover with Parafilm and place at 4 °C for 16 h.
10. Read the IC_{50} value as the inhibitor concentration which gives 50% haemagglutination.[a]

[a] Inhibition of haemagglutination causes the erythrocytes to sediment in the pointed well bottoms forming small and distinct red pellets. 50% inhibition of haemagglutination is seen as incomplete pellet formation.

3.1.2 Scope and limitations

The major advantage of haemagglutination inhibition is that it is a relatively straightforward and rapid assay and it has consequently found widespread use. All the material needed is often readily available and no expensive apparatus is necessary. However, when conducting haemagglutination inhibition experiments one should bear in mind that expression of saccharides and adhesins on erythrocytes and bacteria may vary, thus causing quantitative variations. The relative inhibition results generally do not vary. Consequently, the fine dissection of an epitope is only possible if a library of chemically modified carbohydrate analogues is assayed and compared to a reference compound in one set of experiments using one single batch of erythrocytes and bacteria. Furthermore, the haemagglutination inhibition assay is experimentalist-dependent, since the IC_{50} value is a subjective estimate by the naked eye of the interval between the lowest inhibitor concentration that inhibits and the highest concentration that does not inhibit. Finally, haemagglutination inhibition requires a multivalent presentation of the carbohydrate-binding protein on the carrier (e.g. multivalent expression of adhesins on a bacterial cell surface). This requirement prevents isolated monovalent carbohydrate-binding proteins, such as a purified PapG protein, from being investigated and alternative assays have to be considered.

3.2 Competitive ELISA

3.2.1 Principles

Competitive ELISA (enzyme-linked immunosorbent assay) has also found widespread use mainly due to its generality and simplicity. In ELISA an active carbohydrate receptor is immobilized onto a solid surface, which in most cases is a

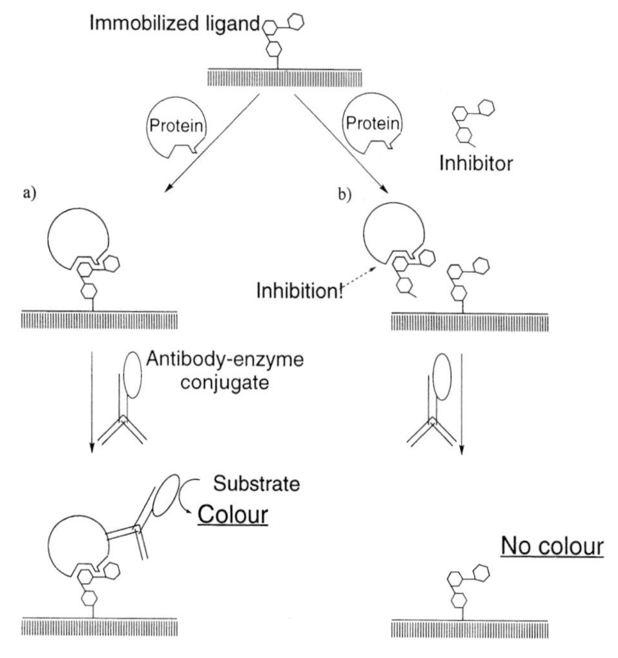

Figure 5 The principles of ELISA and competitive ELISA.

microtitre plate well. Subsequent incubation with a carbohydrate-binding protein results in binding to the immobilized receptor. Detection of bound protein is then done with a protein-specific antibody. This antibody is either directly conjugated with an enzyme or detected with a secondary antibody which is conjugated to an enzyme (e.g. horseradish peroxidase or alkaline phosphatase). A substrate releasing a coloured product upon enzymatic reaction is added and the colour produced is detected in a microplate reader (*Figure 5a*). The amount of colour produced (the optical density) is related to the amount of protein bound to the solid surface. In a competitive experiment, where the carbohydrate-binding protein is incubated in the presence of a soluble inhibitor, less protein will bind to the microtitre plate surface and less colour will consequently be produced upon detection (*Figure 5b*).

3.2.2 Immobilization of carbohydrate receptors

Adsorption of a carbohydrate receptor onto a microtitre well is conveniently performed by evaporation of a glycolipid-containing methanol solution. Microtitre plates are usually manufactured from hydrophobic plastic materials, such as polystyrene, onto which the lipid part of a glycolipid is readily adsorbed (*Figure 6a*). Adsorption of a glycolipid mixed with auxiliary lipids (e.g. cholesterol and phosphatidylcholine) is believed to provide a more 'cell membrane-like' presentation of the glycolipid. In addition, the presence of auxiliary lipids prevents 'clumping' of the glycolipids, making them more evenly distributed in the microtitre well (36). Alternatively, neoglycoproteins, e.g. saccharide–BSA conjugates, can be used to coat microtitre wells (*Figure 6b*). Many proteins, including BSA have a pronounced tendency to adsorb onto hydrophobic plastic materials, such as polystyrene. The use of a neoglycoprotein may sometimes result in a more stable adsorption than the use of a glycolipid. Whole cells presenting the appropriate saccharide structure can also be immobilized onto microtitre wells (37). Nevertheless, leakage of adsorbed glycolipid, neoglycoprotein, or cells into the buffer solution sometimes occurs during an assay, thus affecting sensitivity and reproducibility. The most stable and reproducible means of immobilization is by covalent linking of a carbohydrate receptor to the microtitre plate (*Figure 6c*). Covalent linking in aqueous solution of carboxy-functionalized spacer glycosides to microtitre plates carrying secondary amines, resulting in a well-defined and chemically stable amide bond, has been developed in our laboratory (30) (*Figure 7*). Carbohydrates covalently coupled to a microtitre plate via a long and flexible spacer were hypothesized to allow a homogeneous presentation of all significant epitopes to a protein. The name 'Glycoplate' will be used in the text below for microtitre plates carrying a covalently bound carbohydrate. The covalent linking reaction was found to be reproducible and ELISA assays using 'Glycoplates' were consequently also highly reproducible. *Protocol 3* describes how a carboxy-functionalized spacer galabioside (**37**) is coupled to a microtitre plate carrying secondary amino groups to give a 'Glycoplate' (**38**). *Protocols 4* and *5* describe ELISA experiments using the 'Glycoplate' (**38**).

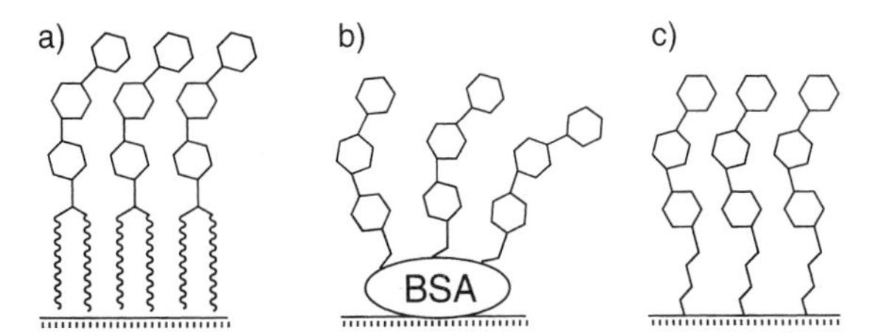

Figure 6 Immobilization of a carbohydrate ligand onto a solid surface by means of physical adsorption of (a) glycolipids or (b) neoglycoproteins, and by means of (c) covalent coupling of a pre-functionalized carbohydrate.

Protocol 3

Preparation of 'Glycoplate' (38): covalent coupling of carboxy-functionalized spacer galabioside (37) to a microtitre plate functionalized with secondary amino groups[a] (ref. 30)

Equipment and reagents

- CovaLink® microtitre plates functionalized with secondary amino groups (A/S Nunc)
- Carboxy-functionalized spacer galabioside (**37**)
- N-Hydroxysuccinimide (NHS)

Protocol 3 continued

- *N*-(3-dimethylaminopropyl)-*N'*-ethylcarbodiimide hydrochloride (EDC)
- Cova buffer: 1 litre PBS (see *Protocol 1*), 116.9 g NaCl, 10.0 g $MgSO_4$, and Tween 20

Method

1. Add to each well of the CovaLink® plate 50 μl of a freshly prepared solution of 4 mM[b] carboxy-functionalized spacer galabioside (**37**) and 8 mM NHS in distilled or deionized water.
2. Add to each well 50 μl of a freshly[c] prepared solution of 8 mM EDC in distilled or deionized water.
3. Shake the plate for 2–6 h.
4. Wash the plate three times with Cova buffer.

[a] Procedure adapted from the suppliers protocol.

[b] The efficiency of the coupling reaction drastically decreases when the concentration of the carboxy-functionalized spacer glycoside is below 1 mM.

[c] The solution has to be absolutely fresh, since EDC slowly decomposes in water.

Unreacted carboxy-functionalized spacer galabioside (**37**) can be recovered after completion of the coupling reaction. NHS is removed from **37** by silica gel chromatography and the by-product *N*-(3-dimethylaminopropyl)-*N'*-ethyl urea is removed by gel filtration (Sephadex G10 or G25) or by ion exchange chromatography (38). The 'Glycoplates' are fully functional after eight weeks storage at 4°C (longer-term stability tests have not been undertaken).

3.2.3 Background controls, blocking of non-specific binding, and incubation times

It is important to include appropriate control wells when conducting competitive ELISA experiments, so that the optical density in a control well reflects the extent of non-specific binding to the microtitre plate. The optical density read in a control well is subtracted from the optical densities read in other wells. The strain HB101/pPAP24, which expresses P pili devoid of the PapG adhesin, was used in control wells when conducting competitive ELISA experiments with HB101/pPAP5 (*Protocol 4*). Wells devoid of a covalently linked galabioside were used as control wells in the competitive ELISA experiments with the $PapD_{J96}PapG_{J96}$ (DG) complex (*Protocol 5*).

Most proteins have a tendency to adsorb non-specifically onto the hydrophobic microtitre plates. Undesirable non-specific binding by a carbohydrate-binding protein can be largely suppressed by pre-treating the microtitre plate with a 'blocking' protein that does not recognize the immobilized carbohydrate but is readily adsorbed to hydrophobic surfaces. Bovine serum albumin (BSA) or milk powder are the most commonly used 'blocking' proteins. Non-specific binding of living bacteria to hydrophobic surfaces on microtitre plates is generally

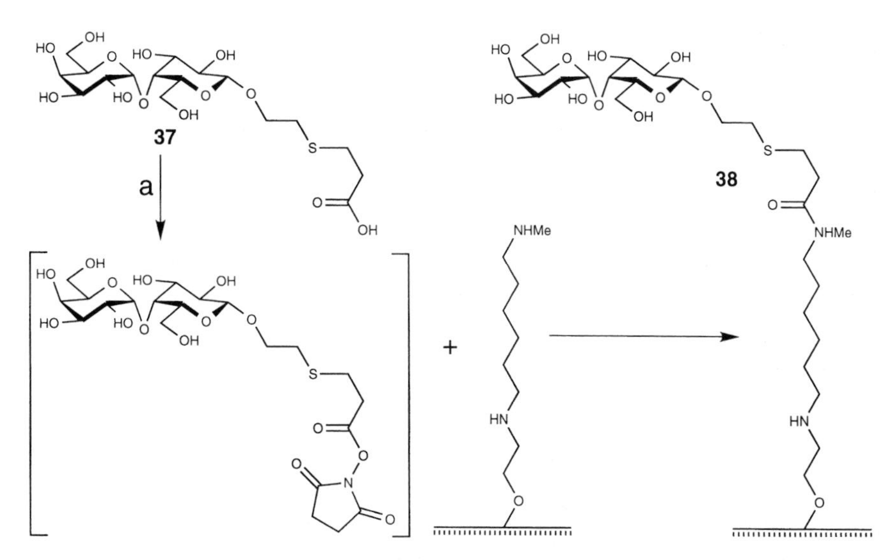

Figure 7 Preparation of a 'Glycoplate' (**38**). Covalent immobilization via amide bond formation between a carboxy-functionalized galabioside (**37**) and an amino-functionalized microtitre plate. (a) *N*-(3-dimethylaminopropyl)-*N'*-ethylcarbodiimide hydrochloride, *N*-hydroxysuccinimide, water.

blocked by pre-treating the plate with bacteria devoid of receptor-recognizing proteins.

Incubation times for the carbohydrate-binding protein and the antisera, as well as the washing times, often affect the signal-to-noise ratio (due to large variations in on-off rates for different protein–ligand complexes). Consequently, incubation and washing times have to be optimized for each new system investigated. It should also be noted that an ELISA experiment as described in *Protocols* 4 and 5, could in principle be performed using a microtitre plate coated via physical adsorption of a glycolipid or a neoglycoprotein.

Protocol 4

Competitive ELISA with HB101/pPAP5 bacteria

Equipment and reagents

- Microtitre plate reader
- HB101 (39) bacteria
- HB101/pPAP5 (34) bacteria (expressing P pili presenting the $PapG_{J96}$ adhesin at the tip, see *Protocol 1*)
- HB101/pPAP24 (40) bacteria (expressing P pili devoid of the $PapG_{J96}$ adhesin)
- TS agar
- Carbenicillin
- Bovine serum albumin (BSA)
- PBS (see *Protocol 1*)
- Carbohydrate analogues as inhibitors
- 'Glycoplate' **38** (see *Protocol 3*)
- Cova buffer (see *Protocol 3*)
- Anti-pili rabbit antiserum (WU93 diluted 1/200 in 2% BSA/PBS)

Protocol 4 continued

- Alkaline phosphatase-conjugated anti-rabbit IgG (Sigma A7539, diluted 1/500)
- Substrate buffer: 10% diethanolamine, 0.5 mM $MgCl_2$ pH 9.4
- Sigma 104 phosphatase substrate (i.e. *p*-nitrophenyl phosphate)
- Substrate solution: 1 mg/ml Sigma 104 phosphatase substrate in substrate buffer

A. Preparation of bacterial strains HB101/pPAP5 and HB101/pPAP24

1. Grow the bacteria on TS agar with 100 μg/ml carbenicillin and passage for three days.
2. Suspend bacteria in PBS to an A_{540} of 1.0.
3. Spin down a known volume of the bacterial suspension and resuspend the pellet in PBS containing 2% BSA (1/20th of the original volume).
4. Use this bacterial suspension in the competitive ELISA experiment described below.

B. Competitive ELISA with HB101/pPAP5 bacteria

1. Block the 'Glycoplate' for non-specific binding by incubation overnight at 4°C with 400 μl of PBS containing 2% BSA and *E. coli* HB101 cells (devoid of P pili) suspended to a klett of 50.
2. Wash three times with PBS and leave the last portion of washing for 15 min in the wells.
3. Dissolve each inhibitor to 18 mM in PBS.
4. Dispense 75 μl of inhibitor solution to the first well.
5. Dispense 50 μl of PBS to all other wells.
6. Make serial threefold dilutions by transferring 25 μl from the first well to the second well, mixing, transferring 25 μl from the second well to the third, and so on. Remove and discard 25 μl from the second last well. The very last well will contain no inhibitor.
7. Add 50 μl HB101/pPAP5 suspension to all wells except control wells.
8. Add 50 μl HB101/pPAP24 suspension to control wells.
9. Shake the plate gently.
10. Incubate for 45 min at room temperature.
11. Wash three times with Cova buffer and leave the last portion of washing for 15 min in the wells.
12. Add 100 μl anti-pili rabbit antiserum solution.
13. Incubate for 60 min at room temperature.
14. Wash three times with Cova buffer and leave the last portion of washing for 15 min in the wells.
15. Add 100 μl alkaline phosphatase-conjugated anti-rabbit IgG solution.
16. Incubate for 60 min at room temperature.

Protocol 4 continued

17 Wash three times with Cova buffer and leave the last portion of washing for 15 min in the wells.

18 Wash once with substrate buffer.

19 Filter substrate solution through 0.22 μm filter.

20 Add 100 μl of substrate solution to each well.

21 Read the plate in the microtitre plate reader at 405 nm.

22 Subtract the optical density reading from control wells (HB101/pPAP24) from the optical density reading from other wells (HB101/pPAP5).

23 Fit the optical density readings to the standard binding function:

$$OD = OD_0[1 - IC / (IC_{50} + IC)]$$

which will give the IC_{50} value. (OD is the optical density at the inhibitor concentration IC. OD_0 is the optical density in the absence of inhibitor and IC_{50} is the inhibitor concentration causing 50% inhibition.)

Protocol 5

Competitive ELISA with the periplasmic $PapD_{J96}PapG_{J96}$ pre-assembly complex

Equipment and reagents

- Microtitre plate reader
- PBS (see *Protocol 1*)
- Carbohydrate analogues as inhibitors
- 'Glycoplate' **38** (see *Protocol 3*)
- Solution of purified $PapD_{J96}PapG_{J96}$ (DG) complex (41): 0.1 mg/ml in 2% BSA/PBS
- Bovine serum albumin (BSA)
- Cova buffer (see *Protocol 3*)
- Alkaline phosphatase-conjugated anti-rabbit IgG (Sigma A7539, diluted 1/500)
- DG rabbit antiserum (diluted 1/500 in 2% BSA/PBS)
- Alkaline phosphatase-conjugated anti-rabbit IgG (Sigma A7539, diluted 1/500)
- Substrate buffer: 10% diethanolamine, 0.5 mM $MgCl_2$ pH 9.4
- Sigma 104 phosphatase substrate (i.e. *p*-nitrophenyl phosphate)
- Substrate solution: 1 mg/ml Sigma 104 phosphatase substrate in substrate buffer

Method

1 Dissolve each inhibitor to 18 mM in PBS.

2 Dispense 75 μl of inhibitor solution to the first well.

3 Dispense 50 μl of PBS to all other wells.

4 Make serial threefold dilutions as described in *Protocol 4*.

5 Add 50 μl DG solution to each well.

6 Incubate for 45 min at room temperature with gentle shaking.

Protocol 5 continued

7. Wash three times with PBS and leave the last portion of washing for 15 min in the wells.
8. Block the 'Glycoplate' for non-specific binding by incubation with 400 μl of 2% BSA in PBS for 1 h at room temperature (or overnight at 4 °C).[a]
9. Wash three times with Cova buffer and leave the last portion of washing for 15 min in the wells.
10. Add 100 μl of DG rabbit antiserum solution to each well.
11. Incubate for 60 min at room temperature with gentle shaking.
12. Wash three times with Cova buffer and leave the last portion of washing for 15 min in the wells.
13. Add 100 μl alkaline phosphatase-conjugated anti-rabbit IgG solution.
14. Incubate for 60 min at room temperature with gentle shaking.
15. Wash three times with Cova buffer and leave the last portion of washing for 15 min in the wells.
16. Wash once with substrate buffer.
17. Filter substrate solution through 0.22 μm filter.
18. Add 100 μl of substrate solution to each well.
19. Read the plate in the microtitre plate reader at 405 nm.
20. Subtract the optical density reading from control wells (devoid of covalently linked galabioside) from the optical density reading from other wells.
21. Fit the optical density readings to the standard binding function:

$$OD = OD_0[1 - IC / (IC_{50} + IC)]$$

which will give the IC_{50} value. (OD is the optical density at the inhibitor concentration IC. OD_0 is the optical density in the absence of inhibitor and IC_{50} is the inhibitor concentration causing 50% inhibition.)

[a] It is sometimes preferable to block for non-specific binding *prior* to incubation with the carbohydrate-binding protein (as described in *Protocol 4*).

3.2.4 Scope and limitations

Competitive ELISA is a general and easy to perform assay, employing standard equipment. As in a HAI experiment (Section 3.1), a competitive ELISA experiment is performed with a dilution series of the inhibitor. The result is a series of data points (optical density versus inhibitor concentration) from which the IC_{50} value can be obtained. An advantage of ELISA over haemagglutination is that the reading of optical densities is spectroscopic and thus experimentalist-independent. Consequently, more accurate IC_{50} values are obtained and a statistical treatment can be performed by fitting a curve to the data points (*Figure 8*). Furthermore, the ELISA assay is suitable for studying purified monovalent carbohydrate-binding

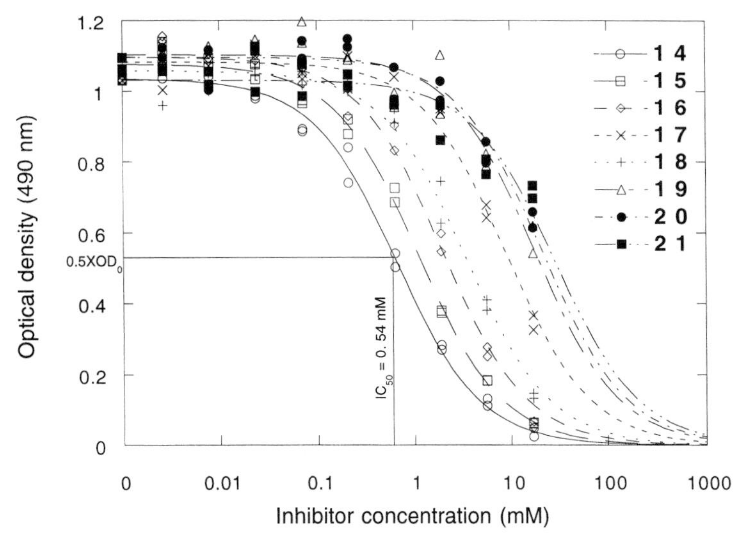

Figure 8 Inhibition of binding of the $PapD_{J96}PapG_{J96}$ complex (DG) to 'Glycoplate' **38**, using methyl β-D-galabioside (**14**) and its deoxy analogues (**15–21**). The IC_{50} value (0.54 mM) of **14** is indicated.

proteins, such as $PapD_{J96}PapG_{J96}$. This is in contrast to haemagglutination, which requires multivalent presentation of the protein on intact bacteria.

The closely related competitive solid phase RIA technique relies on detection of protein bound to an immobilized saccharide using a radiolabelled protein or protein-specific antibody (42) instead of an enzyme-conjugated antibody. In RIA, the radioactivity read in the assay vessel gives a measure of the amount of protein bound to the solid surface.

3.3 Techniques for direct measurement of carbohydrate–protein interaction

X-ray crystallography of a protein–carbohydrate complex (43, 44) has the potential of revealing all submolecular details of recognition. However, crystals of carbohydrate-protein complexes are often difficult to obtain and X-ray crystallography does not provide information of the thermodynamics of the recognition. A number of techniques exist for direct measurement of carbohydrate–protein interaction thermodynamics when a purified protein is available and a selection of such methods are briefly discussed below. Titration microcalorimetry (44) is an attractive technique for studying carbohydrate–protein interactions, since it provides all the thermodynamic parameters for association in solution. A well-defined volume of a carbohydrate solution is repeatedly injected into the microcalorimetric cell, which contains a solution of the carbohydrate-binding protein. The heat of complex formation (ΔH) accompanying each injection is recorded and the data are fitted to a mathematical model of the binding process. One single experiment will provide the ΔH, association constant (K_a), and the stoichiometry of the binding process, which allows the ΔG and ΔS values to be calculated. Furthermore, the Δc_p value of the binding process can be obtained if the ΔH is measured at different temperatures. The Δc_p value provides information concerning hydrophobic contributions to the binding process. However, a drawback of titration microcalorimetry is that rather large amounts and high concentrations of the carbohydrate-binding protein are required.

Changes in the UV (42) or fluorescence (45, 46) spectra of the protein or of the carbohydrate ligand, induced by the binding event, can be used to obtain the association constant of a binding process and thus its ΔG value. Determination of the association constants at different temperatures allows the ΔH and ΔS values to be calculated.

The interaction between a carbohydrate-binding protein and its ligand may sometimes be studied using nuclear magnetic resonance (NMR) spectroscopy (47–49). The association constant can sometimes be estimated from line broadenings and changes in chemical shifts of saccharide atoms induced by the binding process. In addition, NMR spectroscopy provides direct information about how the carbohydrate interacts with protein residues in cases were transfer NOE effects can be detected. Such detailed information is otherwise only possible to obtain from an X-ray structure of the complex.

Surface plasmon resonance has recently emerged as a tool for studying carbohydrate–protein interactions (50, 51). It differs from the techniques described above in that it provides the *kinetic* parameters of a binding process, i.e. the on (k_{on}) and off rates (k_{off}). The association constant is calculated from the on and off rates.

3.4 Enzyme assays

Epitope mapping of one specific class of carbohydrate-recognizing proteins, the carbohydrate-processing enzymes (glycosidases, glycosyl transferases, etc.), is done by evaluating saccharide analogues (deoxy, deoxyfluoro, deoxyamino, etc.) as substrates or inhibitors in appropriate enzyme assays (52). The effect of a chemical modification of a substrate or inhibitor on v_{max} and K_m or K_i can be interpreted in terms of submolecular interactions with the enzyme in the same way as described in Section 4 for other carbohydrate-recognizing proteins.

3.5 Qualitative *in situ* assays

In situ assays are relatively complicated to perform as compared to the *in vitro* assays described above, since they require access to tissue samples. However, *in situ* assays can be used to investigate if the epitope mapping results obtained in an *in vitro* assay (e.g. HAI or competitive ELISA) correlate with binding processes occurring under more physiologically relevant conditions. For example, in order to see if the results of epitope mapping of the $PapG_{AD110}$ adhesin (relevant in human pyelonephritis) (53) by HAI really reflect the binding process in a pyelonephritis patient, we devised an *in situ* assay based on inhibition of fluorescein isothiocyanate (FITC)-labelled bacteria to fixed human kidney sections (29). Synthetic oligosaccharide fragments were evaluated as inhibitors and although exact quantification of inhibitor efficiencies was not possible, the relative inhibitor efficiencies estimated in the *in situ* assay correlated well with results obtained in HAI experiments. The good correlation between the *in situ* assay and the HAI, suggests that results obtained with the less complicated HAI experiments reflect how the bacterial binding to carbohydrate receptors occurs during pyelonephritis and that these results can be used to design and develop efficient inhibitors of the $PapG_{AD110}$ adhesin.

4 Data interpretation

The data output from competitive binding assays is an IC_{50} value for each inhibitor. The relative equilibrium constant (K_{rel}) is calculated as the ratio of IC_{50} values between the inhibitor and a reference compound. Inhibitory power is defined as 100 K_{rel}. The difference in free energy of binding compared to a reference inhibitor is calculated using the expression $\Delta\Delta G = -RT\ln K_{rel}$ (54). HA, HAI, and competitive ELISA should be performed in at least triplicate in order to obtain statistically treatable and reliable data.

The inhibitors are divided into four categories when compared to a reference inhibitor:

(1) $\Delta\Delta G < 0$ kJ/mol (better inhibitor than the reference).

(2) $\Delta\Delta G = 0$–2 kJ/mol (marginally less efficient inhibitor than the reference).

(3) $\Delta\Delta G = 3$–10 kJ/mol (significantly less efficient inhibitor than the reference).

(4) $\Delta\Delta G > 11$ kJ/mol (a non-inhibitor).

It has been reported that deletion (by site-specific mutagenesis) of a hydrogen-bonding amino acid residue in tyrosyl-t-RNA synthetase resulted in decreased binding strength of the ligand by 2.1–6.3 kJ/mol (55). Consequently, the loss of inhibitory power in category 3, $\Delta\Delta G = 3$–10 kJ/mol, suggests the loss of a hydrogen bond (or possibly a severe steric repulsion) upon binding to the protein. The non-inhibitory compounds belonging to category 4 have lost more than one critical interaction with the protein.

4.1 Conformational analysis

Systematic synthesis and biological evaluation of saccharide analogues requires the analogues to have very similar conformations, thus ensuring that any difference in biological activity originates in the chemical modification. Therefore, conformational analysis of synthetic analogues should always be undertaken in conjunction with binding studies. The Forssman pentasaccharide and all its fragments (**1–10**) (56), the deoxy (**15–21**), deoxyfluoro (**22–25**) (15, 16, 57), and the charged (**26–28**) (18) galabiosides, as well as the thiodisaccharide analogue **29** (22) and the conformationally restricted analogue **30** (20) were subjected to rigorous conformational analysis using nuclear magnetic resonance (NMR) spectroscopy and computer calculations. Torsion angles around glycosidic linkages in the fragments (**1–10**) were shown to be very similar to corresponding glycosidic bonds in the parent Forssman pentasaccharide. Furthermore, conformations of the chemically modified galabiosides **15–28** were essentially the same as in the parent galabiose disaccharides (**9** and **14**). The thiodisaccharide analogue **29** adopted similar dihedral angles around the interglycosidic bonds as compared to the parent *O*-disaccharide **9**. However, the relative orientation of the two galactose residues in **29** was slightly altered as compared to **9**, due to different bond lengths and angle of C–S–C as compared to C–O–C. The conformationally restricted analogue **30** adopted a minimum conformation different from that of the parent compound **9**.

4.2 Fragments of oligosaccharides

Probing the importance of individual monosaccharide residues in an oligosaccharide ligand is exemplified below with our investigation of the binding specificities of the adhesins of human uropathogenic *E. coli*. Prior to our work, inhibition of haemagglutination revealed glycolipids of the P blood group system (i.e. P (58) and P^k antigens (59), *Figure 1a*) as receptors for uropathogenic *E. coli*. Screening against a large panel of natural glycolipids in TLC overlay experiments (60) later identified the minimal receptor epitope as galabiose [αGal*p*(1–4)βGal*p*]. Evaluation of the synthetic fragments (**1–10**, *Figure 1b*) as inhibitors of $PapG_{J96}$ (ELISA) (30) and $PapG_{AD110}$ (HAI) (29), confirmed the earlier results obtained by TLC overlay techniques; galabiose was found to constitute an absolute prerequisite for recognition (*Table 1* and *Figure 9*). Only the fragments containing the galabiose disaccharide (**3–6**, **8–9**) were inhibitory. However, the use of synthetic fragments revealed further details that could not be detected by screening against isolated natural glycoconjugates. None of the monosaccharide residues flanking the galabiose disaccharide contributed to the binding by $PapG_{J96}$ (compare **9** with **3**, **6** and **8**). On the contrary, the βGlc*p* residue, which is present in the reducing end of the Forssman pentasaccharide, was detrimental to binding. The $PapG_{J96}$ adhesin instead preferred hydrophobic aglycons, such as the 2-(trimethylsilyl)ethyl group, on the galabiose disaccharide (compare **8** with **9**, **5** with **6**, and **3** with **4**). No significant differences were observed between $PapG_{J96}$ present on whole bacteria (HB101/pPAP5) and in a complex ($PapD_{J96}PapG_{J96}$) with the chaperon $PapD_{J96}$.

The $PapG_{AD110}$ adhesin revealed a different recognition motif when compared to that of the $PapG_{J96}$ adhesin. The two closest residues flanking the galabiose disaccharide, βGalNAc*p* and βGlc*p*, were shown to interact with the $PapG_{AD110}$ adhesin in a HAI experiments (*Table 1* and *Figure 9*). The βGalNAc*p* residue contributed with a binding interaction of approximately $\Delta\Delta G = -1$ kJ/mol (compare **8** with **5** and **9** with **6**), which may be attributed to the *N*-acetyl group. The 4″-deoxy (**11**) and 4″-epi (**12**) analogues of **6** were as efficient inhibitors as **6**, while substitution of the *N*-acetyl group in **6** with a hydroxy group (→**13**) resulted in a decrease in inhibitory power as compared to **6**. Compound **13** was as good an inhibitor as the galabioside **9**. However, binding energy differences of $\Delta\Delta G = -1$ kJ/mol are on the borderline of being statistically significant and conclusive argumentation should be done with care. On the contrary, the βGlc*p* residue

Table 1 Inhibition experiments with fragments and fragment analogues of the Forssman pentasaccharide

Compound	HB101/pPAP5[a] (30)			DG[b] (30)			HB101/pPIL110–35[c] (29)		
	IC_{50} (mM)	100 K_{rel}[d] (%)	ΔΔG[d] (kJ/mol)	IC_{50} (mM)	100 K_{rel}[d] (%)	ΔΔG[d] (kJ/mol)	IC_{50} (mM)	100 K_{rel}[d] (%)	ΔΔG[d] (kJ/mol)
1	n.d.[e]	n.d.	n.d.	n.d.	n.d.	n.d.	> 13.6	< 31	> 2.8
2	> 50	< 1.4	> 10.5	> 10	< 2	> 9.7	> 13.8	< 30	> 2.9
3	0.73	99	0	0.13	146	–0.9	2.5	170	–1.3
4	4.0	18	4.2	> 10	< 2	> 9.7	1.0	430	–3.5
5	2.8	26	3.4	> 10	< 2	> 9.7	0.75	560	–4.2
6	1.3	55	1.5	0.22	86	0.3	2.8	150	–1.0
7	> 50	< 1.4	> 10.5	> 10	< 2	> 9.7	> 13.3	< 32	> 2.7
8	6.4	11	5.4	> 10	< 2	> 9.7	1.0	420	–3.5
9	0.72	100	0	0.19	100	0	4.2	100	0
10	> 50	< 1.4	> 10.5	n.d.	n.d.	n.d.	> 14.2	< 29	> 3.0
11	n.d.	n.d.	n.d.	n.d.	n.d.	n.d.	1.4	310	–2.7
12	n.d.	n.d.	n.d.	n.d.	n.d.	n.d.	2.0	210	–1.4
13	n.d.	n.d.	n.d.	n.d.	n.d.	n.d.	4.7	90	0.3

[a] Bacteria expressing P pili carrying the $PapG_{J96}$ adhesin; ELISA.

[b] The periplasmic $PapD_{J96}PapG_{J96}$ pre-assembly complex; ELISA.

[c] Bacteria expressing P pili carrying the $PapG_{AD110}$ adhesin; HAI.

[d] The 100 K_{rel} (inhibitory power) and ΔΔG values are related to compound **9**.

[e] Not determined.

contributed with a binding interaction to $PapG_{AD110}$ of approximately ΔΔG = −3.5 kJ/mol (compare **5** with **6**, **8** with **9**, and **4** with **3**), which is statistically significant and corresponds to a medium strength hydrogen bond.

4.3 Deoxy and deoxyfluoro analogues

The results from mapping of hydrogen bonding interactions of the $PapG_{J96}$ and $PapG_{AD110}$ adhesins with the monodeoxy, monodeoxyfluoro, and *C*-methyl analogues (**15–25**) of methyl β-D-galabioside (**14**) are shown in *Table 2*. None of the two adhesins formed a hydrogen bond to HO-2 of galabiose, since compound **15** displayed only a marginal loss of inhibitory power as compared to the parent galabioside **14**. The monodeoxy analogues **17–21** showed a significant reduction in inhibitory power with both adhesins, suggesting that the HO-6, -2′, -3′, -4′, and -6′ formed critical hydrogen bonds with the adhesins (*Figure 9*). Compound **16** displayed intermediate inhibitory power (19–33%) against $PapG_{J96}$, which suggests that it is involved in a weak hydrogen bond to the adhesin or that the size/surface of the HO-3 is important for binding. The latter explanation was verified by substituting the HO-3 for a *C*-methyl (→**25**), which resulted in recovery (50%) of some of the inhibitory power lost in **16**. A likely explanation to this observation is that HO-3 is intramolecularly hydrogen bonded to O-5′ in the galabiose disaccharide, thus extending a large hydrophobic patch formed by H-1,

Table 2 Inhibition experiments with monodeoxy and monodeoxyfluoro analogues of methyl β-D-galabioside (**14**)

Compound	HB101/pPAP5[a] (28, 30)			DG[b] (30)			HB101/pPIL110–35[c] (29)		
	IC_{50} (mM)	100 K_{rel}[d] (%)	ΔΔG[d] (kJ/mol)	IC_{50} (mM)	100 K_{rel}[d] (%)	ΔΔG[d] (kJ/mol)	IC_{50} (mM)	100 K_{rel}[d] (%)	ΔΔG[d] (kJ/mol)
14	0.18/4.2[e]	100/100	0/0	0.54	100	0	7.4	100	0
15	0.3/9.0	61/47	1.1/1.9	1.1	60	1.3	17.1	51	1.4
16	0.98/15.0	19/28	3.9/3.2	1.95	33	2.7	4.8	143	–0.8
17	4.2/> 40	4.4/< 9.5	7.2/> 5.8	9.0	7	6.6	> 33.5	< 22	> 3.7
18	2.3/15.8	7.6/27	5.9/3.2	3.4	19	4.1	> 25.9	< 30	–> 2.9
19	6.4/> 40	2.7/< 9.5	8.3/> 5.8	> 20	< 3	> 8.5	> 34.4	< 21	> 3.8
20	10/> 60	1.9/< 7.0	9.1/> 6.6	> 20	< 3	> 8.5	> 33.8	< 21	> 3.8
21	3.6/> 60	5.4/< 7.0	6.7/> 6.6	> 20	< 3	> 8.5	> 35.4	< 21	> 3.8
22	9.2/n.d.[f]	1.9/n.d.	9.1/n.d.	n.d.	n.d.	n.d.	n.d.	n.d.	n.d.
23	3.3/n.d.	6.2/n.d.	6.4/n.d.	n.d.	n.d.	n.d.	n.d.	n.d.	n.d.
24	0.33/n.d.	50/n.d.	1.6/n.d.	n.d.	n.d.	n.d.	n.d.	n.d.	n.d.
25	0.39/n.d.	50/n.d.	1.6/n.d.	n.d.	n.d.	n.d.	3.9	191	–1.4

[a] Bacteria expressing P pili carrying the $PapG_{J96}$ adhesin; HAI and ELISA.

[b] The periplasmic $PapD_{J96}PapG_{J96}$ pre-assembly complex; ELISA.

[c] Bacteria expressing P pili carrying the $PapG_{AD110}$ adhesin; HAI.

[d] The 100 K_{rel} (inhibitory power) and ΔΔG values are related to compound **14**.

[e] HAI/ELISA.

[f] Not determined.

-3, -4, -5, -1′, and -2′, which interacts with a hydrophobic region in the combining site of $PapG_{J96}$ (*Figure 9*). A similar, but even more pronounced effect, was observed with the $PapG_{AD110}$ adhesin. Compound **16** showed a slight (ΔΔG = −0.8 kJ/mol), and compound **25** a moderate, improvement (ΔΔG = −1.4 kJ/mol) as inhibitors of the $PapG_{AD110}$ adhesin when compared to **14**. This strongly suggests that the hydrophobic patch on galabiose is interacting with a hydrophobic region in the combining site of $PapG_{AD110}$ adhesin.

The monodeoxyfluoro analogues **22** and **23** showed significant reduction in inhibitory power suggesting that HO-6 and -4′ donates a proton in a hydrogen bond to the $PapG_{J96}$ adhesin, while the monodeoxyfluoro analogue **24** displayed only a slight reduction of inhibitory power suggesting that HO-6′ accepts a proton in a hydrogen bond to the $PapG_{J96}$ adhesin.

Quite interestingly, the adhesins of the Gram positive pig pathogen *Streptococcus suis* (61) recognize a different epitope on the galabiose disaccharide. Competitive RIA and HAI experiments using the monodeoxy galabiosides **14–21** as inhibitors of *S. suis* identified the HO-2, -3, -4′, and -6′ as the key polar groups.

4.4 Amino and carboxy analogues

The HO-6 and -2′ of galabiose were identified as two of the key polar groups in the interaction with PapG adhesins of uropathogenic *E. coli* from inhibition

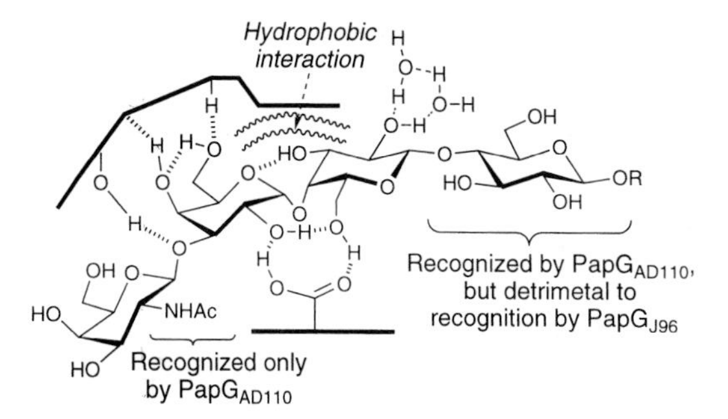

Figure 9 Model of the globoside epitopes recognized by the $PapG_{J96}$ and $PapG_{AD110}$ adhesins. Hydrogen bonds was identified from inhibitions with the monodeoxy galabiosides **15–21** and are indicated with dashed lines. The critical co-operative hydrogen bond network between HO-6 and -2′ and a carboxy group of $PapG_{J96}$ was identified from inhibitions with the amino and carboxy analogues **26–28** and the thiodisaccharide **29**.

experiments with the monodeoxy analogues **14–21**. Furthermore, conformational analysis of galabiosides suggested that HO-6 and -2′ were closely situated, which was supported by the observation in NMR experiments that an intramolecular hydrogen bond exists between HO-6 and -2′ (22, 57). The close proximity of HO-6 and -2′ resembles that of vicinal carbohydrate diols, which have been observed to form co-operative hydrogen bonds with polar amino acids (carboxy, amido, and guanidino groups) in protein combining sites (43). The amino and carboxy galabioside analogues **26–28** (*Figure 3*) were synthesized to investigate if HO-6 and -2′ interact with such polar amino acids in the $PapG_{J96}$ adhesin combining site. Evaluation of **26–28** as inhibitors of $PapD_{J96}PapG_{J96}$ in the competitive ELISA showed that the H_2N-2′ compound **26** was a better inhibitor ($\Delta\Delta G = -1.1$ kJ/mol) than the reference galabioside **9**, while the H_2N-6 compound **26** only retained a fraction of the inhibitor potency ($\Delta\Delta G = 5.0$ kJ/mol) of **9** (*Table 3*). The uronic acid **28** was non-inhibitory. These results are in agreement with the previous results using the monodeoxy analogues **14–21** in that HO-6 and -2′ are involved in co-operative hydrogen bonding to a glutamate or aspartate in the $PapG_{J96}$ adhesin. The increased inhibitory power of compound **26** is consistent with the formation of a salt bridge between a protonated H_2N-2′ and a carboxylic acid in $PapG_{J96}$ without disruption of the intramolecular hydrogen bond to HO-6 (*Figure 9*). The lack of inhibitory power of compound **27** can be interpreted in terms of a precise geometrical requirement for the position of the amino group in order to form a salt bridge. The total lack of activity of the uronic acid **28** also suggests the presence of a carboxylic acid in the combining site of $PapG_{J96}$, since under physiological conditions both carboxylic acids would be charged and thus repel each other.

4.5 Conformationally altered analogues

Replacement of the intersaccharidic oxygen with sulfur (**9**→**29**) induced a slight change in the relative orientation of the two galactose residues as revealed by

Table 3 Competitive ELISA with the $PapD_{J96}PapG_{J96}$ complex using amino and carboxy analogues of 2-(trimethylsilyl)ethyl β-D-galabioside (**9**) as inhibitors (18)

Compound	IC_{50} (mM)	100 K_{rel}[a] (%)	ΔΔG[a] (kJ/mol)
9	2.2	100	0
26	1.4	157	−1.1
27	17.2	13	5.0
28	> 24.9	< 6	> 21

[a] The 100 K_{rel} (inhibitory power) and ΔΔG values are related to compound **9**.

Table 4 Competitive ELISA with the $PapD_{J96}PapG_{J96}$ complex using conformationally altered or restricted analogues of 2-(trimethylsilyl)ethyl β-D-galabioside (**9**) as inhibitors (20, 22, 30)

Compound	IC_{50} (mM)	100 K_{rel}[a] (%)	ΔΔG[a] (kJ/mol)
9	0.16	100	0
29	4.4	4	8.0
30	n.i.[b]	n.i.	n.i.

[a] The 100 K_{rel} (inhibitory power) and ΔΔG values are related to compound **9**.

[b] Non-inhibitory.

conformational analysis. The conformational change was further supported by the fact that an intramolecular hydrogen bond between HO-6 and -2′ could be observed in **9** but not in **29** in NMR experiments. This minor change in relative orientation of the two galactose residues was due to different bond lengths and angle of C–S–C, since the conformational analysis indicated that the dihedral angles of **29** were very similar to those of **9**. The thiodisaccharide **29** displayed a decrease of inhibitory activity against the $PapD_{J96}PapG_{J96}$ adhesin (ΔΔG = 8.0 kJ/mol, *Table 4*) corresponding to the loss of one intermolecular hydrogen bond as compared to the reference **9** in competitive ELISA experiments. This is most likely due to the change in the spatial arrangement of HO-6 relative to HO-2′, -3′, -4′, and -6′, disrupting the intramolecular hydrogen bond between HO-6 and HO-2′, which in turn leads to the loss of one intermolecular hydrogen bond to the adhesin. The extraordinary sensitivity of the $PapD_{J96}PapG_{J96}$ adhesin to the relative orientation of the two galactose residues in galabiose nicely demonstrates that a stunningly precise specificity may be obtained when a protein recognizes hydroxyl groups from more than one monosaccharide residue.

References

1. Varki, A. (1993). *Glycobiology*, **3**, 97.
2. Dwek, R. A. (1996). *Chem. Rev.*, **96**, 683.
3. Karlsson, K.-A. (1995). *Curr. Opin. Struct. Biol.*, **5**, 622.
4. McAuliffe, J. C. and Hindsgaul, O. (1997). *Chem. Ind.*, 170.
5. Lemieux, R. U. (1996). *Acc. Chem. Res.*, **29**, 373.

6. Bundle, D. R. (1997). In *Glycosciences: status and perspectives* (ed. H.-J. Gabius and S. Gabius), p. 311. Chapman & Hall, Weinheim, Germany.
7. Magnusson, G., Chernyak, A. Y., Kihlberg, J., and Kononov, L. O. (1994). In *Neoglycoconjugates: preparation and application* (ed. Y. C. Lee and R. T. Lee), p. 53. Academic Press, San Diego, USA.
8. Toshima, K. and Tatsuta, K. (1993). *Chem. Rev.*, **93**, 1503.
9. Boons, G. J. (1996). *Tetrahedron*, **52**, 1095.
10. Nilsson, U., Ray, A. K., and Magnusson, G. (1994). *Carbohydr. Res.*, **252**, 117.
11. Nilsson, U., Wendler, A., and Magnusson, G. (1994). *Acta Chem. Scand.*, **48**, 356.
12. Lindberg, F., Lund, B., Johansson, L., and Normark, S. (1987). *Nature*, **328**, 84.
13. Kuehn, M. J., Heuser, J., Normark, S., and Hultgren, S. J. (1992). *Nature*, **356**, 252.
14. Kihlberg, J., Frejd, T., Jansson, K., and Magnusson, G. (1986). *Carbohydr. Res.*, **152**, 113.
15. Kihlberg, J., Frejd, T., Jansson, K., Sundin, A., and Magnusson, G. (1988). *Carbohydr. Res.*, **176**, 271.
16. Kihlberg, J., Frejd, T., Jansson, K., Kitzing, S., and Magnusson, G. (1989). *Carbohydr. Res.*, **185**, 171.
17. Nikrad, P. V., Beierbeck, H., and Lemieux, R. U. (1992). *Can. J. Chem.*, **70**, 241.
18. Hansen, H. C. and Magnusson, G. (1998). *Carbohydr. Res.*, **307**, 233.
19. Navarre, N., van Oijen, A. H., and Boons, G. J. (1997). *Tetrahedron Lett.*, **38**, 2023.
20. Wilstermann, M., Balogh, J., and Magnusson, G. (1997). *J. Org. Chem.*, **62**, 3659.
21. Wilstermann, M., Kononov, L. O., Nilsson, U., Ray, A. K., and Magnusson, G. (1995). *J. Am. Chem. Soc.*, **117**, 4742.
22. Nilsson, U., Johansson, R., and Magnusson, G. (1996). *Chem. Eur. J.*, **2**, 295.
23. Thompson, L. A. and Ellman, J. A. (1996). *Chem. Rev.*, **96**, 555.
24. Liang, R., Yan, L., Loebach, J., Ge, M., Uozumi, Y., Sekanina, K., *et al.* (1996). *Science*, **274**, 1520.
25. Nilsson, U. J., Fournier, E., and Hindsgaul, O. (1998). *Bioorg. Med. Chem.*, **6**, 1563.
26. Hultgren, S. J., Abraham, S. N., Caparon, M., Falk, P., St Geme III, J. W., and Normark, S. (1993). *Cell*, **73**, 887.
27. Strömberg, N., Marklund, B.-I., Lund, B., Ilver, D., Hamers, A., Gaastra, W., *et al.* (1990). *EMBO J.*, **9**, 2001.
28. Kihlberg, J., Hultgren, S. J., Normark, S., and Magnusson, G. (1989). *J. Am. Chem. Soc.*, **111**, 6364.
29. Striker, R., Nilsson, U., Stonecipher, A., Magnusson, G., and Hultgren, S. J. (1995). *Mol. Microbiol.*, **16**, 1021.
30. Nilsson, U., Striker, R. T., Hultgren, S. J., and Magnusson, G. (1996). *Bioorg. Med. Chem.*, **4**, 1809.
31. Magnusson, G., Hultgren, S. J., and Kihlberg, J. (1995). In *Methods in enzymology* (ed. R. J. Doyle and I. Ofek), Vol. 253, p. 105. Academic Press, London.
32. de Man, P., Cedergren, B., Enerbäck, S., Larsson, A.-C., Leffler, H., Lundell, A.-L., *et al.* (1987). *J. Clin. Microbiol.*, **25**, 401.
33. Hull, R. A., Gill, R. E., Hsu, P., Minshew, B. H., and Falkow, S. (1981). *Infect. Immun.*, **33**, 933.
34. Lindberg, F., Lund, B., and Normark, S. (1984). *EMBO J.*, **3**, 1167.
35. Hultgren, S. J., Schwan, W. R., Schaeffer, A. S., and Duncan, J. L. (1986). *Infect. Immun.*, **54**, 613.
36. Pellizzari, A., Pang, H., and Lingwood, C. A. (1992). *Biochemistry*, **31**, 1363.
37. Ofek, I. (1995). In *Methods in enzymology* (ed. R. J. Doyle and I. Ofek), Vol. 253, p. 528. Academic Press, London.
38. Lawrence, R. M., Biller, S. A., Fryszman, O. M., and Poss, M. A. (1997). *Synthesis*, 553.
39. Boyer, H. W. and Roulland-Dussoix, D. (1969). *J. Mol. Biol.*, **41**, 459.

40. Lindberg, F., Lund, B., and Normark, S. (1986). *Proc. Natl. Acad. Sci. USA*, **83**, 1891.
41. Hultgren, S. J., Lindberg, F., Magnusson, G., Kihlberg, J., Tennent, J. M., and Normark, S. (1989). *Proc. Natl. Acad. Sci. USA*, **86**, 4357.
42. Spohr, U., Hindsgaul, O., and Lemieux, R. U. (1985). *Can. J. Chem.*, **63**, 2644.
43. Quiocho, F. (1989). *Pure Appl. Chem.*, **61**, 1293.
44. Toone, E. J. (1994). *Curr. Opin. Struct. Biol.*, **4**, 719.
45. Vermersch, P. S., Tesmer, J. J. G., Lemon, D. D., and Quiocho, F. A. (1990). *J. Biol. Chem.*, **265**, 16592.
46. Ziegler, T., Pavliak, V., Lin, T.-H., Kovác, P., and Glaudemans, C. P. J. (1990). *Carbohydr. Res.*, **204**, 167.
47. Sauter, N. K., Bednarski, M. D., Wurzburg, B. A., Hanson, J. E., Whitesides, G. M., Skehel, J. J., *et al.* (1989). *Biochemistry*, **28**, 8388.
48. Glaudemans, C. P. J., Lerner, L., Dvaes Jr., G. D., Kovác, P., Venable, R., and Bax, A. (1990). *Biochemistry*, **29**, 10906.
49. Bundle, D. R., Bauman, H., Brisson, J.-M., Gagné, S. M., Zdanov, A., and Cygler, M. (1994). *Biochemistry*, **33**, 5183.
50. MacKenzie, C. R., Hirama, T., Deng, S., Bundle, D. R., Narang, S. A., and Young, N. M. (1996). *J. Biol. Chem.*, **271**, 1527.
51. MacKenzie, C. R., Hirama, T., Lee, K. K., Altman, E., and Young, N. M. (1997). *J. Biol. Chem.*, **272**, 5533.
52. Palcic, M. M. and Hindsgaul, O. (1996). *Trends Glycosci. Glycotechn.*, **8**, 37.
53. Marklund, B.-I., Tennent, J. M., Garcia, E., Hamers, A., Båga, M., Lindberg, F., *et al.* (1992). *Mol. Microbiol.*, **6**, 2225.
54. Pressman, D. and Grossberg, A. L. (1968). In *The structural basis of antibody specificity*, p. 16. W. A. Benjamin Inc., New York, USA.
55. Fersht, A. R., Shi, J.-P., Knill-Jones, J., Lowe, D. M., Wilkinson, A. J., Blow, D. M., *et al.* (1985). *Nature*, **314**, 235.
56. Grönberg, G., Nilsson, U., Bock, K., and Magnusson, G. (1994). *Carbohydr. Res.*, **257**, 35.
57. Bock, K., Frejd, T., Kihlberg, J., and Magnusson, G. (1988). *Carbohydr. Res.*, **176**, 253.
58. Leffler, H. and Svanborg-Edén, C. (1980). *FEMS Lett.*, **8**, 127.
59. Källenius, G., Möllby, R., Svensson, S. B., Winberg, J., Lundblad, A., Svensson, S., *et al.* (1980). *FEMS Lett.*, **7**, 297.
60. Bock, K., Breimer, M. E., Brignole, A., Hansson, G. C., Karlsson, K.-A., Larsson, G., *et al.* (1985). *J. Biol. Chem.*, **260**, 8545.
61. Haataja, S., Tikkanen, K., Nilsson, U., Magnusson, G., Karlsson, K.-A., and Finne, J. (1994). *J. Biol. Chem.*, **269**, 27466.

Chapter 9
Phage display libraries

Samantha Williams

Paul van der Logt, and Volker Germaschewski
Unilever Research, Colworth Laboratory, Colworth House, Sharnbrook, Beds MK44 1LQ, UK.

1 Introduction

1.1 General

Two independent developments were brought together in the late 1980s to create a technique which since then has been shown to prove extremely powerful for the determination of amino acid sequences recognized by antibodies. First, the construction of synthetic peptide libraries on solid matrices, for example plastic pins (1) offered the possibility to screen for peptides that were bound by an antibody of interest (see also Chapters 2 and 3). Although peptides were isolated on the basis of binding the antibody used for screening the library, in many cases the amino acid sequences of the peptides bore no resemblance to any part of the native sequence of the antigen against which the antibody was originally raised. Rather, these sequences mimicked native epitopes and Geysen consequently called them 'mimotope libraries'.

Secondly, the discovery by George Smith (2) that filamentous bacteriophage can tolerate foreign peptides as fusions to their coat proteins and consequently display them on the surface of the phage particle paved the way for a new concept of biological peptide display libraries. In this pioneering work, Smith constructed fusion phage which displayed fragments of the *Eco*RI restriction endonuclease (2). Using Petri dishes with adsorbed anti-*Eco*RI antibodies he was able to demonstrate the feasibility of an affinity purification that achieved the isolation of these fusion-phage from a large background of unspecific phage. This affinity selection principle sparked off the idea of using a library of random peptides instead of defined protein domains. These random peptide libraries could then be used to select for peptides that bind a specific target molecule like for instance an antibody (3). It is possible to screen an extremely large pool of peptides which is one of the crucial benefits of using these libraries. One further significant advantage of these biological libraries is that once phage particles have been isolated on the basis of binding to a particular antibody the phage can be readily amplified in bacterial culture. This not only allows the isolation of the phage DNA genome and consequently the amino acid sequence of the displayed peptide, but also offers the option for further analysis being carried out on individual phage clones such as phage ELISAs. This is only possible since the binding activity displayed on the phage particle (in this case the random peptide) and the genetic information which codes for the amino acid sequence of the peptide are physically linked (in the phage particle).

In this chapter we will demonstrate how a peptide library comprising nine random amino acids was used for determining the epitopes recognized by antibodies against beta-lactoglobulin. (BLG) Initially most studies involving epitope mapping using phage display were looking at monoclonal antibodies and a whole range of epitopes recognized by monoclonal antibodies have since been elucidated this way (3, 4). However, recently the study of polyclonal antisera and the determination of immunodominant epitopes has also received attention (5, 6). In this chapter we demonstrate that polyclonal serum IgG can be used to screen phage display libraries and search for immunodominant epitopes represented in the immune repertoire of an immunized donor. In Section 2.2.7 these results are directly compared with results obtained using a synthetic peptide library on plastic pins (Pepscan) for screening the same antiserum.

Foreign peptides or protein domains of fusion phage can be linked to the pIII or the pVIII coat protein. Recently fusion to pVI, another minor coat protein, has also been reported (7). In theory any fusion to pVIII, the major coat protein, allows a high degree of multivalency and also circumvents potential infectivity problems since, in contrast to pIII fusions, pVIII is not involved in the infection process. However, although it is possible to fuse peptides to each of the about 2700 copies of the pVIII coat protein, pVIII fusion phage libraries that exceed six amino acids or so have to be based on so-called phagemid vectors since inserts longer than a few amino acids seem to interfere with assembly of the phage particle. In this system wild-type pVIII and indeed all other phage gene products are provided by an external helper phage. The manipulated copies of pVIII harbouring the fused peptides are encoded by the phagemid which is a type of plasmid that carries a phage replication origin and packaging signal. The superinfection of bacterial cells containing the phagemid DNA with helper phage then results in a competition between wild-type pVIII and manipulated pVIII (encoded by the phagemid) for packaging into new phage particles during assembly. The generated hybrid phage particles will carry varying numbers of peptide fusions and wild-type pVIII but the presence of wild-type coat protein assures correct assembly of particles and increases the tolerance for larger fusions (8).

It is also possible to use this phagemid system for pIII fusion phage production and if the wild-type pIII encoded by the helper phage is in vast excess over the manipulated copy of pIII a near monovalency of the displayed peptide or domain can be achieved which may have advantages for certain applications (9). In most cases however, fusions of peptides and proteins to pIII do not interfere with the infection process of the *E. coli* host cell and therefore libraries based on pIII fusions do usually not require the phagemid system. In this case every copy of pIII carries the fusion peptide on the surface of the phage whilst maintaining infectivity. Since there is no involvement of helper phage and no superinfection, phage recovered from the selection procedure can be readily reintroduced into new host cells reducing the time required between repeated selection cycles.

Although not directly relevant for epitope mapping it ought to be mentioned that phage display as such is not restricted to peptides alone and a number of different proteins or protein domains have been displayed on filamentous phage particles. They include enzymes such as alkaline phosphatase (10), trypsin (11), and *Staphylococcus* nuclease (12), growth factors and cytokines such as human growth hormone hGH (13) and IL-3 (14), and receptors like the IgE binding extracellular domain of Fc R1 (15). The display of antibody fragments (Fab) or single chain Fv fragments (scFv) has been extremely successful since they show full antigen binding activity and can be used essentially like free antibodies (16). Some phagemid vectors contain amber stop codons between the foreign gene and the gene coding for the phage coat protein which subsequently allows the production of soluble fusion fragments in non-suppressor strains (9, 15, 16). This facilitates the further analysis of the fused domain or peptide.

The display of antibody specificities as Fab or scFv fragments resulted in a whole new branch of phage display technology and today whole biotech companies are based on this technology. The principle behind this area of phage display is to create huge libraries consisting of fusion phage particles that aim at a complete representation of the entire pool of antibody specificities of an immunized animal, human, or even unimmunized donor. For these so-called 'naive' libraries a further degree of complexity is introduced by replacing the natural V genes of the heavy and light chains with synthetic amino acid sequences which are free of any bias. Screening of these libraries by affinity selection allows the researcher to search for completely new 'monoclonal' antibody specificities that bind to any target of interest including self-antigens and not naturally occurring antigens. This technique also offers the possibility of generating what effectively are human monoclonal antibodies which may become important in a pharmaceutical context. Moreover, it has been demonstrated that the affinity maturation process that usually occurs during an immune response *in vivo* to generate high affinity antibodies by somatic mutation can be mimicked in this *in vitro* system by introducing random mutations and subsequent screening (14, 17).

In this chapter, we introduce the use of a pIII scFv antibody fragment library for the selection of antibody fragments which bind beta-lactoglobulin. It is demonstrated that the antibody specificities isolated from the phage displayed antibody library recognized the same main epitopes as the natural immune response of the animal.

1.2 Overview of phage display libraries

After it had been demonstrated by George Smith that fusion phage can be isolated from large backgrounds of unspecific phage on the basis of using their displayed protein moiety for affinity selection, the next breakthrough was to display random peptides and create large pools of such fusion phage in order to search for phage that bind specifically to any molecule of interest. Since these libraries were initially almost exclusively used in connection with antibodies in order to perform epitope mapping studies the term 'epitope library' was introduced (3) but strictly speaking this term does not apply if other interactions are studied. Therefore the more general term 'random peptide library' is preferable and widely used now.

As already mentioned above biological libraries have distinct advantages over chemical libraries. First, sequence analysis of the isolated ligands (phage particles in this case) is much easier since the selection for binding activity also means selecting for the encoding gene. Secondly, biological libraries are cheaper to make and can be propagated indefinitely. Thirdly, multiple rounds of selection followed by amplification enhance the chances of finding very rare ligands that, in theory, may have existed only once in the original library. On the other hand synthetic libraries offer the advantage of not having a biological bias towards certain amino acid sequences since there is no *in vivo* selection pressure. Further, the presented peptide sequences can almost be seen as 'naked' epitopes since they are not presented in the context of any carrier protein, like for example, a phage coat protein.

Most random peptide libraries comprise between 6–15 amino acids but larger libraries have been reported (18). For reasons explained above pVIII libraries are usually phagemid libraries and therefore require superinfection with helper phage. Both pIII and pVIII fusion libraries offer distinct advantages depending on the application. pVIII libraries have a higher degree of multivalency which results in phage particles capable of interacting with multiple binding sites. This may be helpful if interactions with weak affinities are studied which otherwise would be lost during the selection procedure. In the same way a pIII fusion library can have an advantage if stronger interactions are pursued since too many interactions per phage particle could result in too tight binding and failure to recover the bound phage during the elution step of the selection procedure.

Ideally, a random library covers all possible amino acid sequences using all 20 amino acids. For a random hexapeptide library this means that there have to be at least 20^6 or 6.4×10^7 different transformants to have every possible amino acid combination represented in theory. In reality this will of course not be precisely the case due to biological preference for some amino acid sequences during the infection–amplification cycle of the phage and the abortive effects of others. Further, some amino acid sequences are likely to be thermodynamically less favourable than others and therefore less likely to be present. It is also probable that some sequences will occur more than once further reducing the overall complexity. Consequently, even higher numbers of transformants are desirable, in practice about 2.5×10^9 transformants if one would like to achieve 90% confidence for having all possible peptides represented (19). However, the transformation efficiency of the *E. coli* host cells limits library sizes to around 10^7 to 10^8 transformants if still manageable amounts of competent cells are to be handled. Additionally the transformation efficiency of RF DNA (replicative form;

= double-stranded phage genome intermediate), which is the manipulated vector DNA in the case of non-phagemid libraries, is notoriously poor. In summary this means that all libraries consisting of more than five random residues will only contain a fraction of the theoretically possible amino acid sequences (19). Although it seems contradictive, it has been proposed to increase the number of possible short peptides by cloning larger random peptides. The idea is that, for example, a 38 residue random peptide offers 33 'windows' for unique hexapeptides thus significantly reducing the number of independent transformants required for a hexapeptide library (18).

Several cloning strategies have been developed for the construction of phage display libraries, but essentially they all use chemically synthesized random oligonucleotides which are converted into double-stranded fragments by annealing or polymerase chain reaction (PCR) and ligated into the linearized vector DNA whilst maintaining the reading frame for the phage fusion protein. Often the third base of the triplets coding for the random amino acids are restricted to G or T (3). This still provides all 20 amino acids but reduces the degeneracy of triplets used to code for a single amino acid since several amino acids are represented by more codons than others in the genetic code. Furthermore two of the three stop codons are avoided which would lead to non-functional nonsense fusions. After ligation, *E. coli* cells are transformed with the ligation mixture, usually by electroporation in order to obtain higher transformation efficiencies. To increase the library size several independent transformations are performed and the transformants pooled. Aliquots of the transformed cells are plated out prior to amplification to check transformation efficiency and provide colonies for sequence analysis which is an indicator of the quality of the library. It is important for the integrity of the library that there is no preference for certain amino acids or positions within the displayed peptide. The library is then amplified by growing up the transformed host cells in large scale and if a phagemid-based library is used phage have to be rescued at this stage by superinfection with helper phage. The library consisting of the produced phage particles are then purified by precipitation with polyethylene glycol.

1.3 Applications of phage display libraries

As mentioned before, phage display libraries have, until recently, mainly been used for epitope mapping purposes. Many examples of this have been reported, but the number of completely new defined epitopes that can be located precisely on the antigen remains low (4). Most of the successes are confined to continuous or linear epitopes where the contributing amino acids follow each other in the same sequence as in the primary sequence of the antigen against which the antibodies were raised thereby making identification of the epitope straightforward (3, 4, 8).

One of the key advantages of phage display libraries is that no structural information about the protein of interest is required prior to biopanning. Since the random population will theoretically cover all possible binding sites it was assumed that random peptide libraries might also allow the isolation of sequences which mimic conformational epitopes or binding sites consisting of amino acids that are far apart in the primary sequence of the protein but in close proximity in the folded form. In many cases phage binding such sequences could be isolated but it was difficult or impossible to identify the corresponding residues, i.e. the epitope, in the antigen unless the structure of the native protein was known. As a result, numerous attempts with antibodies recognizing discontinuous epitopes have been unsuccessful in terms of epitope mapping purposes but often successful in identifying sequences which mimic the binding site. These so-called mimotopes can prove to be of some value if they specifically recognize a ligand about which very little prior information was available. In particular, if they recognize molecules of biological or medical interest, and bear the potential of being lead compounds for the development of drugs (20). For the identification of lead compounds for drug discovery binding affinity and high specificity are paramount, and not necessarily the homology of the amino acid sequence with the native antigen as in epitope mapping. There are examples where structural information about the antigen was available and consequently helped the interpretation of screening results (21). Here we report an example where phage display was used for screening antibodies directed against an antigen with known three-dimensional structure. The peptides that bound antibodies against beta-lactoglobulin showed strong consensus motifs which could be modelled on the three-dimensional structure, clearly indicating the true epitopes (5).

Since the introduction of the phage display system and the first demonstrations of its capabilities by George Smith, many researchers have worked with the system and sometimes added or altered it according to their specific needs. Today there are many different types of libraries in use and although the general principle remains the same, the design and methodology vary considerably. Alternatives to bacteriophage display have also been developed which are based on the same idea of complex molecular libraries and screening but use entirely different display vehicles and host organisms. Some of the variations and alternatives are as follows.

For '*constrained phage display libraries*' the conformation of the displayed peptides is constrained by, for example, two cysteine residues flanking the random sequence which then form a disulfide bond with each other (21). This forces the peptide in a cyclic form and limits the range of possible conformations the peptide can assume on the phage particle. It has been found that in many cases peptides that are displayed in a constrained conformation are better recognized since often the target sequence of a ligand also occurs in a more or less constrained environment. The danger is however, that the range of restricted conformations engineered during library construction might not include the conformation suitable for interaction with the molecule, in which case it may be impossible to select for a ligand with this specific library.

'*Synthetic peptide libraries*' have been mentioned earlier and remain important through their advantages already discussed but have in many instances been replaced by the cheaper biological libraries. In fact the identification of reactive peptides with these libraries is often a mass screen rather than a selection procedure and requires high throughput screening technology (22).

Peptides have also been expressed as *fusions to surface proteins of E. coli cells* (23), *yeast cells, and viruses*. In the 'peptide-on-plasmid approach' the peptides are even directly attached to the plasmids that encode them (24). Some of these techniques may offer advantages in very specific situations but will probably not gain the universal success of bacteriophage display or the synthetic peptide libraries. An interesting variation of the pIII fusion phage system has been described recently where the coat-anchored and the exposed infection-mediating domain of pIII are split into two independent molecules and infectivity is only restored in cases where the two domains interact via protein–protein interactions; 'direct interaction rescue' (25) or 'selectively infective phage' (26).

2 Case study: identification of dominant peptide epitopes

2.1 Project strategy

We have used phage display of peptides to characterize the main immunoreactive epitopes within the bovine milk protein beta-lactoglobulin (BLG) (5) as part of a larger project funded by the Agro and Agro-Industrial research program of the European Community. As a whole, the project was set up to increase the overall understanding of defined allergenic epitopes within milk proteins. Our experience of phage display of peptides for epitope mapping of polyclonal antibodies against BLG will be used in this chapter as an example of how to approach an epitope mapping study. We have also made a phage displayed antibody library derived from an immunized mouse spleen using BLG as an antigen and tested the 'monoclonal' antibody fragments for their antigen specificity.

Beta-lactoglobulin (BLG) is the major whey protein in the milk of various species including cows, sheep, goats, horses, and pigs but is not present in humans or rodents (for review see ref. 27). BLG is generally regarded as one of the major allergens in cow milk allergy which affects between 2–7.5% of infants (for review see ref. 28). We have studied in detail defined antigenic determinants (epitopes) within BLG which are recognized by polyclonal antibodies from different immunized species using both phage display of peptides and the Pepscan technique.

Although the Pepscan and phage display epitope mapping techniques are often used to study proteins recognized by monoclonal antibodies we have also shown that existing methodology works equally well for the study of polyclonal antibodies.

2.2 Procedure for the identification of peptide epitopes by phage display

2.2.1 Early considerations for screening phage peptide libraries

i. Choice of phage display peptide library

At the start of the reported study we obtained a sample of the pVIII9aa-cys library from Dr Alfredo Nicosia of the Institute of Molecular Biology, Pomezia, Rome. This library is described by Felici *et al.* (8) and Luzzago *et al.* (21) and consists of random nonapeptides fused to the major coat protein pVIII so that several hundred peptides are displayed on each phage particle. In addition, each peptide is flanked by two cysteines so that at least some of these cysteines form disulfide bonds (21). The presence of constrained peptides within the library may give it significant advantages over other libraries as the conformations which the peptides can adopt are limited (21, 29). We were particularly interested in using this phage library as it had already been used previously to study epitopes recognized by polyclonal antibodies (30).

ii. Purification of polyclonal antibodies from immune sera

For screening of the peptide phage library polyclonal antibodies from rabbits which had been immunized with BLG the antibodies were first affinity purified (see *Protocol 1*). The affinity purification process eliminates the possibility of isolating any other peptides within the library which may bind to antibodies recognizing antigens other than BLG. There are however, examples in the literature of procedures for screening antibodies from whole sera containing polyclonal antibodies of interest (usually patient sera where detailed information about the antigen is unknown) (30, 31).

Protocol 1

Affinity purification of antibodies from immunized rabbits for screening of phage peptide library

Reagents

- Two eight-week-old New Zealand White rabbits
- CNBr-activated Sepharose 4B (Amersham Pharmacia Biotech)
- Tris-buffered saline (TBS): 50 mM Tris–HCl, 140 mM NaCl pH 7.4

Method

1. Immunize rabbits with antigen of interest (in this case beta-lactoglobulin B (BLG) from Sigma) using the immunization schedule of choice.
2. Couple the antigen of interest to approx. 2 ml of CNBr-activated Sepharose using the manufacturer's instructions.
3. Pre-equilibrate a glass column (8 × 1 cm containing 2 ml of CNBr-activated Sepharose coupled to approx. 1–6 mg of antigen) with 10 ml of PBS at room temperature.
4. Load the rabbit antiserum containing specific antibodies (2 ml) onto the column and incubate for 20 min at room temperature.
5. Wash the resin with 10 column volumes of PBS followed by 2 column volumes of 10 mM sodium phosphate, 0.5 M NaCl pH 7.4.
6. Carry out a further wash (5 ml) with Milli Q water to remove traces of sodium phosphate buffer before eluting the antibodies with 3 M $MgCl_2$ (pH 7.4, adjusted with 1 M Tris).

Protocol 1 continued

7 Collect fractions of 1 ml and analyse samples for binding to antigen-coated microtitre plates by ELISA.

8 Pool the purified fractions and dialyse extensively against TBS. Store the pooled fractions in aliquots at −20 °C and use to coat Immunotubes (Nunc) for panning of the phage library, and for phage ELISAs (see *Protocols 2* and *3*).

2.2.2 Panning of phage displayed peptide library

The panning procedure is shown schematically in *Figure 1* and described in detail in *Protocol 2*. The *E. coli* strain chosen for panning is important as it must express an F-pillus so that it can be infected by filamentous phage. We have used the strain XL-1 Blue as this strain has been used routinely in phage work.

i. Immobilization of antibodies and blocking conditions

We panned the pVIII9aa-cys library with antibodies immobilized directly on polystyrene tubes. Other methods for panning involve immobilization of biotinylated antibody to streptavidin-coated plates or to streptavidin-coated magnetic beads. We have found direct immobilization of antibody to be perfectly adequate for panning although it may be advisable to check the activity of the immobilized antibody before commencing, e.g. by detecting activity using labelled antigen if the antigen is known. Blocking of the panning tubes is an important step as the bacteriophage are inherently 'sticky' and minimization of non-specific binding will aid the discovery of specific phage clones. Routinely, the blocking buffer used would contain 1% (w/v) Marvel and 1% (w/v) BSA, but for the BLG epitope mapping study the blocking buffer contained ovalbumin (see *Protocol 2*) instead as BLG is a bovine milk protein.

ii. Phage numbers during panning

We always carry out panning with two tubes, an antibody-coated tube and an uncoated tube (a new uncoated tube is introduced at each panning round). The number of phage added to each tube is usually about 1×10^{11} as this allows approximately $\times$ 1000 copies of the library (the complexity of the library used in this study is 1×10^{8} (8). We usually obtain recovery of 10^{5} infective phage from the uncoated polystyrene panning tube as assessed by titering infective phage at each round of panning. These background results are consistent with previous findings (30). The phage numbers eluted off the uncoated panning tube are compared with the phage numbers eluted off the antigen-coated panning tube after each round by titering the infective phage at each round (see *Protocol 2*). This gives an indication of whether the panning procedure has resulted in enrichment of specific phage displayed peptides which bind to the antibody of interest. After three rounds of panning with rabbit anti-BLG coated tubes we noted that approx. 500 times more phage were eluted off the antibody-coated tube as compared to the uncoated tube.

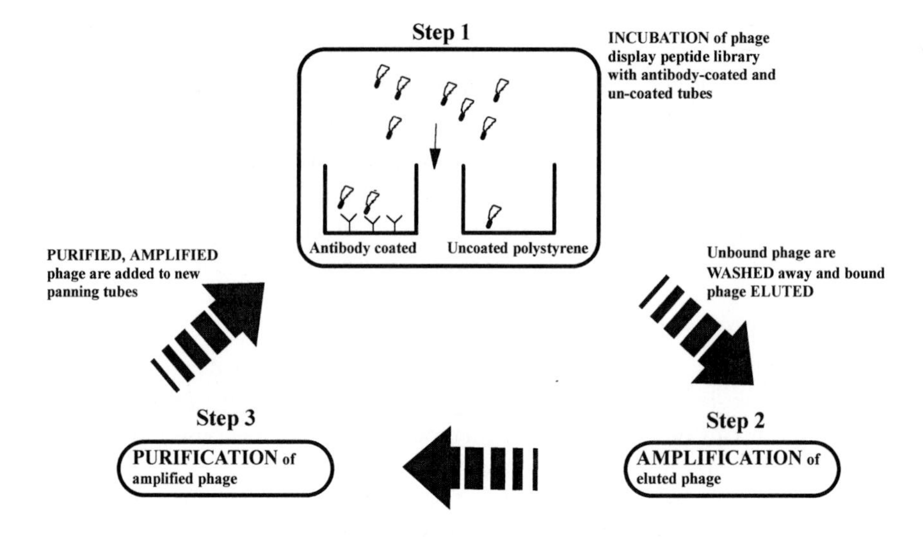

Figure 1 Schematic diagram to show panning of a phage displayed peptide library.

iii. Differing conditions for panning

We carried out two separate screens of the phage library with the same purified antibodies. In the first screen we kept the antibody concentration the same at each round of panning and in the second screen we lowered the coating concentration × 10 at each round of panning. Since the two separate screens gave different peptide epitopes we recommend carrying out a number of separate screens when studying a polyclonal antibody preparation.

Protocol 2

Panning of phage display peptide library

Equipment and reagents

- Polystyrene panning tubes (Immunotubes from Nunc)
- Ampicillin and kanamycin (Sigma)
- M13K07 helper phage (Gibco BRL Life Technologies)
- Coating buffer: 0.1 M $NaHCO_3$ pH 9.0
- pVIII9aa-cys phage library (Institute of Molecular Biology, Pomezia, Rome) shown to have a titre of 1×10^{13} infective phage by infection of logarithmic XL-1 Blue bacteria
- TBS: 50 mM Tris–HCl, 140 mM NaCl pH 7.4
- TTBS: TBS, 0.5% (v/v) Tween 20
- Blocking buffer: TBS containing 10 mg/ml ovalbumin
- Elution buffer: 0.1 M HCl pH 2.2, adjust with glycine, containing 1 mg/ml ovalbumin
- SOBAG agar: 20 g Bacto tryptone, 5 g Bacto yeast, 0.5 g NaCl, 15 g Bacto agar, 10 ml of 1 M $MgCl_2$, 1% glucose, 100 μg/ml ampicillin, make up to 1 litre in distilled water

Protocol 2 continued

- 2TY/glucose/ampicillin: 20 g Bacto tryptone, 5 g Bacto yeast, 0.5 g NaCl, make up to 1 litre with distilled water and autoclave; after cooling, add glucose to 1% and ampicillin to 100 μg/ml
- PEG/NaCl: 2.5 M NaCl containing 20% (w/v) polyethylene glycol 8000

A. Panning on immobilized antibody and amplification of eluted phage[a]

1. Coat polystyrene tubes used for panning with either affinity purified antibodies (20 μg) in 2 ml of coating buffer or with coating buffer only overnight at 4 °C.
2. Wash tubes three times with TTBS and incubate both tubes with 4 ml of blocking buffer for 4 h at room temperature.
3. Add aliquots (10 μl; 1×10^{11} infective phage) of pVIII9aa-cys phage library stock to the antibody-coated and uncoated polystyrene tubes each containing 1 ml of TBS and 1 mg/ml ovalbumin, and incubate overnight at 4 °C.
4. Remove unbound phage by 15 washes (each of 4 ml) with TTBS followed by 5 washes with TBS at room temperature.
5. Elute bound phage by incubation of washed panning tubes with 1 ml of elution buffer for 12 min at room temperature.
6. Transfer eluted phage to 1 ml polypropylene tubes and neutralize with 60 μl of 2 M Tris (pH not adjusted). Also add aliquots (200 μl) of 1 M Tris–HCl pH 7.4 to the panning tubes for neutralization.
7. Add the eluted neutralized phage particles to 9 ml of logarithmic XL-1 Blue bacteria (OD_{600} = 0.2–0.5) growing in 2TY containing 1% (w/v) glucose. Also add logarithmic XL-1 Blue bacteria (4 ml) directly to the neutralized panning tubes.
8. Carry out infection of XL-1 Blue with the eluted phage for 30 min at 37 °C with no shaking.
9. Pool the infected bacteria (total volume 14 ml), add ampicillin (to 100 μg/ml), and incubate the cultures overnight with shaking at 37 °C.
10. Remove a small aliquot (10 μl) of infected bacterial cells prior to overnight incubation for titration (dilute 10^{-2} to 10^{-6} in 2TY/amp/glucose) and plate out 100 μl on SOBAG agar plates.
11. The next day remove an aliquot (150 μl) of the overnight XL-1 Blue culture infected with phage eluted from the panning tube coated with antibodies, add to 15 ml of 2TY/glucose/ampicillin, and grow to logarithmic phase.
12. Superinfect the logarithmic culture with M13K07 helper phage (1×10^{11} phage/ml) by incubation for 30 min without shaking at 37 °C.
13. Centrifuge the bacterial culture for 20 min at 3000 g and drain the cell pellet on tissue. Resuspend the cell pellet in 200 ml of 2TY containing 100 μg/ml ampicillin

Protocol 2 continued

and 50 μg/ml kanamycin and no glucose, and incubate the bacterial cell culture overnight at 37 °C with shaking.

B. Purification of phage

1. Pellet the bacteria grown in part A, step 13 by centrifugation (3000 g, 15 min). Precipitate the phage particles in the supernatant by adding 40 ml of PEG/NaCl and incubate on ice for 1 h.
2. Centrifuge the phage suspension at 15 000 g for 20 min at 4 °C and resuspend the resulting pellet in 20 ml of TBS.
3. Carry out a further PEG/NaCl precipitation by adding 4 ml of PEG/NaCl and incubate on ice for a further 20 min.
4. Resuspend the final phage pellet in 2 ml of TBS which usually results in phage titres of the order of 1×10^{13} infective phage/ml. Calculate this number accurately by infection of logarithmic XL-1 Blue bacteria.
5. Add aliquots of this phage suspension (1×10^{11} phage) directly to new coated or uncoated panning tubes.
6. Repeat the entire panning procedure (parts A and B) a further two times.

[a] Affinity selection of phage was performed by a combination of the methods of Folgori *et al.* (30) and Parmley and Smith (32).

2.2.3 Screening of panned library by phage ELISA

The next step involves screening of individual phage clones (usually about 200) from the third round of panning for their ability to bind to the antibodies of interest as described in *Protocol 3*. If enrichment is obvious after pan 2 then we also screen 200 colonies from this round of panning. Phage rescue is carried out for each individual phage clone as described in *Protocol 3* so that individual phage supernatants displaying only one peptide sequence on their surface can be assessed for binding to the antibody of interest. The assay we employ (*Protocol 3*) involves capture of phage particles by sheep anti-M13 antibodies immobilized on polystyrene microtitre plates, followed by addition of the rabbit anti-BLG antibodies, and then detection of bound antibodies using an anti-rabbit IgG alkaline phosphatase conjugate. We have found this assay to be sensitive and to give low background. It is also possible to carry out phage ELISAs by screening the positive clones by immobilizing the affinity purified antibodies directly on the polystyrene surface, followed by phage supernatants, then sheep anti-M13 antibodies, and then detection by a rabbit anti-sheep IgG alkaline phosphatase conjugate.

Protocol 3

Phage ELISA

Equipment and reagents

- See *Protocol 2* for other common reagents
- 96-well polystyrene microtitre plates (Sterilin and Greiner)
- Sterile wooden cocktail sticks
- Affinity purified sheep anti-M13 bacteriophage antibodies (C. P. Laboratories)
- Anti-rabbit IgG alkaline phosphatase conjugate (Zymed)
- M13K07 helper phage (Life Technologies)
- Substrate solution: *p*-nitrophenyl phosphate (1 mg/ml; Sigma) in 1 M diethanolamine, 1 mM $MgCl_2$ pH 9.8

Method

1. Plate out the output from the third round of panning (XL-1 Blue infected with phage eluted off the antibody-coated plate and the uncoated control) on SOBAG agar plates and incubate overnight at 37 °C.
2. Pick random individual bacterial colonies (~ 200) with toothpicks and add each to the wells of 96-well microtitre plates (Sterilin) each containing 100 μl of 2TY/glucose/ampicillin.
3. Incubate the microtitre plates overnight with shaking at 37 °C.
4. The following day remove aliquots from each well (20 μl) and add to the wells of fresh microtitre plates each containing 200 μl of 2TY/glucose/ampicillin, and incubate with shaking for 1 h at 37 °C.
5. At the next stage, add 25 μl of 2TY/glucose/ampicillin and 10^9 M13KO7 helper phage to each well and incubate for 30 min at 37 °C without shaking.
6. Follow this step by a further incubation for 1 h at 37 °C with shaking.
7. Centrifuge the plates at 800 *g* for 20 min at room temperature, remove the supernatant, and drain the cell pellets. Resuspend each pellet in 200 μl of 2TY containing ampicillin (100 μg/ml) and kanamycin (50 μg/ml) and no glucose, and incubate with shaking at 37 °C overnight.
8. Centrifuge the microtitre plates containing the overnight bacterial cultures at 800 *g* for 20 min and add the phage-containing supernatants to sheep anti-M13 bacteriophage-coated microtitre plates (100 μl/well). The anti-M13 plates are prepared as follows:-
   (a) Coat the wells of Greiner high bind plates with sheep anti-M13 antibodies by overnight incubation (100 μl/well; 10 μg/ml) at 4 °C in binding buffer (see *Protocol 2*).
   (b) Block the plates with blocking buffer (see *Protocol 2*) (200 μl/well) for 1 h at room temperature.
9. Remove unbound phage from the sheep anti-M13-coated plates by five washes with PBST.

Protocol 3 continued

10 Add affinity purified rabbit antibodies purified by the method described in *Protocol 1* (2 μg/ml in blocking buffer; 100 μl/well) and incubate for 2 h at room temperature.

11 Add alkaline phosphatase-conjugated goat anti-rabbit immunoglobulin (100 μl/well) at a dilution of 1/1000 (in blocking buffer) and incubate for a further 2 h at room temperature.

12 Develop the assay with 100 μl/well of substrate solution for 10–20 min and read the optical densities at 410 nm.

2.2.4 Results of phage ELISA

After three rounds of panning with anti-BLG antibodies, approx. 10% of the individual phage clones screened by phage ELISA were considered to be positive. Positive signals were judged as those that gave ELISA signals which were four to five times higher than the background, i.e. in wells with no phage present or in wells with no added antibody conjugate.

A number of assays can be designed to study the specificity of the putative positives in more detail. The individual phage clones which gave higher signals were assessed in assays which tested the inhibitive effect of intact BLG on the binding of purified anti-BLG antibodies to individual phage (see *Figure 2*). In the presence of increasing concentrations of intact BLG, dose-dependent inhibition of antibody binding to individual phage clones was observed. An irrelevant competitor (BSA) had very little effect on binding (see *Figure 2*).

2.2.5 Analysis of peptide sequences

i. Sequencing of random peptides

In order to analyse the sequences of the peptides which are positive in phage ELISA double-stranded phagemid DNA can be purified from bacterial cultures infected with individual phage clones and used for DNA sequencing. We used the oligonucleotide primer:- TTT CCC AGT CAC GAC GTT G so that analysis of the inserted DNA sequence encoding the random nonamer could be obtained by automated sequencing. Sequences in the region encoding the random nonapeptide can then be assessed for homology to the primary sequence of the antigen of interest (if known).

ii. Linear versus conformational epitopes

The sequences we obtained during two separate screens of the pVIII9aa-cys library resulted in 19 different peptides which could be divided into groups which had homology with four different regions of BLG (see *Table 1*). The peptides were aligned with the BLG sequence and similarities between the peptide and whole antigen assessed by eye. All identified peptides were found to have similarities with the linear sequence of BLG indicating that there were no conformational epitopes represented in the peptides isolated. In a separate study on antibodies

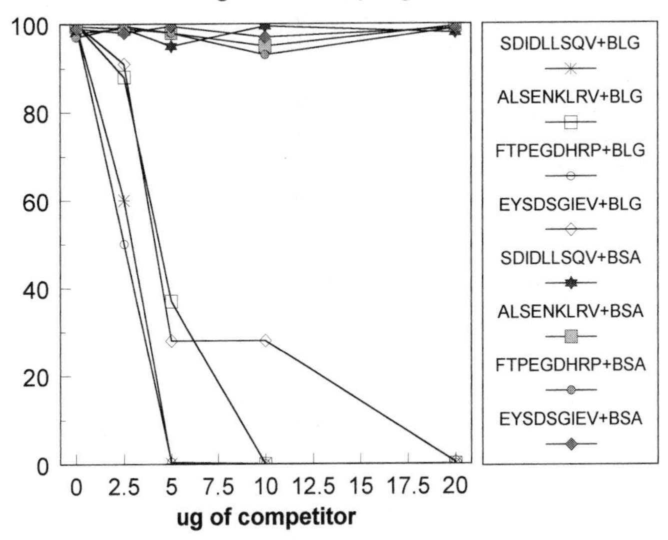

Figure 2 Effect of intact BLG on the binding of rabbit anti-BLG IgG to individual phage clones. The phage ELISA was carried out as described in Protocol 3 except that serial twofold dilutions of BLG (100 μl; maximum conc. 200 μg/ml), BSA, or PBST only were added to wells containing positive phage clones captured by immobilized sheep anti-M13. Affinity purified rabbit anti-BLG antibodies were then added immediately to each well (2.5 μl from purified stock; 130 μg/ml in TBS). Detection of rabbit antibody bound to individual phage clones was then made using the standard alkaline phosphatase anti-rabbit IgG conjugate used for all other phage ELISAs. Phage clones were chosen so that one peptide would be tested from each of the peptide groups shown in *Table 1*.

against heat denatured BLG we did find peptides which were convincing positives in the specificity assays but when sequenced had no obvious sequence similarity to the linear sequence of the BLG. These peptides may represent conformational epitopes rather than linear ones which was an unexpected finding when screening antibodies against denatured antigen.

iii. Detailed analysis of positive phage peptides obtained with anti-BLG antibodies

Some peptides were isolated more than once in one screen, e.g. RWWQETDDS which was identified six times in the first screen and falls into Group I in *Table 1*. This group of peptides is homologous to residues 126–134 in BLG. Group II was made up of peptides which were found to mimic residues 86–94 in BLG. In this group clone 6 showed a striking resemblance to BLG as its peptide sequence (ALSENKLRV) contained six exact matches to the BLG sequence and one residue which was a conservative replacement. Groups I and II contained phage peptides which were isolated only in the first screen or second screen as well as some which were isolated after both screens. The phage peptides which formed Groups III and IV had sequence homology to regions 49–53 and 26–35 in BLG respectively and were all isolated in the second screen of the library only.

Table 1 Alignment of BLG sequence with deduced nonapeptide sequences of the phage peptides isolated by two separate screens of the library[a]

Group I		
BLG	123 VRT**PEVDD**EAL**EK** 135	No. sequenced
1	**P**WW**DD**PQDI	1 (1)
2	RWWQ**ET**D**DS**	6 (1)
3	**E**A**DD**SSPGR	1 (1)
4	NIYA**E***L***DD**S	2 (1)
5	**E**YS**D**SGI**EV**	6 (1 + 2)
Group II		
BLG	83 KID**ALNENK***V***LV** 94	
6	**AL**S**ENK***L***RV**	2 (1 + 2)
7	**A**QS**EN**ARP**V**	1 (1)
8	HSS**EN***R*SS	3 (1 + 2)
9	FPS**A**FS**EN**Q	1 (1)
10	LPFS**EN***R*T	2 (1)
11	PTS**EN***R*AHS	2 (1 + 2)
12	LPSP**LLEN***R*	5 (1 + 2)
13	**SL**Y**EN***RRV*	1 (2)
14	IKNPSS**ENA**	1 (2)
Group III		
BLG	47 KP**TPEGD**LEILL 58	
15	F**TPEG**HRP	1 (2)
16	QFF**PEGD**YL	1 (2)
Group IV		
BLG	26 **ASDISLLD**AQ 35	
17	**SDI**D**LL**SQV	1 (2)
18	RQ**D**L**S**T**LD**	1 (2)
19	R**AS**VDRD**L**A	1 (2)

[a] The numbers either side of the BLG sequence denote amino acid sequence numbers. Letters in bold denote exact matches of phage sequences with the BLG sequence and italicized bold letters denote conservative substitutions. Each peptide is flanked by two cysteines which have not been shown in this table. The numbers on the right indicate the number of times the corresponding peptide was isolated and the numbers in brackets refer to the screen at which the peptide was isolated, i.e. the first screen, the second screen or both.

iv. Effect of linear synthetic peptides on the binding of polyclonal antibodies to individual phage clones

We assessed the effect of two synthetic peptides mimicking two individual regions of the antigen on the binding of purified antibodies to the individual phage clones (see *Figure 3*). Peptide 1 corresponding to amino acids 120–135 (QCLVRTPEVDDEALEK) of BLG was synthesized by Ms L. Law at the Institute of Food Research, Norwich and peptide 2 corresponding to amino acids 81–91 (CFKIDALNENKV) of BLG was synthesized by Dr M. Munns (Peptide and Protein Research Consultants) at the University of Exeter, UK.

Somewhat surprisingly, the presence of the linear synthetic peptide resulted in complete blocking of antibody binding to phage clones displaying the constrained peptide which mimicked the same region of BLG. In the case of the BLG

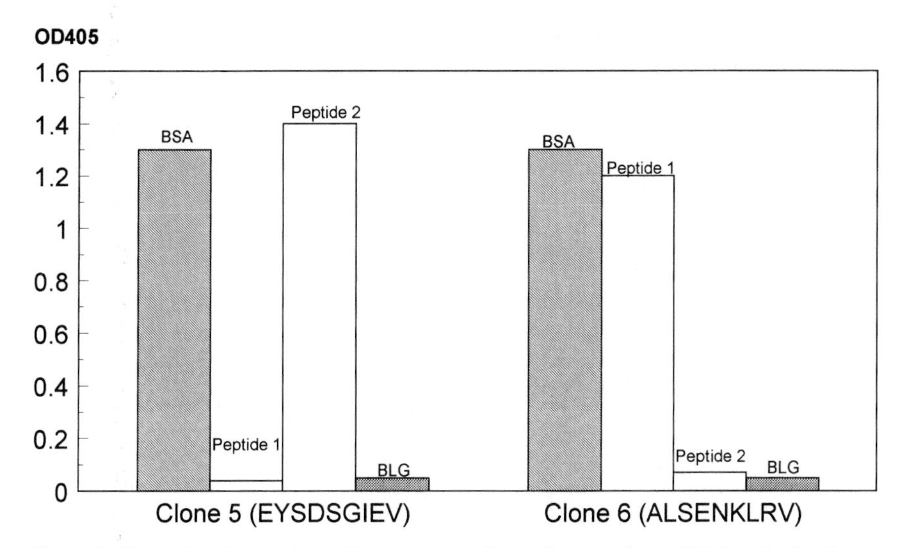

Figure 3 Effect of synthesized peptides corresponding to linear regions of BLG on the binding of rabbit anti-BLG IgG to individual phage clones. Inhibition experiments were carried out as for *Figure 2* except that antibody binding to individual phage clones (clone 5 [EYSDSGIEV] and clone 6 [ALSENKLRV]) was tested in the presence of 250 μg/ml (100 μl, in PBS) of peptide 1 (CLVRTPEVDDEALEK), peptide 2 (CFKIDALNENKV), BLG, or BSA.

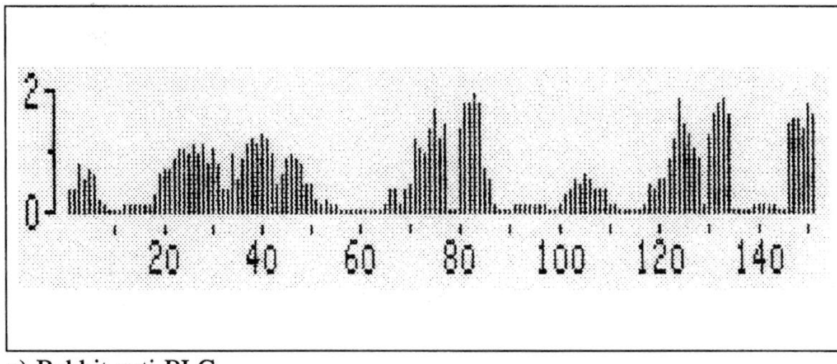

a) Rabbit anti-BLG

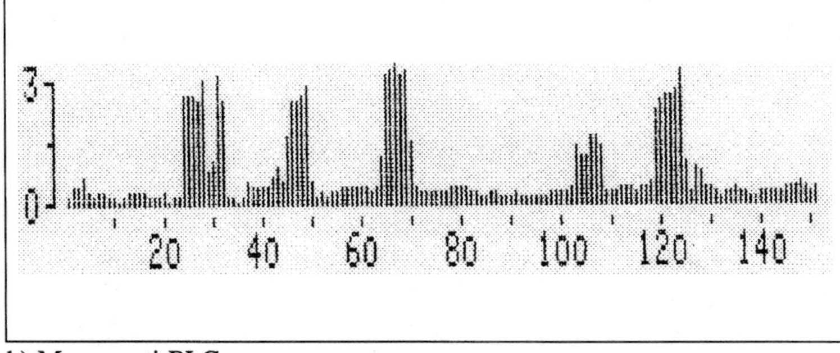

b) Mouse anti-BLG

Figure 4 Pepscan analysis of (a) anti-BLG rabbit antisera, and (b) mouse antisera.

epitopes isolated it would therefore seem that the constraint introduced by loop formation within the peptide is not necessary for antibody binding.

2.2.6 Pepscan assays

The same rabbit antibodies used to screen the phage peptide library were also analysed using the alternative technique of Pepscan in order to be able to compare the two techniques (*Figure 4a*). We also used Pepscan to characterize epitopes recognized by an antiserum raised in a mouse, the spleen of which was subsequently used to construct an antibody scFv antibody library using phage technology (*Figure 4b*, also see Section 2.3). A set of 151 peptides corresponding to 12-mers spanning the amino acid sequence of BLG (each peptide overlapping with the previous peptide by 11 amino acids) were synthesized on polyethylene pins using established Pepscan methodology at ID-DLO, Lelystad, The Netherlands (22). The polyethylene pins were held on racks which had a 96-well format compatible with conventional ELISA plates. The peptides were tested for their ability to bind rabbit and mouse anti-BLG polyclonal antibodies using ELISA as described in *Protocol 4*.

Protocol 4

Pepscan ELISA

Reagents

- Blocking buffer: PBS containing 5% ovalbumin, 5% horse serum, and 1% Tween 80
- Substrate solution: 50 mg ABTS and 20 μl of 30% (w/w) H_2O_2 solution in 100 ml of 0.1 M sodium phosphate/0.08 M citrate buffer pH 4.0
- Wash buffer: PBS containing 1% (v/v) Tween 80
- Sonicating buffer: 10 mM sodium phosphate buffer, 1% (w/v) SDS, and 0.1% (w/v) 2-mercaptoethanol (added just prior to sonication)

Method

1. Block non-specific binding sites on the pins by incubation in blocking buffer in the wells of a Sterilin microtitre plate (150 μl/well) for 30 min at room temperature.
2. Dilute antiserum in blocking buffer (100 μl/well) and incubate overnight at 4 °C.
3. Wash the pins carefully (× 3) in wash buffer at room temperature.
4. Transfer the pins to the wells of a microtitre plate containing 100 μl of the relevant anti-immunoglobulin horseradish peroxidase conjugate and incubate for 1 h at 37 °C.
5. Carry out further washes and detect the presence of antibody by incubation in fresh substrate solution. Develop the plate for 10–20 min.
6. Read plates at 405 nm.

Protocol 4 continued

7. In order to remove bound antibody from pins for reuse sonicate the pins in sonicating buffer in a polyethylene container (wall approx. 2 mm thick) in a 60 litre 2000 watt sonicating water-bath for 60 min at 70 °C.
8. Rinse the pins with Milli Q water and then re-sonicate in Milli Q water for a further 1 h at 70 °C.
9. Store the pins in a plastic container at −20 °C.

2.2.7 Comparison of epitopes recognized by rabbit anti-BLG antibodies discovered by Pepscan and by phage display

Two screens of the phage library provided peptide sequences which had sequence similarity with four regions within BLG. The Pepscan analysis for the same antibody preparation exhibited nine positive regions (*Figure 4a*). There may be several explanations for the fact that the technique of phage display resulted in isolation of a subset of the epitopes that are recognized by Pepscan. It is possible that screening of larger numbers of phage clones would have resulted in the isolation of additional phage peptides which had sequence homology to the extra BLG epitope regions found by Pepscan.

The results obtained may also point to the fact that the use of different epitope mapping techniques may give different peptides as epitopes. The environment in which the peptide is displayed by the two techniques is very different, i.e. the bacteriophage surface compared to the polypropylene of the Pepscan pins. It is therefore perhaps not surprising that the epitopes identified differ slightly. The screening of several different phage peptide libraries may also result in the accumulation of more epitope data especially for a polyclonal antibody preparation, e.g. libraries displaying peptides of differing lengths or linear as well as constrained, etc. The epitopes identified in this study have also been mapped onto the three-dimensional structure of BLG and they have all been shown to be clustered on the surface of the molecule (5).

2.3 Deriving peptide binding antibody fragments

Pepscan and random peptide library screening/analysis of polyclonal sera has shown to be a very powerful tool for the identification of those regions of a protein which are most antigenic. Pepscan and random peptide phage display techniques in the BLG study described in Sections 2.1 and 2.2, identified nine or four regions in the BLG molecule respectively, to be most antigenic. However, this still does not say anything about the total response, because especially the Pepscan analysis only identifies the linear epitopes that are recognized by antibodies produced by the immune response. We therefore decided to isolate monoclonal anti-BLG antibodies and screen these for fine antigen recognition using Pepscan and phage display. It was anticipated that among the panel of monoclonal antibodies a significant portion would recognize non-linear and/or hidden epitopes. These results, together with the data obtained with the analysis of the polyclonal response, would then give us a complete picture of antibody response against BLG.

In addition to the above there is a second argument that can be raised for combining Pepscan analysis and the isolation of monoclonal antibodies. Microbial and plant production system have the potential to allow the industrial exploitation of antibody fragments as fine chemicals. It is implicit that during the processing, product recovery is attained with full bioactivity. The route to ensure this is achieved through affinity purification processes. However, basic research models of antigen affinity reactions using the whole antigen as the affinity target are generally not suited for scale up to allow industrial application. One solution to this problem is the use of small peptide analogues of the antigen which can be covalently coupled to an inert affinity matrix material. These peptides can be consistent with epitopes of the antigen, or unrelated sequences which are shown to be recognized by the same antibody.

Table 2 Primer sequences

Code	Target	Sequence
MK09	VL cDNA	5′ AGATGGATACAGTTGGTGCAGCAT 3′
MG008	VH cDNA	5′ GATAGACAGATGGGGGTGTCGTTT 3′
PCR.116	3′ VL	5′ GTTAGATCTCGAGCTTGGTCCC 3′
PCR.154	3′ VL	5′ CCGTTTTATTTCCAACTTTGTCCC 3′
PCR.322	3′ VL	5′ GAGTCATTCTGCGGCCGCCCGTTT(C/T)A(G/T)CTCGAGCTT(G/T)GTCCC 3′
PCR.90	5′ VL	5′ GACATTGAGCTCACCCAGTCTCCA 3′
PCR.175	5′ VL	5′ GAAATTGT(G/T)CTCAC(C/A)CA(G/A)TCTCC 3′
PCR.176	5′ VL	5′ GACATCCAGATGAC(C/A)CAG(T/A)CT(C/A)C 3′
PCR.177	5′ VL	5′ GATATTGTGATGATGAC(C/A)CAGG(C/A)T 3′
PCR.178	5′ VL	5′ GATGTTGTGATGACCCAAACTCC 3′
PCR.179	5′ VL	5′ A(A/G) (T/C)ATTGTGATGACCCAG(A/T)CTC 3′
PCR.191	5′ VL	5′ GCCATCGAGCTCAC(C/T)CA(A/G)AT(T/C)AC 3′
PCR.185	5′ VL	5′ GACGTGGTGATGACCCAG(T/A) (C/G) (T/C)CC 3′
PCR.186	5′ VL	5′ GACAT(T/C)CAGATGAC(C/T)CAGAC(T/C)AC 3′
PCR.226	5′ VL	5′ GACATTGTGCTGAC(T/C)CA(G/A)T(C/T)TCC 3′
PCR.227	5′ VL	5′ CAAATTGTTCTC(A/T) (C/T) (C/A)CAGTCTCC 3′
PCR.228	5′ VL	5′ GACATTGTGATG(A/T)CACA(G/A)TC(T/G)CCA 3′
PCR.229	5′ VL	5′ GATATTGTGATGAC(G/T)CAGG(C/A)T(G/A) (C/A)A 3′
PCR.358	5′ VL	5′ GAT(G/A)TT(G/T)TGATGACCCA(G/A)AC(G/T)GCA 3′
PCR.359	5′ VL	5′ GATATTGTGATGACCCA(G/A) (C/G) (A/C)TG 3′
PCR.89	3′ VH	5′ TGAGGAGACGGTGACCGTGGTCCCTTGGCCCC 3′
PCR.51	5′ VH	5′ AGGT(C/G) (A/C)A(A/G)CTGCAG(C/G)AGTC(A/T)GG 3′
PCR.162	5′ VH	5′ CATGCCATGACTCGCGGCCCAGCCGGCCATGGCC(C/G)AGGT(C/G) (A/C)A(A/G)CTGCAG(C/G)AGTC(A/T)GG 3′

2.3.1 Construction of an antibody fragment library

For reasons outlined above we set out to isolate monoclonal antibodies from a mouse whose anti-BLG serum was analysed by Pepscan and was shown to be very similar to the pattern observed for the rabbit anti-BLG response illustrated in *Figure 4*. Instead of following the traditional route to isolate monoclonal antibodies we decided to use phage display technology to allow the direct isolation of BLG-specific antibodies from the spleen of an immunized mouse.

Following the method outlined in *Protocol 5* and using the primers detailed in *Table 2*, a VH library of 6×10^5 individual members and a final scFv.BLG library of 4×10^6 individual clones in pHEN.1 (16) was constructed.

Protocol 5

Construction of a scFv phage display library

Reagents

- mRNA isolation kit (Amersham Pharmacia Biotech "Quickprep")
- SOBAG agar: 20 g Bacto tryptone, 5 g Bacto yeast, 0.5 g NaCl, 15 g Bacto agar, make up to 1 litre with distilled water and autoclave; add 10 mM $MgCl_2$, 1% glucose, 100 μg/ml ampicillin
- XL-1 Blue electrocompetent *E. coli* (Stratagene)
- Restriction enzymes: *Sfi*I, *Bst*EII, *Not*I, and *Sac*I (New England Biolabs)
- Restriction enzyme buffer (10 ×): 500 mM Tris–HCl pH 8.0, 100 mM $MgCl_2$, 500 mM NaCl (Gibco BRL-2)

A. Isolation of mRNA

1. Isolate spleen cells from one whole spleen by perfusion with PBS.
2. Isolate the mRNA from these spleen cells (we used the Pharmacia Quickprep mRNA isolation kit according to the manufacturer's instructions).
3. Store the purified mRNA in 70% ethanol.

B. cDNA synthesis

1. Dissolve approx. 5 μg mRNA in 22 μl distilled H_2O and denature for 10 min at 65 °C.
2. Place on ice and add 5 μl mRNA solution to each of the following mixes. For sequences of primers used see *Table 2*.

Code	Mix	Primer	H_2O	$RT^{Superscript}$
L+	10 μl	2 μl (5 pmol) MK09	2 μl	1 μl (10 U)
L-	10 μl	2 μl (5 pmol) MK09	3 μl	-
H+	10 μl	2 μl (5 pmol) MG008	2 μl	1 μl (10 U)
H-	10 μl	2 μl (5 pmol) MG008	3 μl	-

Mix:

- 20 μl of 5 × RT buffer (Gibco BRL Superscript)
- 2.5 μl dNTPs (20 mM each)
- 10 μl DTT solution (Gibco BRL Superscript)

Protocol 5 continued

- 4 μl RNasin (40 U/μl)
- 13.5 μl H_2O

3 Incubate reaction mixtures for 2 h at 45 °C.

C. Isolation of VH and VL gene fragments by PCR

1 PCR reactions:
- 10 μl of 10 × Vent DNA polymerase buffer
- 2 μl dNTPs (20 mM stock)
- 30 pmol 5′ primer
- 30 pmol 3′ primer
- 1 μl (1 U) Vent DNA polymerase
- dH_2O to a final volume of 100 μl

2 The reaction mixtures were overlaid with 100 μl of mineral oil. The reaction conditions were as followed:

(a) 1 × 94 °C for 4 min.

(b) 33 × 94 °C for 1 min, 55 °C for 1 min, and 72 °C 1 min.

3 Use 1 μl of the cDNA reactions (L± and H±) for each amplification reaction. For the isolation of the VH genes primers we used PCR.89 and PCR.51 (*Table 2*). For isolation of the VL genes we used all combinations of 5' VL primers (*Table 2*) with PCR.116 and PCR.154.

D. Construction of scFv library in pHEN.1

1 Introduce the 5′ *Sfi*I restriction site necessary for cloning the VH fragments as *Sfi*I/*Bst*EII fragments into pHEN.1 (16) by a second PCR reaction using the initial PCR product as template using primers PCR.162 and PCR.89.

2 Pool the reaction products, purify from agarose gel using DEAE membranes (33), and digest overnight at 50 °C in:
- 20 μl of 10 × restriction enzyme buffer
- 8 μl spermidine (100 mM)
- 8 μl BSA (10 mg/ml)
- 4 μl *Sfi*I (40 U)
- 4 μl *Bst*EII (40 U)
- 154 μl distilled H_2O

3 Purify the *Sfi*I/*Bst*EII digested VH fragments from an agarose gel using DEAE membranes and insert into pHEN.1 *Sfi*I/*Bst*EII digested vector. Each reaction contains:
- 100 ng pHEN.1 *Sfi*I/*Bst*EII
- 50 ng insert DNA
- 2 μl of 10 × ligation buffer
- 1 μl DNA ligase
- H_2O up to a final volume of 20 μl

Protocol 5 continued

4. Incubate the ligation reactions at room temperature for 4 h.
5. Transform electrocompetent *E. coli* XL-1 Blue with the purified ligation reactions (purified by phenol:chloroform extraction, followed by chloroform extraction, and then ethanol precipitation).
6. Select transformed *E. coli* on SOBAG plates (overnight 37 °C).
7. After counting the colonies to estimate the VH library size, scrape the plates and isolate the DNA.
8. Introduce the 5′ *Sac*I and the 3′ *Not*I restriction sites necessary for cloning the VL fragments into the VH library in pHEN1 by a second PCR reaction using primers PCR.90 and PCR.322 and 1 μl of the initial PCR products as template.
9. Pool the PCR products and purify from an agarose gel using DEAE membranes (33) and digest overnight at 37 °C in:
   - 20 μl of 10 × restriction enzyme buffer
   - 8 μl spermidine (100 mM)
   - 8 μl BSA (10 mg/ml)
   - 8 μl *Sac*I (80 U)
   - 154 μl dH_2O
10. Next, digest the VL fragments with *Not*I by overnight incubation at 37 °C after adding:
    - 10 μl of 10 × restriction enzyme buffer
    - 3.75 μl of 4 M NaCl
    - 78.25 μl dH_2O
    - 8 μl *Not*I (80 U)
11. Purify the *Sac*I/*Not*I digested VL fragments from an agarose gel using DEAE membranes and insert into the *Sac*I/*Not*I digested pHEN.VH library. Each reaction contains:
    - 100 ng pHEN.VH (*Sac*I/*Not*I)
    - 50 ng insert DNA
    - 2 μl of 10 × ligation buffer
    - 1 μl DNA ligase
    - H_2O up to a final volume of 20 μl
12. Incubate the ligation reactions at room temperature for 4 h.
13. Transform electrocompetent *E. coli* XL-1 Blue with the purified ligation reactions.

2.3.2 Screening the scFv anti-BLG antibody fragment library

The anti-BLG mouse scFv library was screened for antibodies that bind:

(a) Whole BLG.

(b) Peptide BLG-2: C-LVRTPEVDDEALEK (position 121–135).

(c) Peptide BLG-3: C-AQKKIIAEKTK (position 67–77).

(d) Peptide BLG-4: C-MENSAEPEQSL (position 107–117).

(e) Peptide BLG-5: C-PTPEGDLEILL (position 48–58) using the method outlined in *Protocol 6*.

Protocol 6

Screening of a scFv phage display library

Reagents

- FCS-T: fetal calf serum containing 0.15% (v/v) Tween 20
- Elution buffer: 0.1 M HCl–glycine pH 2.2, 1 mg/ml BSA
- See *Protocol 2* for other common reagents
- The peptides were coupled to maleimide activated BSA via the N-terminal cysteine (protocol as described by the manufacturer, Perbio Science UK)

Method

1. Inoculate 15 ml 2TY/amp/glucose with 50 μl of overnight culture of the anti-BLG library in XL-1 Blue and grow until the culture has reached log phase (A_{600} = 0.3–0.5).
2. Add 4.5 × 10^9 M13K07 helper phage and incubate for 30 min at 37 °C without shaking.
3. Spin the infected cells at 4000 *g* for 10 min. Resuspend the pellet in 200 ml 2TY containing ampicillin (100 μg/ml) and kanamycin (50 μg/ml), and incubate at 37 °C overnight (200 r.p.m.).
4. Spin the overnight culture at 4000 g for 10 min and add 1/5 volume PEG/NaCl to the supernatant, mix well, and leave in ice-water for 1 h.
5. Pellet the phage particles by centrifugation at 13 000 g for 30 min and resuspend the phage pellet in 20 ml water and add 4 ml PEG/NaCl solution. Mix and leave for 15 min in ice-water.
6. Pellet the phage particles by centrifugation at 4000 g for 15 min and resuspend the phage pellet in 2 ml FCS-T.
7. Add 1 ml phage suspension to Nunc Immunotubes (5 ml) which are pre-coated overnight at 37 °C with 1 ml target antigen in carbonate buffer (10 μg/ml) followed by a blocking with 5 ml FCS-T for 1 h at 37 °C.
8. Incubate phages for 2–4 h at room temperature with occasional shaking.
9. Remove unbound phage by washing the tube 20 times with PBS-T followed by 20 washes with PBS.
10. Elute phage from tube by incubating the tubes with 1 ml elution buffer for 15 min at room temperature.
11. Neutralize the eluted phage by adding 60 μl of 2 M Tris.

Protocol 6 continued

12 Reinfect 9 ml log-phase *E. coli* XL-1 Blue with the eluted phages. Also add 4 ml log-phase *E. coli* XL-1 Blue to the Immunotube. Incubate both cultures for 30 min at 37 °C without shaking to allow for infection.

13 Pool the 10 ml and 4 ml of the infected XL-1 Blue bacteria and make serial dilutions of 10^{-1} to 10^{-6}. Plate these dilutions on SOBAG agar plates and grow overnight at 37 °C.

14 Take the remaining infected XL-1 Blue culture and spin at 4000 g for 10 min. Resuspend the pelleted bacteria in 1 ml of 2TY and plate 100 μl aliquots on SOBAG agar plates. Grow overnight at 37 °C. The selected clones can then be harvested and pooled to repeat the panning process or alternatively, individual clones can be assayed for antigen binding activity.

Screening was performed as described in *Protocol 6*. Selection of binding clones (panning) was always carried out with two tubes, an antigen coated tube and an uncoated tube (Ctrl). The results of the panning reactions are outlined in *Table 3*. As with the screening of the phage displayed peptide library the phage numbers eluted off the coated plate are compared to the phage numbers eluted off the uncoated tube. This gives an indication of whether the selection has resulted in an enrichment of specific antigen binding antibody-phage clones. A complicating factor in the screening of the library is the fact that it was not possible to use the preferred blocking agent 'Marvel' because of its high BLG content. This inability to use Marvel resulted in an increase of the background binding or stickiness of the phages which is reflected in the high background observed in the uncoated tubes. Nevertheless, enrichment was observed for whole BLG and peptide BLG-2 after two and three rounds of panning. Because of the high background it was not possible to isolate antibody fragments recognizing the peptides BLG-3 to -5.

Table 3 Panning results[a]

Antigen	Panning I	Panning II	Panning III
Whole BLG	4×10^5	4×10^6 ****	2×10^8 ****
Ctrl	5×10^5	5×10^5	5×10^7
Peptide BLG-2	4×10^5	3×10^6 ****	4×10^8 ****
Ctrl	5×10^5	6×10^5	4×10^7
Peptide BLG-3	2×10^5	1.5×10^6	$> 10^8$
Ctrl	5×10^5	4×10^5	$> 10^8$
Peptide BLG-4	3×10^5	1.3×10^6	5×10^7
Ctrl	5×10^5	10^6 5×10^7	
Peptide BLG-5	6×10^4	9×10^5	$> 10^8$
Ctrl	5×10^5	2×10^6	$> 10^8$

[a] Numbers are total colony counts, e.g. total number of bound phages recovered. Asterisks indicate enrichment over background.

2.3.3 Identification of individual BLG binding antibody fragments

96 individual clones/colonies of libraries panned on whole BLG and the peptide BLG-2 were analysed for BLG binding and (if appropriate) peptide binding following the method outlined in *Protocol 7*.

Protocol 7

ScFv phage ELISA

Reagents

- Blocking buffer: fetal calf serum containing 0.15% (v/v) Tween 20
- See *Protocol 3* for other common reagents
- Rabbit anti-M13 polyclonal serum (in-house reagent)

Method

1. Inoculate 100 μl of 2TY/amp/glucose in 96-well microtitre plates with individual, well-isolated colonies using sterile toothpicks, and grow the cultures overnight at 37 °C with shaking at 100–150 r.p.m. ('master plate').
2. For each microtitre plate, prepare 21 ml of 2TY/amp/glucose medium containing 10^{10} M13K07 helper phage, and pipette 200 μl into each well of a sterile V-bottom microtitre plate.
3. Transfer 20 μl of the culture from the master plate to a corresponding well in the second plate and incubate the V-bottom microtitre plate for 2× h at 37 °C with shaking at 150 r.p.m.
4. Centrifuge the plate at 800 g for 10 min and remove the supernatant from each well by inverting the plate on tissues and tapping the plate dry on tissues.
5. Add 200 μl 2TY/glucose containing 50 μg/ml kanamycin (with no glucose) to each well of the microtitre plate from the previous step and incubate the plate overnight at 37 °C with shaking at 150 r.p.m.
6. Centrifuge the plate as described above and mix the phage supernatants with an equal volume of blocking buffer and incubate at room temperature for 30 min.
7. Add 100 μl phage supernatant/blocking buffer to pre-blocked Greiner microtitre plates which have been sensitized with the antigen of interest which in our case was BLG or BLG peptides (10 μg/ml in carbonate buffer, 100 μl/well, overnight at 37°C).
8. Allow the phages to bind to the antigen for 1–2 h at 37 °C. Remove unbound phage by four washes with PBS-T.
9. Add 100 μl of an appropriate dilution of a polyclonal rabbit anti-M13 antibody (we have an in-house reagent) in blocking buffer to each well and incubate at 37 °C for 1 h.
10. Remove unbound antibody by four washes with PBS-T.

Protocol 7 continued

11. Add 100 μl of an appropriate dilution of an alkaline phosphatase-conjugated anti-rabbit antibody (in blocking buffer) to each well and incubate at 37 °C for 1 h.
12. Remove unbound antibody by four washes with PBS-T and detect alkaline phosphatase activity by adding 100 μl substrate solution to each well.
13. When colour development is clearly visible (usually within × h), take an absorbance reading at 405 nm.

Table 4 Screening results: the number of clones that specifically recognized and bound the target antigen(s) after I, II, and III rounds of panning

Library panned on:	Antigen(s) used in phage ELISA	Panning I	Panning II	Panning III
BLG	BLG	0/96	4/96	1/96
Peptide BLG-2	BLG + Peptide BLG-2	4/96	21/96	11/96

The results of this analysis as outlined in *Table 4* indicate that by panning on BLG we isolated antibody fragments that specifically recognize BLG and by panning on peptide BLG-2 we isolated antibody fragments that specifically bind peptide BLG-2 *and* whole BLG. However, compared to results obtained with many other targets, the overall results are disappointing. Only low numbers of specific clones could be identified from a large background of 'sticky' clones.

2.3.4 Detailed analysis of individual BLG binding antibody fragments

Detailed analysis of the clones isolated via panning on BLG and peptide BLG-2 clearly showed that they all bound specifically to BLG. In addition, both clones isolated via panning on whole BLG and peptide BLG-2 bound peptide BLG-2, suggesting that we isolated antibodies recognizing the same epitope by panning on whole BLG and the peptide BLG-2. DNA sequence analysis of all positive clones confirmed that the same clones were isolated on both target antigens. Only two DNA sequences with only minor differences, most of these located in the CDR-3 of the heavy chain (see *Figure 5*), could be identified. This lack of diversity is very surprising since previous experiments using the same phage display technology to isolate antibodies against different targets usually yielded between 8–16 different DNA/antibody sequences. The fine antigen binding specificity of both scFv fragments was confirmed by Pepscan analysis (*Protocol 4*) of soluble scFv fragments (production of which is outlined in *Protocol 8*), clearly demonstrating the specificity of the antibody binding activity (see *Figure 6*).

In Section 2.3 we have described the use of phage display technology to isolate monoclonal antibody fragments with predetermined antigen binding specificity. Despite the specific problems linked to the isolation of anti-BLG antibodies we have isolated antibody fragments together with the relevant peptide epitope and showed that it is possible to identify the most immunogenic regions of a target protein and use this information to construct small peptides for use as affinity selection ligands. The fact that the antibody fragments isolated via this route also bind the whole intact antigen makes this combined epitope discovery/ antibody phage display approach the preferred route for identifying epitope/ antibody pairs.

Protocol 8

Production of soluble scFv fragments from pHEN.1

Reagents

- *E. coli* D29A1 (or any other non-suppresser strains like HB501 or TOP 10F)
- IPTG (Sigma)

Method

1. Inoculate 2.5 ml SOBAG medium with an individual well-isolated colony from a plate with *freshly transformed* D29A1.
2. Incubate at 37 °C at 200 r.p.m. until the culture reaches an A_{600} of 0.6–1.0.
3. Centrifuge the culture at 1000 g for 10 min at room temperature.
4. Carefully remove all the supernatant from the centrifuge tube.
5. Resuspend the cell pellet in 5 ml SOBAG without glucose plus 1 mM IPTG.
6. Incubate the culture overnight at 25 °C with shaking at 150–200 r.p.m.
7. Centrifuge the overnight culture and use the *E. coli* supernatant for analysis by Pepscan as described in *Protocol 4*, using alkaline phosphatase-conjugated anti-myc antibody as detection reagent.

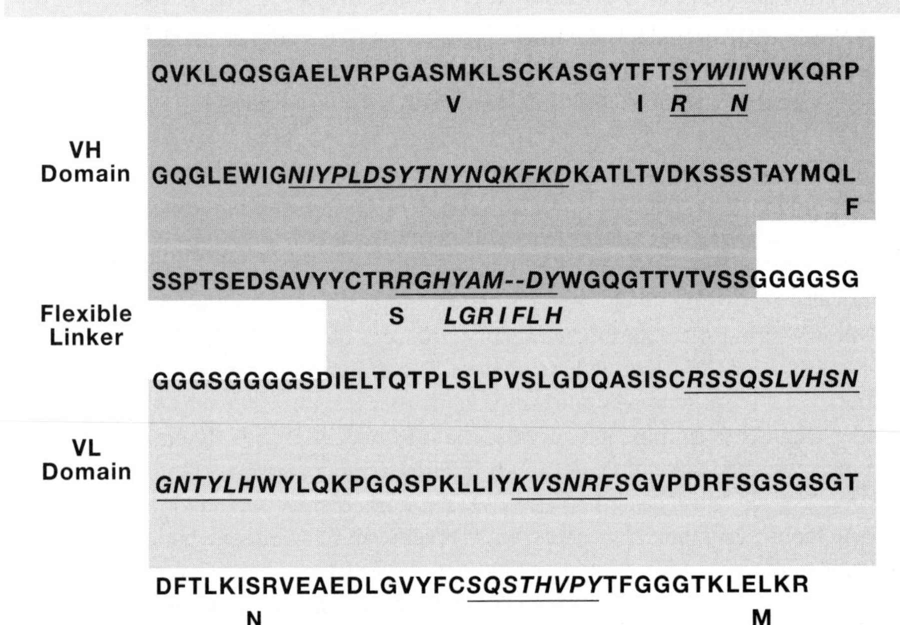

Figure 5 Amino acid sequence of anti-BLG scFv1 (upper line) and scFv.2 (lower line). The differences between scFv.1 and 2 are indicated. The CDR regions are shown underlined and in italics.

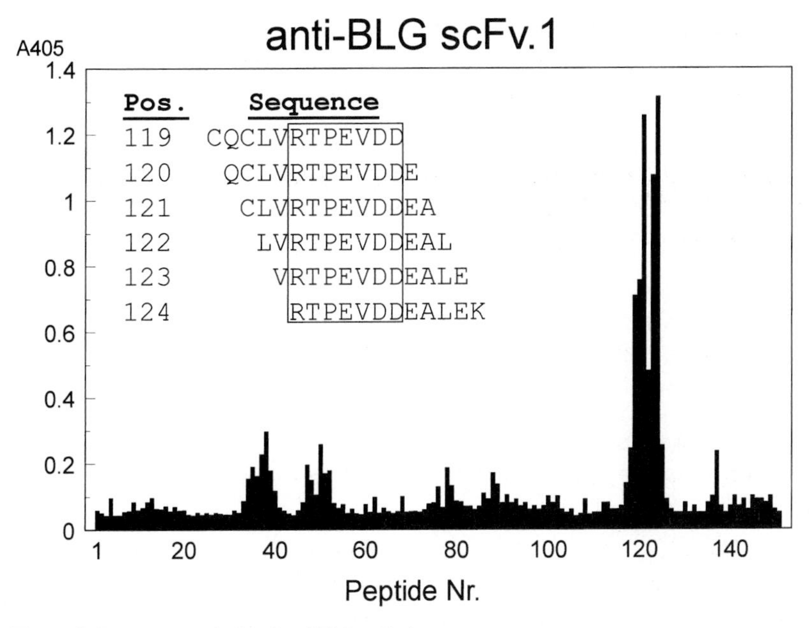

Figure 6 Pepscan analysis of anti-BLG scFv.1.

3 Conclusions

Phage display is a powerful technique which allows several selection cycles to be carried out on large libraries of different peptides or larger proteins in a relatively short length of time. In this chapter we have illustrated the many advantages of the phage display technique with our own experience of the uses of peptide display and also of antibody fragment display.

Acknowledgements

We wish to thank Alfredo Nicosia and Riccardo Cortese for donation of the pVIII9aa-cys phage display peptide library. The Pepscan work was carried out at ID-DLO, Lelystad, The Netherlands and we wish to thank Rob Meloen and Wouter Puijk for their input into this project. The beta-lactoglobulin epitope mapping work was funded by the Agro and Agro-Industrial Research program of the European Community (Grant No. CT94-0970) and we wish to thank the other partners in the program for useful discussions especially Mike Morgan and his group at the Institute of Food Research, Norwich for donation of some of the synthetic peptides used in this work.

References

1. Geysen, H. M., Rodda, S. J., and Mason, T. J. (1986). In *Synthetic peptides as antigens* (ed. R. Porter and J. Wheelan), Ciba Foundation Symposium **119**, p. 131. Wiley Press.
2. Smith, G. P. (1985). *Science*, **228**, 1315.

3. Scott, J. K. and Smith, G. P. (1990). *Science*, **249**, 386.
4. Stephen, C. W. and Lane, D. P. (1992). *J. Mol. Biol.*, **225**, 577.
5. Williams, S. C., Badley, R. A., Davis, P. J., Puijk, W. C., and Meloen, R. H (1998). *J. Immunol. Methods*, **213**, 1.
6. Germaschewski, V. and Murray, K. (1996). *J. Virol. Methods*, **58**, 21.
7. Jespers, L. S., Messens, J. H., De Keyser, A., Eeckhout, D., Van Den Brande, I., Gansemans, T. G., *et al.* (1995). *Biotechnology*, **13**, 378.
8. Felici, F., Castagnoli, L., Musacchio, A., Jappelli, R., and Cesareni, G. (1991). *J. Mol. Biol.*, **222**, 301.
9. Lowman, H. B. and Wells, J. A. (1993). *J. Mol. Biol.*, **234**, 564.
10. McCafferty, J., Griffith, A. D., Winter, G., and Chiswell, D. J. (1991). *Protein Eng.*, **4**, 955.
11. Corey, D. R., Shiau, A. K., Yang, Q., Janowski, B. A., and Craik, C. S. (1993). *Gene*, **128**, 129.
12. Chiswell, D. J. and McCafferty, J. (1992). *TIBTECH*, **10**, 80.
13. Bass, S., Greene, R., and Wells, J. A. (1990). *Proteins*, **8**, 309.
14. Gram, H., Marconi, L.-A., Barbas III, C. F., Collet, T. A., Lerner, R. A., and Kang, A. S. (1992). *Proc. Natl. Acad. Sci. USA*, **89**, 3576.
15. Robertson, M. W. (1993). *J. Biol. Chem.*, **268**, 12736.
16. Hoogenboom, H. R., Griffiths, A. D., Johnson, K. S., Chiswell, D. J., Hudson, P., and Winter, G. (1991). *Nucleic Acids Res.*, **19**, 4133.
17. Hawkins, R. E., Russell, S. J., and Winter, G. (1992). *J. Mol. Biol.*, **226**, 880.
18. Kay, B. K., Adey, N. B., He, Y.-S., Manfredi, J. P., Mantaragnon, A. H., and Folkes, D. M. (1993). *Gene*, **128**, 59.
19. Hoess, R. H. (1993). *Curr. Opin. Struct. Biol.*, **3**, 572.
20. Balass, M., Heldman, Y., Cabilly, S., Givol, D., Katchalski-Katzir, E., and Fuchs, S. (1993). *Proc. Natl. Acad. Sci. USA*, **90**, 10638.
21. Luzzago, A., Felici, F., Tramontano, A., Pessi, A., and Cortese, R. (1993). *Gene*, **128**, 51.
22. Geysen, H. M., Meloen, R. H., and Barteling, S. J. (1984). *Proc. Natl. Acad. Sci. USA*, **81**, 3998.
23. Little, M., Fuchs, P., Breitling, F., and Dubel, S. (1993). *TIBTECH*, **11**, 3.
24. Cull, M. G., Miller, J. F., and Schatz, P. J. (1992). *Proc. Natl. Acad. Sci. USA*, **89**, 1865.
25. Gramatikoff, K., Georgiev, O., and Schaffner, W. (1994). *Nucleic Acids Res.*, **22**, 5761.
26. Krebber, C., Spada, S., Desplanq, D., and Plückthun, A. (1995). *FEBS Lett.*, **377**, 227.
27. Perez, M. D. and Calvo, M. (1995). *J. Dairy Sci.*, **78**, 978.
28. Schwartz, R. H. (1991). *Immunol. Allergy Clinics N. Am.*, **11**, 717.
29. McLafferty, M. A., Kent, R. B., Ladner, R. C., and Markland, W. (1993). *Gene*, **128**, 29.
30. Folgori, A., Tafi, R., Meola, A., Felici, F., Galfre, G., Cortese, R., *et al.* (1994). *EMBO J.*, **13**, 2236.
31. Dybwad, A., Forre, O., Kjeldsen-Kragh, J., Natvig, J. B., and Sioud, M. (1993). *Eur. J. Immunol.*, **23**, 3189.
32. Parmley, S. F. and Smith, G. P. (1988). *Gene*, **73**, 305.
33. Sambrook, I., Fritsch, E. F., and Maniatis, T. (ed.) (1989). In *Molecular cloning: a laboratory manual* (2nd edn), p. 6.25. Cold Spring Harbor Laboratory Press, NY.

Chapter 10
Site-directed mutagenesis in epitope mapping

Samuel S. Perdue

Beirne Carter Centre for Immunology Research, University of Virginia School of Medicine, Charlottesville, Virginia, USA

1 Introduction

The utility of amino acid variants in epitope mapping originated with studies of naturally occurring evolutionary variants of hen egg white lysozyme (HEL) (1, 2) and myoglobin (3, 4). Seven avian lysozymes have been isolated that differ from HEL at 3–10 amino acids (human milk lysozyme differs at 52 positions). The epitopes for several anti-HEL antibodies have been crudely mapped to the lysozyme surface based on their differential recognition of these evolutionary variants (1, 2, 5). Similar studies were also performed using monoclonal antibodies specific for either human or sperm whale myoglobin and a panel of myoglobin variants from a number of vertebrates (3, 4).

Subsequent investigations also analysed the abilities of naturally occurring 'escape' mutants to avoid antibody detection. Single site variants of influenza neuraminidase have been isolated that are not recognized by the anti-neuraminidase antibodies S10/1 (6) and NC41 (7). Similar mutants have been described for influenza haemagglutinin (8), the spirochete bacterium *Borrelia hermsii* (9, 10), and African trypanosomes (11).

While the study of natural variants demonstrated some success in mapping epitopes on protein antigens, the technique had its limitations. First, the number of proteins available for study was restricted to those for which multiple species isolates were available. Secondly, investigators were limited to studying the few amino acid variations that nature had provided, and therefore were unable to look at large numbers of changes in a protein's primary sequence. Thirdly, individual amino acid substitutions rarely occur in isolation, and functional analysis of antigen–antibody binding was not always able to clearly determine the contribution to binding made by a single residue on the antigen surface.

Site-directed mutagenesis is a powerful tool that solves all of these problems. This strategy allows any amino acid within a protein sequence to be substituted with another. When a single residue is changed, the effect of this change on antibody-binding can be assessed by a variety of techniques. Unlike the case with natural variants, the researcher may select different amino acids at a given position and subsequently determine the role of various physiochemical properties in immune complex formation. For example, a glutamine residue may be replaced with a conservative substitution (asparagine), a charge reversal (glutamate), a larger side chain (tryptophan), or a smaller amino acid (alanine). This allows an in-depth analysis of the contribution of that amino acid to the energy of complex formation. Certainly amino acid substitutions can have a marked effect on antigenic structure and thus antibody binding. A recent study amply demonstrated this, showing that mutations for amino acid substitutions in the hepatitis B virus envelope could lead to loss of binding to neutralizing antibodies (12). Likewise site-directed mutagenesis has proved to be a useful technique for mapping the epitope involved in CD46–measles virus H protein interactions (13).

Several studies followed the use of evolutionary variants with site-directed mutagenesis of model antigen–antibody systems. Kam-Morgan and colleagues (1993) were able to functionally map a portion of the monoclonal antibody HyHEL-10 epitope on HEL by analysing 20 single amino acid substitutions at four lysozyme positions—Asn-19, Arg-21, Asp-101, and Gly-102. Substitutions at positions 19 and 102 had no effect on binding to HyHEL-10, while several substitutions at 21 and 101 led to a significant decrease in binding (14). Site-directed mutagenesis has also been used to confirm the importance of two single residues, Arg-45 and Arg-68, to the HyHEL-5 epitope on HEL (15, 16). Later, a more extensive mutagenesis of the antibody D1.3 epitope on HEL was also conducted and compared with identical studies on the D1.3/anti-D1.3 complex (17).

Mutagenesis of 34 surface residues on the 85-residue *Escherichia coli* phosphocarrier protein HPr has also been reported (18, 19), and these mutants were used to map the functional epitopes for three anti-HPr monoclonal antibodies. Following solution of the crystal structure of the HPr:Jel42 antigen–antibody complex (19), the functionally-determined epitope was compared to that determined by structural analysis. This study represented the first direct comparison of a structural and functional epitope. The structural epitope was shown to consist of 14 residues, while mutagenesis had suggested an epitope of at least 12 amino acids. Nine residues were shared between the two epitopes.

Site-directed mutagenesis has also been used in other antigen systems. Three critical residues on influenza neuraminidase (NA) which are recognized by the monoclonal antibody NC41 have been determined to be Ser-368, Asn-400, and Lys-434 (20). This is a finding that corresponds with the previous report of natural escape mutants involving amino acid substitutions at these positions (21). These studies, like the others mentioned previously, also permitted some analysis of the fine details of antigenicity by utilizing multiple substitutions at a single position.

The interactions between mutants of staphylococcal nuclease (SNase) and a panel of monoclonal antibodies has also proven effective at mapping epitopes on this antigen. Smith and co-workers (22) used competitive binding data in conjunction with limited site-directed mutagenesis in order to place 12 antibodies in complementation groups according to their relative ability to block binding by one another to SNase. In addition, two monoclonal antibodies were mapped to distinct positions on SNase based on their sensitivities to specific single amino acid substitutions. Subsequent mutagenesis was then used to expand the epitope of one of these antibodies (23).

We later expanded on the work of Smith and co-workers by analysing the antibody-binding properties of single alanine substitution mutants at 68 positions on the surface of SNase (Perdue and Benjamin, unpublished data). Alanine substitution offers a unique opportunity to assess individual amino acids at the antigen–antibody interface. Because the side chain of alanine consists of a single methyl group (which corresponds to the carbon in the side chains of all amino acids except glycine and proline) alanine substitution allows the removal of most atoms from a given position without the introduction of new atoms. Obviously, glycine possesses an even simpler side chain, but due to the large number of permissible ψ and ϕ angles observed for glycine, the potential effects on gross protein conformation often make this amino acid a poor choice for mutagenesis when precise epitope analysis is one's goal. In the search for putative epitope for use as a subunit vaccine, alanine substitution was used by Gogolak *et al.* (2000) to define the epitope within the 317–329 region of human influenza A virus (H1N1 subtype) haemagglutinin that could produce helper T cell and antibody responses. A helper T cell epitope was identified as being the 320–326 core sequence that could interact with the major histocompatibility class II peptide-binding groove (24) see also Chapter 4).

The replacement of native amino acids with alanine has the advantage of effectively removing from the antigen–antibody interface the energetic contributions of all side chain atoms beyond the Cβ position, including all side chain hydrogen bonds and salt bridges. Alanine-scanning mutagenesis, whereby single point mutants of a protein are systematically generated in order to determine the residues involved in complex formation, has therefore been used to assess the molecular interactions at several protein–protein interfaces.

The most extensive use of this technique in epitope mapping has involved the model antigens SNase (25) and human growth hormone (hGH). In the latter studies, alanine- and homologue-scanning mutagenesis were used to map the functional epitopes of 21 different monoclonal antibodies (mAbs) directed against hGH (26). Homologue-scanning mutagenesis (27) was first used to determine the residues on hGH (cysteine, glycine, and alanine residues which were not mutated). Mutants were then tested for their ability to bind selected mAbs—each mAb being tested against mutants containing substitutions at or near the relevant regions as previously determined by homologue scanning. Reductions in binding affinity upon single alanine substitution ranged from 2-fold to 1000-fold. These studies clearly showed that the entire surface of hGH is antigenic, as previously suggested (28). Also, the epitope for each of the 21 mAbs was discontinuous in nature, and each epitope was dominated by a small subset of residues with a hierarchy of functional importance (by single letter amino acid code) determined as R > P > E > D, F, I. Yet the identification of binding domains is not restricted to antigen–antibody interactions. Site-directed mutagenesis also may be used to identify other types of molecular interactions, e.g. alanine-scanning for evaluating between Factor V C2-domain for binding to phospholipid membranes for blood coagulation (29), homologue-scanning for adhesion molecule–leukocyte interactions (30). Thus it should be emphasized that the studies described here do not represent a complete and definitive list of site-directed mutagenesis in epitope mapping, but do provide the researcher with a general body of work from which to obtain further information.

2 General approaches to mutagenesis in epitope mapping techniques in site-directed mutagenesis

2.1 Traditional approach

Traditionally, site-directed mutagenesis protocols adhered to subtle variations on the following procedure. The gene of interest is first inserted into an M13 phage vector whose DNA is then isolated in single-stranded form (ssDNA). An oligonucleotide differing from the gene sequence at one or a few nucleotides (corresponding to the desired mutation) is then annealed to the ssDNA, and this oligonucleotide subsequently serves as a primer for directing second strand synthesis by the Klenow fragment of DNA polymerase. DNA ligase is used to repair the final nick, and the resulting double-stranded (dsDNA) is used to transform *E. coli*. The M13 phages ultimately lyse the cells and produce plaques, from which the DNA can be isolated. The mutated DNA may then be removed from the M13 vector by endonuclease digestion and cloned into a plasmid for expression or further analysis.

The yield of desired product, i.e. the mutated versus the wild-type strand, depends on a number of factors and is often considerably lower than the 50% yield expected from the semi-conservative replication of DNA alone. Initial methods for improving the efficiency of site-directed mutagenesis focused on improving plaque screening. One effective method of selection for positive (mutant) phage is the use of the original mutagenic primer as a probe for the clones of interest. However, improved screening does not address the underlying problem of low mutagenesis efficiencies. Consequently, methods were developed to enable selective destruction of the original template strand along with the preservation of the mutated gene. A much-sited protocol for increased efficiency was reported by Kunkel and co-workers (1985) and is described in detail below (31).

The Kunkel method takes advantage of the ability of cells to incorporate dUTP in place of dTTP at sites throughout the DNA. Although dUTP can compete with dTTP in all *E. coli* strains, deficiencies in the *dut* and *ung* genes in selected strains leads to elevated dUTP levels. The non-functional *dut* gene results in no dUTPase being produced and the subsequent accumulation of dUTP. Mutants lacking *ung* cannot synthesize the 'proofing' enzyme uracil *N*-glycosylase and therefore are unable to remove uracil from DNA strands. Therefore, *dut ung* strains incorporate dUTP into newly synthesized strands but are unable to correct this error of replication. However, uracilated DNA remains fully capable of serving as a template for subsequent replicative events, an important feature in its usefulness for site-directed mutagenesis.

Once the uracilated template DNA is produced, the traditional steps in mutagenesis are followed. An oligonucleotide containing the mutation of interest is annealed to the ssDNA at the appropriate site and used as a primer for second strand synthesis. However, the subsequent transformation step now utilizes an *ung*$^{+}$ *E. coli* strain that removes uracils from the template strand containing the wild-type gene, producing nicks. These nicks can then serve as sites for specific endonucleases that degrade the template strand, leaving the desired daughter strand intact. In this manner, selection for the mutated DNA strand is greatly enhanced.

The Kunkel method of site-directed mutagenesis has remained a mainstay of many molecular biology laboratories, and serves as the basis for a number of commercially available mutagenesis kits.

Protocol 1

The Kunkel method of site-directed mutagenesis: template preparation

Reagents, Vectors and Bacterial Strains

- Variants of filamentous M13 phage (M13mp) that permit isolation of DNA as either single- or double-stranded closed circles
- Phagemid: pTZ18/19U/R
- Helper phage, required for phagemid production: M13K07
- *E. coli* CJ236: a *dut ung* strain used for preparing uracilated template
- *E. coli* MV1190: an *ung*$^{+}$ strain used for destruction of template DNA
- TE: 10 mM Tris–HCl pH 8.0
- 10 × annealing buffer: 200 mM Tris–HCl pH 7.4, 20 mM $MgCl_2$, 500 mM NaCl
- 10 × hybridization buffer: 20 mM Tris–HCl pH 7.5, 10 mM $MgCl_2$, 2 mM DTT, 500 mM each dNTP, 0.5 mM ATP
- CIAA: chloroform/isoamyl alcohol (24:1)
- Phenol/CIAA: phenol/chloroform/isoamyl alcohol (25:24:1)
- 5 × PEG/NaCl: 150 g PEG 8000, 150 g NaCl, H_2O to 1 litre
- 200 mM EDTA
- LB top agar: 10 g tryptone, 5 g yeast extract, 5 g NaCl, 7 g agar (1 litre)
- LB: 10 g tryptone, 5 g yeast extract, 5 g NaCl (1 litre)
- 2 × YT: 16 g tryptone, 10 g yeast extract, 5 g NaCl (1 litre)
- T4 polynucleotide kinase
- T7 DNA polymerase
- T4 DNA ligase
- RNase A

A. Preparation of phagemid

1. Incubate *E. coli* MV1190 containing phagemid at 37 °C in 5 ml LB (50 μg/ml ampicillin) to an OD_{600} of 0.15–0.3.
2. Add 10^9 pfu M13K07.

Protocol 1 continued

3 Continue incubation for 1–2 h.

4 Pellet 1 ml of cells by centrifugation in a microcentrifuge for 3 min and save supernatant.

5 Incubate 20 ml *E. coli* CJ236 at 37 °C in 20 ml LB (50 μg/ml ampicillin) to an OD_{600} of 3.0.

6 Add 10 μl of supernatant from step 4 and continue to incubate for 2 h.

7 Conduct serial tenfold dilutions (to 10 000) and plate 100 μl onto LB plates (50 μg/ml ampicillin, 20 μg/ml chloramphenicol).

B. Preparation of M13 phage

1 Grow a 3 ml culture of *E. coli* CJ236 to an OD_{600} of 0.2–0.3.

2 Transfer one M13mp plaque to 1 ml of 2 × YT in a 1.5 ml microcentrifuge tube and incubate for 5 min at 60 °C to kill cells.

3 Vortex to disrupt the agar and pellet contents by centrifugation at top speed in a microcentrifuge for 3 min.

4 Transfer 50 μl supernatant to a flask containing 50 ml of 2 × YT (20 μg/ml chloramphenicol) plus the 3 ml CJ236 culture from step 1.

5 Incubate with shaking at 37 °C for 8 h or overnight. If you have reason to believe that your plasmid is unstable, shorter incubation times are generally preferred.

6 Pellet cells by centrifugation at 6000 g for 30 min.

C. Template preparation

1 Precipitate DNA with 0.25 parts 5 × PEG/NaCl (1 part PEG/NaCl to 4 parts supernatant) for 1 h at 0 °C.

2 Pellet DNA by centrifugation at 5000 g for 15 min at 4 °C and allow to dry.

3 Resuspend pellet in 10 × TE pH 8.0 plus 300 mM NaCl.

4 Treat suspension with RNase A to remove RNA contaminants.

5 Vortex thoroughly and place on ice for 1 h.

6 Spin at 5000 g for 15 min at 4 °C. Decant or draw off supernatant and discard solid matter.

7 Extract twice with phenol/CIAA then once with CIAA alone.

8 Add 2 vol. absolute ethanol and 0.1 vol. 3 M Na acetate pH 5.0.

9 Precipitate at −20 °C for 2 h.

10 Isolate DNA by centrifugation, and resuspend pellet containing ssDNA in minimal volume TE.

11 Quantitate DNA by spectrophotometry (OD_{260}) and determine the presence of contaminants by visualization on agarose gel.

2.1.1 Preparation of oligonucleotides

Proper oligonucleotide design is critical for successful use of any site-directed mutagenesis protocol. However, in our experience problems are minimal if a few simple rules are followed:

(a) Design oligonucleotides such that minimal changes to the sequence are made (i.e. few substitutions along the target DNA). Due to the nature of the genetic code, a single or double substitution within a DNA triplet will often allow the desired change in the coding sequence without requiring large scale alteration of the underlying nucleotide order. For example, a change from Ser to Ala, where the Ser is initially encoded by the triplet, TCT, can be accomplished with a single base substitution to GCT.

(b) When possible, design primers such that the substituted bases are placed near the centre of the oligonucleotide sequence. This provides maximum ends for annealing on either side of the site of substitution.

(c) Primers of various lengths are effective, but ideally should contain at least six bases on either side of the substitution(s). As a general rule, longer primers tend to be more reliable in that they provide sufficient end-lengths for annealing and minimize non-specific priming.

(d) Some researchers have suggested that oligonucleotide termini should consist of one or two Gs or Cs, presumably because the stronger electrostatic interactions between these residues make the termini more 'sticky' during annealing. We have adopted this convention when possible, but have had equal success with primers ending in A or T.

(e) Examine all oligonucleotides against the sequence of the target DNA (GenBank) to eliminate primers containing strong sequence homology with other regions of the DNA. Undesired homologies can be reduced by shifting the oligonucleotide sequence a few residues upstream or downstream of the site of mutation.

(f) Examine all oligonucleotides for possible secondary structure. Inverted repeats can be identified using a number of commercially available computer programs, and predicted secondary structures assessed. It is also important to check the ends of the primers to ensure that they are not 'sticky' and therefore likely to form primer dimers. Again, undesired secondary structure and sticky ends can be reduced by shifting the position of the primer.

Protocol 2

The Kunkel method of site-directed mutagenesis: primer extension

Reagents

- See *Protocol 1*

Protocol 2 continued

A. Oligonucleotide annealing to uracilated template DNA

Before the mutagenesis reaction can begin, the oligonucleotide must be phosphorylated at its 5′ hydroxyl group. Most oligonucleotides can now be synthesized commercially with the appropriate phosphorylated end. For non-phosphorylated primers, the phosphate can be added easily using T4 polynucleotide kinase from most distributors of molecular biology reagents.

1. Primer annealing is conducted in a 10 μl reaction in a microcentrifuge tube as follows:
   - 200 ng uracilated template DNA[a]
   - 10–100 ng phosphorylated oligonucleotide[a]
   - 1 ml of 10 × annealing buffer
   - H_2O to 10 μl
2. The optimal reaction temperature for annealing varies depending on the oligonucleotide used. High-homology oligonucleotides can anneal at higher temperatures, while primers containing multiple substitutions require lower temperatures. A good general approach is to incubate the reaction mixture in 65 °C water and allow the reaction to cool naturally to room temperature.
3. The microcentrifuge tube should then be place on ice for 15 min prior to primer extension.

B. Primer extension

1. The product of the annealed primer:template reaction is then added directly to the following reagents for production of non-uracilated daughter DNA:
   - 10 μl annealing reaction mix
   - 10 μl of 10 × hybridization buffer
   - 1 U of T7 DNA polymerase[b]
   - 3 U of T4 DNA ligase
   - H_2O to 100 μl
2. Incubate the reaction mixture for 1 h at 37 °C.[c]
3. Stop the reaction with 8 μl of 200 mM EDTA.
4. Confirm successful completion of mutagenesis reaction by agarose gel electrophoresis using dsDNA and template DNA (without oligonucleotide) as controls.

C. Transfection/transformation

1. The mutated DNA is transformed or transfected into the appropriate cells using any of a number of techniques.
2. Success is determined by analysing plaques/colonies for the desired DNA.

[a] The ratio of oligonucleotide to template DNA can vary from 5:1 to over 20:1, and needs to be determined on a case by case basis. As a general rule, researchers should use the lowest ratio possible in order to minimize adverse reaction events such as mispriming, formation of primer

Protocol 2 continued

dimers, and inhibition of ligation. However, larger ratios (20:1 or higher) may be necessary when there is relatively low sequence identity between oligonucleotide and template or when using primers with exceptionally high A-T content.

[b] T7 DNA polymerase is the enzyme of choice. However, other enzymes are also available and capable of generating complementary strand DNA. The Klenow fragment of *E. coli* DNA polymerase has been used, but has the distinct disadvantage of displacing the annealed oligonucleotide from the DNA template. T4 DNA polymerase overcomes this disadvantage, but is less efficient than T7 polymerase in completing the primer extension event and therefore results in fewer complete DNA sequences.

[c] When oligonucleotides of low homology are used, the extension reaction may first need to be conducted at room temperature for 15 min, followed by 1 h of 37 °C incubation, in order to optimize production. However, this should be avoided except when conditions require it.

2.2 PCR approach

While the Kunkel method of site-directed mutagenesis is a highly effective technique for producing point mutants in desired genes, it requires multiple manipulations of both cells and DNA for effective completion. The development of the polymerase chain reaction (PCR) has provided researchers with a powerful and rapid tool for producing mutants that does not require the large number of steps seen with mutant production from uracilated templates. Many reports in the literature describe methods for PCR-mediated site-directed mutagenesis, and of these three general approaches are described here.

The reagents and time/temperature cycles used in PCR vary according to the enzyme used, nature of the primer/template pair, and 'laboratory tradition'. Researchers are urged to use the polymerase enzyme with the highest fidelity possible in order to avoid undesired point mutations, and DNA sequencing should always be used to confirm sequences following mutagenesis. Furthermore, the concerns involved in oligonucleotide primer design are essentially the same as those discussed for the Kunkel method above. Since this chapter is about general approaches to mutagenesis, not PCR, the reader is referred to a number of excellent texts (32, 33), or manufacturers' directions for detailed analysis of PCR reactions. Rather, this chapter will describe several general approaches to the use of PCR in site-directed mutagenesis, with the fine details such as buffer constituents and cycle times left to the individual researcher to determine according to available reagents and personal preferences.

2.2.1 Overlapping primer mutagenesis

As shown in *Figure 1*, two overlapping mutagenic primers and two flanking primers can be utilized for site-directed mutagenesis. Ideally, 10–15 bases of overlap should be used in the mutagenic primers for optimal effectiveness. The DNA of interest is first cloned into a convenient plasmid vector. While any vector can be used, it is convenient to select one that has commercially available forward and reverse sequencing primers. As always, the cloned DNA fragment

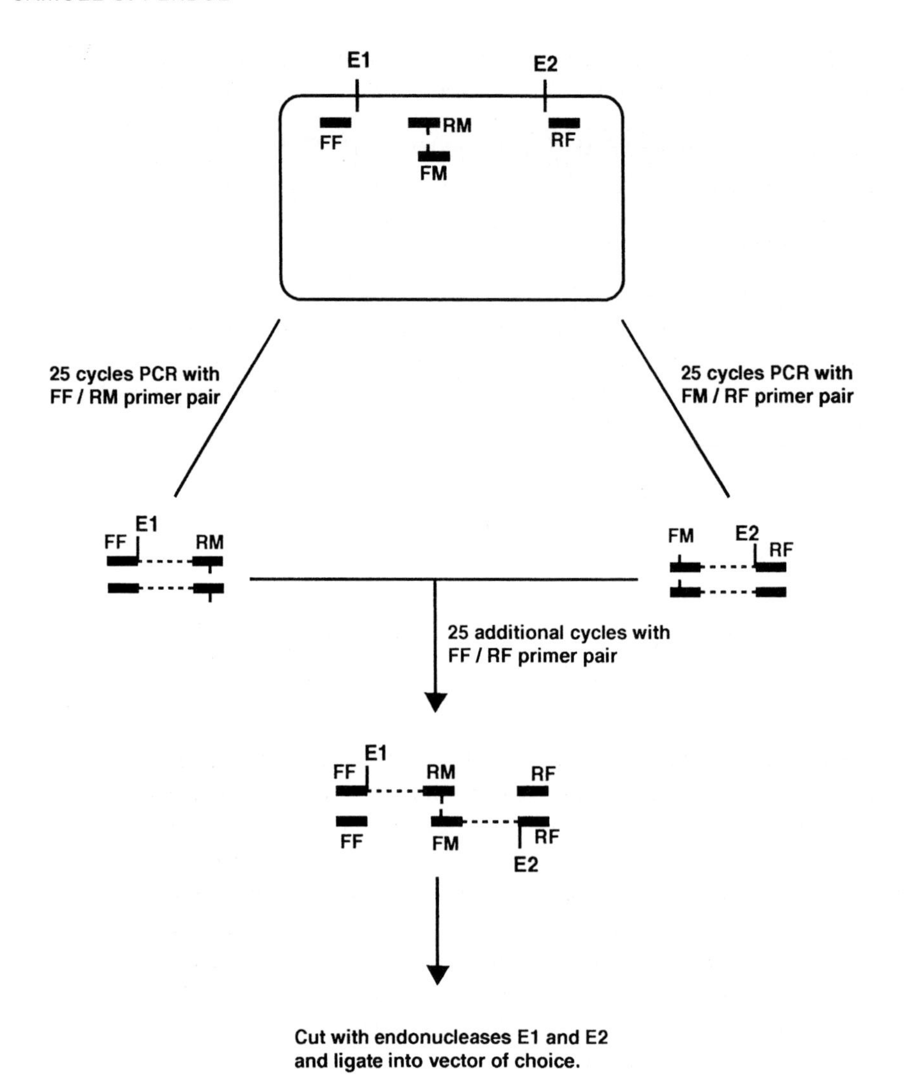

Figure 1 Overlapping primer mutagenesis.

must be free of the restriction sites used in the subsequent mutagenesis reactions.

The reaction begins with two independent 25-cycle reactions. One utilizes the forward flanking (FF) and reverse mutagenic (RM) primers, while the other uses the forward mutagenic (FM) and reverse flanking (RF) primers. These mutagenic oligonucleotides are designed as follows (site of point mutation is indicated in lowercase letters):

```
                      GGCGTGGCATgcTGGAGCTG 5′ (RM primer)
5′ GAGATAACTTCATTCGATTCCGCACCGTATTACCTCGACGTGCATC
   3′CTCTATTGAAGTAAGCTAAGGCGTGGCATAATGGAGCTGCACGTAG
                     5′ATTCCGCACCGTAcgACCT (FM primer)
```

This results in two predominant dsDNA species that have complementary ends due to the overlapping nature of the RM and FM oligonucleotides. The two species are then joined by a subsequent 25-cycle PCR using the RF and FF oligonucleotides as primers. The resulting PCR product can then be cut with the appropriate restriction endonucleases (E1 and E2) and the mutated DNA cloned into the desired vector(s) for sequencing and expression.

2.2.2 Non-overlapping PCR mutagenesis

Should the investigator wish to avoid using overlapping primers, non-overlapping oligonucleotides may be used. However, this protocol involves additional molecular manipulations such as blunt-end ligation that are unnecessary when using the previous approach. *Figure 2* diagrams the steps involved in non-overlapping PCR mutagenesis. In this protocol, only one of the internal primers is mutagenic (M), while the other is simply an internal flanking (IF) primer. These primers are designed as demonstrated below (site of point mutation indicated with lowercase letters):

```
                   GTAAGCTAAGGCGTGGCATgc 5′ (M primer)
5′ GAGATAACTTCATTCGATTCCGCACCGTATTACCTCGACGTGCATC
3′ CTCTATTGAAGTAAGCTAAGGCGTGGCATAATGGAGCTGCACGTAG
                              5′ ACCTCGACGTGCATC (IF primer)
```

The same forward flanking (FF) and reverse flanking (RF) oligonucleotides can be used as were utilized in the previous protocol. Again two independent 25-cycle PCRs are conducted, one using the FF/M primer pair and one with the IF/RF pair. However, the resulting PCR products do not have cohesive termini, and therefore cannot be 'attached' through an addition PCR step as was described above. Rather the PCR fragments are digested with the appropriate restriction endonucleases (E1 and E2), blunt-ended, and ligated into a vector of choice. Obviously, care must be taken to select clones that contain the two PCR fragments in the proper orientation, and DNA sequencing to confirm sequence integrity is critical when using this approach.

2.2.3 Megaprimer synthesis

A variation on the non-overlapping PCR approach is the megaprimer method introduced by Kammann and colleagues (1989) (34), and later modified by several groups (35, 36). In fact, this procedure adopts features of both overlapping and non-overlapping primer mutagenesis. Megaprimer synthesis does not require overlapping primers, but also does not require blunt-end ligation of DNA when constructing the final product.

The procedure utilizes a two-step PCR, as diagrammed in *Figure 3*. In the first step, a mutagenic forward primer (M) and flanking reverse primer (RF) are used in a 25-cycle reaction to generate a 'megaprimer' (MP). The megaprimer contains the mutation of interest, and therefore can serve as the new reverse prime for a second PCR reaction utilizing the forward flanking (FF) oligonucleotide. Megaprimer PCR is a fast, efficient method for obtaining site-directed mutants, and

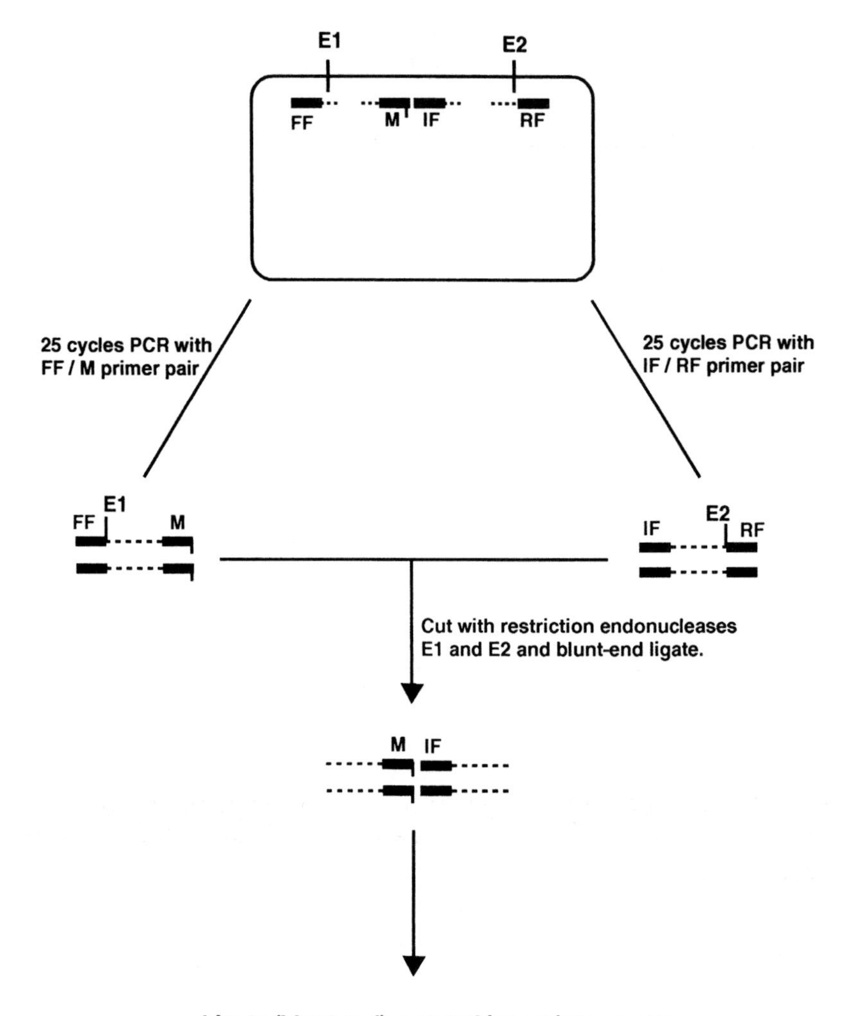

Figure 2 Non-overlapping PCR mutagenesis.

has the added advantage of requiring fewer oligonucleotides than the first two protocols. However, unlike more traditional PCR approaches, megaprimer PCR typically involves careful attention to protocol for optimal success. Therefore, a sample protocol is provided here.

While this protocol is effective for mutagenesis through megaprimer, very large megaprimers may result in very poor yields due to inhibition of annealing caused by primer dimerization and self-annealing. This problem can often be overcome by using the 5-step pre-extension described here, and by increasing the amount of megaprimer relative to template in the second PCR reaction. Megaprimer quantities as high as 6 μg (to 5 ng template) have been reported to greatly enhance yields of megaprimer PCR (38).

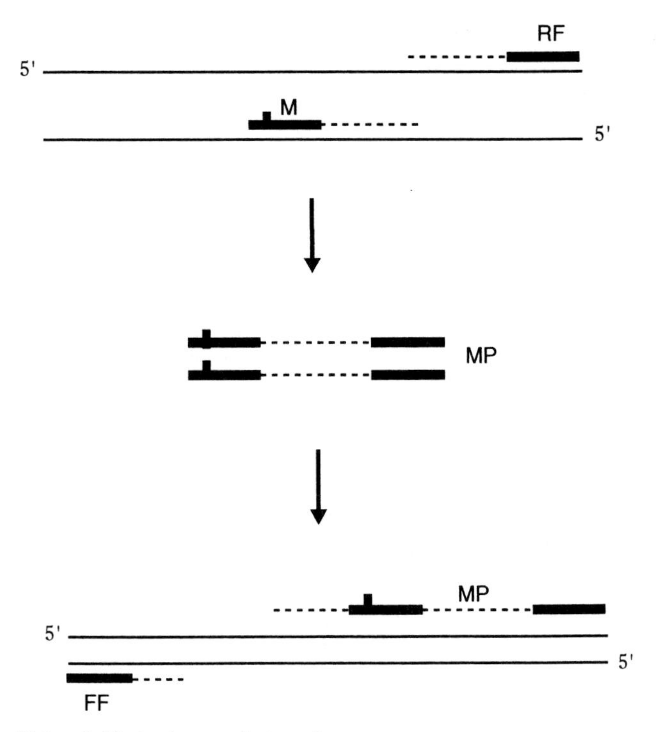

Figure 3 Megaprimer mutagenesis.

Protocol 3

Megaprimer PCR

Reagents

- See *Protocol 1*

A Initial PCR step

1. In a 50 μl reaction mix:
   - 2 ng DNA template
   - 1 μM mutagenic primer (M)
   - 1 μM reverse flanking primer (RF)
   - 1 × polymerase buffer (provided by enzyme supplier)
   - 200 μM each dNTP
   - DNA polymerase (*Taq*, *Vent*, *Pfu*, *Pwo*, or other)
   - H_2O to 50 μl
2. Conduct 25 cycles of PCR using the following temperatures as a guide:
   - 93 °C for 1 min (denaturation)
   - 60 °C for 1 min (annealing)[a]
   - 72 °C for 2 min (extension)[b]

Protocol 3 continued

B Second PCR step

1 Purify megaprimer (PCR product) by extraction from agarose gel and add to second PCR reaction:
- 5 ng template DNA
- 50–100 ng megaprimer DNA
- 1 × PCR buffer (supplied by enzyme provider)
- 200 μM each dNTP
- DNA polymerase
- H_2O to 50 μl

Note that the forward primer (FF) has not been added at this point. Several laboratories have reported improved yields when an initial 5-cycle asymmetrical PCR is first conducted to permit full extension of the megaprimer along the template (37).

2 Conduct 5 cycles using the following temperature settings:
- 93 °C for 1 min (denaturation)
- 72 °C for 2 min (extension)

3 Add forward primer (FF) to 1 μM.

4 Repeat 25-cycle program as described for first PCR step.

[a] As with all PCR reactions, optimal annealing temperatures vary depending on the stability of the primer/template complex. Less stable or lower homology primers require lower annealing temperatures, but such temperatures also increase the risk of mispriming.

[b] If desired, a final 3–5 min extension step can be added at the end of the 25 cycles to ensure full extension of all DNA products.

3 Conclusions

Site-directed mutagenesis is a powerful tool for specific mapping of antibody epitopes. Moreover, while mutagenesis used to require multiple manipulations of different vectors and cell lines, PCR techniques have greatly simplified the process and now allow for rapid and efficient generation of mutant proteins. By permitting virtually limitless changes to the amino acids on the antigen surface, including the substitution of residues that do not produce gross conformational changes, oligonucleotide-directed mutagenesis allows a comprehensive examination of the nature of antigen–antibody interactions.

References

1. Smith-Gill, S. J., Wilson, A. C., Potter, M., Prager, E. M., Feldmann, R. J., and Mainhart, C. R. (1982). *J. Immunol.*, **128**, 314.
2. Smith-Gill, S. J., Lavoie, T. B., and Mainhart, C. R. (1984). *J. Immunol.*, **133**, 384.
3. Berzofsky, J. A., Buckenmeyer, G. K., Hicks, G., Gurd, F. R. N., Feldmann, R. J., and Minna, J. (1982). *J. Biol. Chem.*, **257**, 3189.

4. East, I. J., Hurrell, J. G. R., Todd, P. E. E., and Leach, S. J. (1982). *J. Biol. Chem.*, **257**, 3199.
5. Harper, M., Lema, F., Boulot, G., and Poljak, R. J. (1987). *Mol. Immunol.*, **24**, 9.
6. Varghese, J. N., Webster, R. G., Laver, W. G., and Colman, P. M. (1988). *J. Mol. Biol.*, **200**, 201.
7. Tulip, W. R., Varghese, J. N., Webster, R. G., and Colman, P. M. (1992). *J. Mol. Biol.*, **227**, 149.
8. Knossow, M., Daniels, R. S., Douglas, A. R., Skehel, J. J., and Wiley, D. C. (1984). *Nature (London)*, **311**, 678.
9. Plasterk, R. H., Simon, M. I., and Barbour, A. G. (1985). *Nature (London)*, **318**, 257.
10. Stoenner, H. G., Dodd, T., and Larsen, C. (1982). *J. Exp. Med.*, **156**, 1297.
11. Donelson, J. E. (1995). *J. Biol. Chem.*, **270**, 7783.
12. Seddigh-Tonekaboni, S., Waters, J. A., Jeffers, S., Gehrke, R., Ofenloch, B., Horsch, A., *et al.* (2000). *J. Med. Virol.*, **60**, 113.
13. Hsu, E. C., Sabatinos, S., Hoedemaeker, F. J., Rose, D. R., and Richardson, C. D. (1999). *Virology*, **258**, 314.
14. Kam-Morgan, L. N. W., Smith-Gill, S. J., Taylor, M. G., Zhang, L., Wilson, A. C., and Kirsch, J. F. (1993). *Proc. Natl. Acad. Sci. USA*, **90**, 3958.
15. Chacko, S., Silverton, E., Kam-Morgan, L., Smith-Gill, S., Cohen, G., and Davies, D. (1996). *J. Mol. Biol.*, **245**, 261.
16. Lavoie, T. B., Kam-Morgan, L. N. W., Hartman, A. B., Mallett, C. P., Sheriff, S., Saroff, D. A., *et al.* (1990). In *The immune response to structurally defined proteins: the lysozyme model* (ed. S. J. Smith-Gill and E. E. Sercarz), p. 151. Adenine Press, Schenectady, NY.
17. Dall'Acqua, W., Goldman, E. R., Eisenstein, E., and Mariuzza, R. A. (1996). *Biochemistry*, **35**, 9667.
18. Sharma, S., Georges, F., Delbaere, L. T. J., Lee, J. S., Klevit, R. E., and Waygood, E. B. (1991). *Proc. Natl. Acad. Sci. USA*, **88**, 4877.
19. Prasad, L., Sharma, S., Vandonselaar, M., Quail, J. W., Lee, J. S., Waygood, E. B., *et al.* (1993). *J. Biol. Chem.*, **268**, 10705.
20. Nuss, J. M., Whitaker, P. B., and Air, G. M. (1993). *Protein*, **15**, 121.
21. Webster, R. G., Air, G. M., Metzger, D. W., Colman, P. M., Varghese, J. N., Baker, A. T., *et al.* (1987). *J. Virol.*, **61**, 2910.
22. Smith, A. M., Woodward, M. P., Hershey, C. W., Hershey, E. D., and Benjamin, D. C. (1991). *J. Immunol.*, **146**, 1254.
23. Smith, A. M. and Benjamin, D. C. (1991). *J. Immunol.*, **146**, 1259.
24. Gogolak, P., Simon, A., Horvath, A., Rethi, B., Simon, I., Berkics, K., *et al.* (2000). *Biochem. Biophys. Res. Commun.*, **270**, 190.
25. Benjamin, D. C. and Perdue, S. S. (1996). *Methods*, **9**, 508.
26. Jin, L., Fendly, B. M., and Wells, J. A. (1992). *J. Mol. Biol.*, **226**, 851.
27. Cunningham, B. C., Jhurani, P., Ng, P., and Wells, J. A. (1989). *Science*, **243**, 1330.
28. Benjamin, D. C., Berzofsky, J. A., East, I. J., Gurd, F. R. N., Hannum, C., Leach, S. J., *et al.* (1984). *Annu. Rev. Immunol.*, **2**, 67.
29. Kim, S. W., Quinn-Allen, M. A., Camp, J. T., Macedo-Ribeiro, S., Fuentes-Prior, P., Bode, W., *et al.* (2000). *Biochemistry*, **39**, 1951.
30. Ruchaud-Sparagano, M. H., Malaud, E., Gayet, O., Chignier, E., Buckland, R., and McGregor, J. L. (2000). *Biochem. J.*, **332**, 309.
31. Kunkel, T. A. (1985). *Proc. Natl. Acad. Sci. USA*, **82**, 488.
32. Innis, M. A., Gelfand, D. H., Sninsky, J. J., and White, T. J. (1990). *PCR protocols: a guide to methods and applications.* Academic Press, San Diego, California.
33. Ausubel, F., Brent, R., Kingston, R. E., Moore, D. E., Seidman, J. G., Smith, J. A., *et al.* (1997). *Short protocols in molecular biology*, 3rd edn. Wiley, New York.

34. Kammann, M., Laufs, J., Schell, J., and Gronenborn, B. (1989). *Nucleic Acids Res.*, **17**, 5404.
35. Sarkar, G. and Sommer, S. S. (1990). *BioTechniques*, **8**, 404.
36. Landt, O., Grunert, H. P., and Hahn, U. (1990). *Gene*, **96**, 125.
37. Datta, A. K. (1995). *Nucleic Acids Res.*, **23**, 4530.
38. Smith, A. M. and Klugman, K. P. (1997). *BioTechniques*, **22**, 438.

List of suppliers

Abbott Diagnostics Ltd, Abbotts House Norden Rd., Maidenhead, Berks SL6 4XF
Tel: 01628 784041
Abbott Laboratories Limited, Diagnostics Division, 7115 Mill Creek Drive (2nd Floor) Mississauga, Ontario L5N 3R3
Tel: (905) 858 2450

Actigen Ltd, Signet Court, Swanns Road, Cambridge CB5 8LA
Tel: 01223 319101

Alt Bioscience, School of Biochemistry, The University of Birmingham, Edgbaston, Birmingham
Tel: 0121 414 3376

Amersham Pharmacia Biotech UK Ltd, Amersham Place, Little Chalfont, Buckinghamshire HP7 9NA, UK (see also Nycomed Amersham Imaging UK; Pharmacia)
Tel: 0800 515313
Fax: 0800 616927
URL: http//www.apbiotech.com/

Anachem Ltd, Anachem House, 20 Charles St., Luton, Bedfordshire LU2 OEB
Tel: 01582 745000

Anderman and Co. Ltd, 145 London Road, Kingston-upon-Thames, Surrey KT2 6NH, UK
Tel: 0181 5410035
Fax: 0181 5410623

BD Bioscience, 1 Becton Drive, Franklin Lakes NJ 07417-1880
Tel: (201) 847 6800

Beckman Coulter (UK) Ltd, Oakley Court, Kingsmead Business Park, London Road, High Wycombe, Buckinghamshire HP11 1JU, UK
Tel: 01494 441181
Fax: 01494 447558
URL: http://www.beckman.com/
Beckman Coulter Inc., 4300 N. Harbor Boulevard, PO Box 3100, Fullerton, CA 92834–3100, USA
Tel: (714) 871 4848
Fax: (714) 773 8283
URL: http://www.beckman.com/

Becton Dickinson and Co., 21 Between Towns Road, Cowley, Oxford OX4 3LY, UK
Tel: 01865 748844
Fax: 01865 781627
URL: http://www.bd.com/
B. D. Bioscience, 1 Becton Drive, Franklin Lakes, NJ 07417–1883, USA
Tel: (201) 847 6800
URL: http://www.bd.com/

Bibby-Sterilin Ltd., Tilling Drive, Stone, Staffs. ST15 OSA
Tel: 01785 812121

Bio-Rad Laboratories, Bio-Rad House, Maylands Avenue, Hemel Hempstead, Hertfordshire HP2 7TD
Tel: 0800 181134

Bio-Rad Chemical Division, Group Headquarter, Bio-Rad Laboratories, 2000 Alfred Nobel Drive, Hercules CA 94547
Tel: 1-(800) 224 6723

BioWhittaker House, 1 Ashville Way, Wokingham, RG41 2PL
Tel: 0118-979-5234
BioWhittaker, Inc. USA, 8830 Biggs Ford Road, Walkersville, Maryland 21793-0127, USA
Tel: 1-(800) 638-8174
Fax: (301) 898 7025

Boehringer-Mannheim (see Roche Molecular Biochemicals)

C. A. Hendley, Oakwood Hill Industrial Estate, Loughton, Essex
Tel: 020 8502 1821

Calbiochem-Novabiochem UK Ltd., Boulevard Industrial Park, Padge Rd., Beeston, Nottingham NG9 2JR
Tel: 01159 430840
Calbiochem-Novabiochem Corp., 10394 Pacific Centre Court, San Diego CA 9212
Tel: (858) 450 9600

Cambridge Bioscience, 25 Signet Court, Stourbridge Common Business Park, Swann's Rd., Cambridge CB1 2BL
Tel: 01223 316855

Carl Zeiss Ltd., P.O.Box 78, Woodfield Road, Welwyn Garden City Herts AL7 1LU,
Tel: 01707 871200
Carl Zeiss, Inc., Microscopy and Imaging Systems, One Zeiss Drive, Thornwood, NY 10594
Tel: 1-(800) 233 2343

CP Instrument Company Ltd, Unit 3, The Shires, Shire Hill Industrial Estate, Saffron Walden, Essex, CB11 3AN.
Tel: 01799 581321

DAKO Ltd., Denmark House, Angel Drove, Ely, Cambs. CB7 4ET
Tel: 01353 669911
DAKO Corp. 6392 Via Real, Carpinteria CA 93013
Tel: (805) 235 5743

Difco Laboratories Ltd., PO Box 14b Central Avenue, West Molesey KT8 2SE
Tel: 020 8979 9951

DuPont (UK) Ltd., Industrial Products Division, Wedgewood Way, Stevenage CM23 3DX
Tel: 01279 757711

Dynal (UK) Ltd, 10 Thursby Rd., Croft Business Park, Bromborough, Wirral, Merseyside L62 3PW
Tel: 0151 346 1234
Dynal Inc., 5 Delaware Drive, Lake Success NY 11042
Tel: 1-(800) 638 9416

Dynex Technologies Ltd., Action Court, Ashford Road, Ashford TW15 1XB
Tel: 01784 251225
Dynex Technologies Inc., 14340 Sullyfield Circle, Chantilly, VA 22021
Tel: (703) 631 7800

Eastman Chemical Co., 100 North Eastman Road, PO Box 511, Kingsport, TN 37662–5075, USA
Tel: 001 423 2292000
URL: http//:www.eastman.com/

E-C Apparatus Corp, ISC House, Progress Business Centre, 5 Whittle Parkway, Slough SL1 6DQ
Tel: 01628-668881
E-C Apparatus Corp, 30 Controls Drive, P.O. Box 870, Shelton, CT 06484-0870, U.S.A.
Tel: (203) 926 9300

European Collection of Animal Cultures, PHLS Centre for Applied Microbiology and Research, Porton Down, Salisbury, Wiltshire SP4 0JG
Tel: 01980 612100

Fisher Scientific UK Ltd, Bishop Meadow Road, Loughborough, Leicestershire LE11 5RG, UK
Tel: 01509 231166
Fax: 01509 231893
URL: http://www.fisher.co.uk/
Fisher Scientific, Fisher Research, 2761 Walnut Avenue, Tustin, CA 92780, USA
Tel: (714) 669 4600
Fax: (714) 669 1613
URL: http://www.fishersci.com/

Gibco-BRL, (see Life Technologies)

GlaxoSmithKlein, Research and Development, Langley Court, Beckenham, Kent BR3 3BS
Tel: 020 8658 2211
Glaxo-Wellcome Inc., PO Box 13398, Research Triangle Park, North Carolina 27709
Tel: (919) 483 2100

HyClone Laboratories, 1725 South HyClone Road, Logan, UT 84321, USA
Tel: 001 435 7534584
Fax: 001 435 7534589
URL: http//:www.hyclone.com/

HyClone Laboratories, South HyClone Road, Logan UT 84321
Tel: (435) 753 4584

ICN Pharmaceuticals, Cedarwood, Chineham Business Park, Crockford Lane, Basingstoke, Hants. RG24 8WG
Tel: 01256 374620
ICN Pharmaceuticals Inc., ICN Plaza, 3300 Hyland Avenue, Costa Mesa CA 92626
Tel: (714) 545 0100

Ilford Imaging UK Ltd, Town Lane, Mobberley, Nutsford, Cheshire WA16 7JL
Tel: 01565 650000

Immune Systems Ltd, PO Box 120, Paignton, TQ4 7XD
Tel: 01803 526 556
Fax: 01803 526 776

INTEGRA Biosciences, Unit 9, Industrial Estate Icknield Way, Letchworth SG6 1TD
Tel: 01462 48 65 48

Jencons (Scientific) Ltd., Cherrycourt Way Industrial Estate, Stanbridge Road, Leighton Buzzard, Bedfordshire LU7 8UA
Tel: 01525 372010
Jencons Inc., 800 Bursca Drive, Suite 801, Bridgeville PA15017
Tel: (412) 257 8861

Jones Chromatography Ltd, New Road, Hengoed, Mid Glamorgan CF82 8AU
Tel: 01443 816991

LabSystems, St. Georges Court, Hanover Business Park, Altrincham WA14 5TP
Tel: 0161 942 300

Leica Microsystems (UK) Ltd, Davy Avenue, Knowlhill, Milton Keynes MK5 8LB
Tel: 01908 666663
Leica Microsystems Inc., 111 Deerlake Rd., Deerfield, IL 60015
Tel: (847) 405 0123

Life Sciences International UK Ltd., Unit 5, Ringway Centre, Edison Rd., Basingstoke, Hants RG21 6YH
01256 817282

Life Technologies, 3 Fountain Road, Inchinnan Business Park, Paisley
Tel: 0141 814 6100
Life Technologies Inc, 9800 Medical Centre Drive, Rockville, Maryland 20850
Tel: 1-(800) 338 5772

Mabtech, Gamla Värmdöv 2, SE-131 37 Nacka, SWEDEN
Tel: (46) 8 716 27 00

Marathon Laboratories Supplies, Unit 6, 55–57 Park Royal Road, London NW10 7TJ
Tel: 020 8965 6865

Medicell, 239 Liverpool Road, London
N1 1LX
Tel: 020 7607 2295

Merck Laboratory Supplies, Hunter Boulevard, Magna Park, Lutterworth, Leicester LE17 4XN
Tel: 01455 558600

Millipore UK Ltd., The Boulevard, Blackmoor Lane, Watford, Herts. WD1 8YW
Tel: 01923 816375
Millipore Corporation, Corporate Headquarters, 80 Ashby Rd, Bedford, MA, 01730
Tel: 1-(800) 645 5476

Mimotopes, PO Box 13, Heswall, CH61 5WA
Tel: 0151 648 3343
Mimotopes, 1073 Bullard Court, Suite B, Raleigh NC 27615
Tel: (919) 873 1123

National Diagnostics, Unit 4, Fleet Business Park, Itlings Lane, Hessle, Kingston-upon-Hull, HU13 9LX
Tel: 01482 646022
National Diagnostics Inc., 305 Pattern Drive, Atlanta, Georgia 30336
Tel: (404) 699 2121

National Institute for Biological Standards and Controls, Blanche Lane, South Mimms, Potters Bar, Herts EN6 3QG
01707 654753

NBS Biologicals Ltd, 14 Tower Square, Huntingdon PE18 7DT
Tel: 01480-433875

New England Biolabs (UK) Ltd., 73 Knowl Piece, Wilbury Way, Hitchin, Herts, SG4 0TY
Tel: 0800 318486
Fax: 01462 420616
New England Biolabs Ltd., 3397 American Drive, Unit 12, Mississauga, Ontario, Canada L4V 1T8
Tel: 1-800-387-1095
Fax: (905) 672-3370

Nikon Inc., 1300 Walt Whitman Road, Melville, NY 11747-3064
Tel: (631) 547-4200
Nikon Corporation, Fuji Building, 2-3, 3-chome, Marunouchi, Chiyoda-ku, Tokyo 100, Japan
Tel: 011 81 03 3214 5311

Nycomed-Amersham plc, Amersham Place, Little Chalfont, Bucks HP7 9NA
Tel: 01494 544000
Nycomed Amersham Inc, 2636 South Clearbrook Drive, Arlington Heights IL 60005
Tel: (847) 593 6300

Organon Laboratories Ltd., Cambridge Science Park, Milton Road, Cambridge
Tel: 01223 432700

Pall Gelman Laboratories, Europa House, Havant Street, Portsmouth PO1 3PD
Tel: 02392 302600
Pall Gelman Laboratories, 600 South Wagner Road, Ann Arbor, MI 48103-9019
Tel: (734) 665-0651

Perbio Science UK. Ltd, 44 Upper Northgate Street, Chester, CH1 4EF
Tel: 01244 382525

Perkin Elmer Life Sciences, 204 Cambridge Science park, Milton Road, Cambridge CB4 0GZ
Tel: 01223 437400
Perkin Elmer Instruments, Inc., Princeton Applied Research, 801 South Illinois Ave., Oak Ridge, TN 37831
Tel: 1-(800) 366 2741
Fax: (865) 481 2442

Phillip Harris Scientific, Novara House, Excelsior Road, Ashby Park, Ashby-de-la Zouch LE65 1NG

PHLS Centre for Applied Microbiology and Research, Porton Down, Salisbury, Wiltshire SP4 0JG
Tel: 01980 612100

Promega UK, Delta House, Chilworth Research Centre, Southhampton SO16 7NS
Tel: 0800 378994
Promega Corp, 2800 Woods Hollow Road, Madison WI 53711
Tel: (608) 274 4330

Qiagen, Boundary Court, Gatwick Road, Crawley RH10 2 AX
Tel: 01293 422911
Qiagen Inc, 28159 Avenue Stanford, Valencia CA 91355
Tel: 1-(800) 426 8157

Roche Molecular Biochemicals, Roche Diagnostics Ltd., Bell Lane, Lewes, BN7 1LG
Tel: 01273 484644
Roche Diagnostics Corp., 9115 Hague Road, PO Box 50457, Indianapolis IN 46256
Tel: (317) 845 2358

Sanyo-Gallenkamp plc, Park House, Meridian East, Meridian Business Park, Leicester LE3 2UZ
Tel: 01509 265 265
Fax: 0116 263 0530
Sanyo Science USA, 900 North Arlington Heights Rd., Suite 320, Itasca, IL60143
Tel: 1-(800) 858 8442

Sarstedt Ltd., 68 Boston Road, Leicester LE4 1AW
Tel: 0116 2359023

Sartorius Instruments Ltd., 18 Avenue Road, Belmont, Surrey
Tel: 020 8642 8691

Schleicher & Schuell UK Ltd., Unit 11, Brunswick Park Industrial Estate, London, N11 1 JL
Tel: 020 8361 3111
Schleicher & Schuell Inc., 10, Optical Avenue, Keene N.H. 03431 USA
Tel: (603) 352 3810

Serotec, 22 Bankside Station Approach, Kidlington, Oxford OX5 1JE
Tel: 01865 852700

Shandon Southern Products, Chadwick Road, Astmoor, Runcorn, Cheshire WA7 1PR
Tel: 01928 566611

Sigma-Aldrich, Fancy Road, Poole, Dorset BH17 7NH
Tel: 01202 733114
Sigma Chemical Co., PO Box 14508, St Louis, MO 63178, USA
Tel: 001 314 7715765
Fax: 001 314 7715757
URL: http://www.sigma-aldrich.com/

Stratagene Inc., 11011 North Torrey Pines Road, La Jolla CA 92037
Tel: (858) 535 5400

Vector Laboratories, Unit 3 Accent Park, Bakewell Street, Orton, Southgate Peterborough
Tel: 01733 237999
Vector Laboratories, 30 Ingold Road, Burlingame, CA 94010
Tel: 1-(800) 227 6666

Whatman Lab Sales Ltd., Whatman House, St. Leonards Rd., Maidstone, Kent ME16 OLS
Tel: 01622 676670
Whatman Inc. (Whatman), 9 Bridewell Pl., Clifton, NJ 07014
Tel: (201) 773 5800

Wolf Laboratories Limited, Unit J, Lancaster Road, Pocklington Industrial Estate, York YO42 2NR
Tel: 01759 301142

X O-GRAPH Imaging Systems, X O-GRAPH House, Hampton Street, Tetbury GL8 8LD
Tel: 01666 501501

Index

Note: pages with Protocols are indicated in **bold**.